SIMULATION AND CHAOTIC BEHAVIOR OF α-STABLE STOCHASTIC PROCESSES

PURE AND APPLIED MATHEMATICS

A Program of Monographs, Textbooks, and Lecture Notes

MONOGRAPHS AND TEXTBOOKS IN PURE AND APPLIED MATHEMATICS

1. *K. Yano,* Integral Formulas in Riemannian Geometry (1970)
2. *S. Kobayashi,* Hyperbolic Manifolds and Holomorphic Mappings (1970)
3. *V. S. Vladimirov,* Equations of Mathematical Physics (A. Jeffrey, ed.; A. Littlewood, trans.) (1970)
4. *B. N. Pshenichnyi,* Necessary Conditions for an Extremum (L. Neustadt, translation ed.; K. Makowski, trans.) (1971)
5. *L. Narici et al.,* Functional Analysis and Valuation Theory (1971)
6. *S. S. Passman,* Infinite Group Rings (1971)
7. *L. Dornhoff,* Group Representation Theory. Part A: Ordinary Representation Theory. Part B: Modular Representation Theory (1971, 1972)
8. *W. Boothby and G. L. Weiss, eds.,* Symmetric Spaces (1972)
9. *Y. Matsushima,* Differentiable Manifolds (E. T. Kobayashi, trans.) (1972)
10. *L. E. Ward, Jr.,* Topology (1972)
11. *A. Babakhanian,* Cohomological Methods in Group Theory (1972)
12. *R. Gilmer,* Multiplicative Ideal Theory (1972)
13. *J. Yeh,* Stochastic Processes and the Wiener Integral (1973)
14. *J. Barros-Neto,* Introduction to the Theory of Distributions (1973)
15. *R. Larsen,* Functional Analysis (1973)
16. *K. Yano and S. Ishihara,* Tangent and Cotangent Bundles (1973)
17. *C. Procesi,* Rings with Polynomial Identities (1973)
18. *R. Hermann,* Geometry, Physics, and Systems (1973)
19. *N. R. Wallach,* Harmonic Analysis on Homogeneous Spaces (1973)
20. *J. Dieudonné,* Introduction to the Theory of Formal Groups (1973)
21. *I. Vaisman,* Cohomology and Differential Forms (1973)
22. *B.-Y. Chen,* Geometry of Submanifolds (1973)
23. *M. Marcus,* Finite Dimensional Multilinear Algebra (in two parts) (1973, 1975)
24. *R. Larsen,* Banach Algebras (1973)
25. *R. O. Kujala and A. L. Vitter, eds.,* Value Distribution Theory: Part A; Part B: Deficit and Bezout Estimates by Wilhelm Stoll (1973)
26. *K. B. Stolarsky,* Algebraic Numbers and Diophantine Approximation (1974)
27. *A. R. Magid,* The Separable Galois Theory of Commutative Rings (1974)
28. *B. R. McDonald,* Finite Rings with Identity (1974)
29. *J. Satake,* Linear Algebra (S. Koh et al., trans.) (1975)
30. *J. S. Golan,* Localization of Noncommutative Rings (1975)
31. *G. Klambauer,* Mathematical Analysis (1975)
32. *M. K. Agoston,* Algebraic Topology (1976)
33. *K. R. Goodearl,* Ring Theory (1976)
34. *L. E. Mansfield,* Linear Algebra with Geometric Applications (1976)
35. *N. J. Pullman,* Matrix Theory and Its Applications (1976)
36. *B. R. McDonald,* Geometric Algebra Over Local Rings (1976)
37. *C. W. Groetsch,* Generalized Inverses of Linear Operators (1977)
38. *J. E. Kuczkowski and J. L. Gersting,* Abstract Algebra (1977)
39. *C. O. Christenson and W. L. Voxman,* Aspects of Topology (1977)
40. *M. Nagata,* Field Theory (1977)
41. *R. L. Long,* Algebraic Number Theory (1977)
42. *W. F. Pfeffer,* Integrals and Measures (1977)
43. *R. L. Wheeden and A. Zygmund,* Measure and Integral (1977)
44. *J. H. Curtiss,* Introduction to Functions of a Complex Variable (1978)
45. *K. Hrbacek and T. Jech,* Introduction to Set Theory (1978)
46. *W. S. Massey,* Homology and Cohomology Theory (1978)
47. *M. Marcus,* Introduction to Modern Algebra (1978)
48. *E. C. Young,* Vector and Tensor Analysis (1978)
49. *S. B. Nadler, Jr.,* Hyperspaces of Sets (1978)
50. *S. K. Segal,* Topics in Group Kings (1978)
51. *A. C. M. van Rooij,* Non-Archimedean Functional Analysis (1978)
52. *L. Corwin and R. Szczarba,* Calculus in Vector Spaces (1979)

53. *C. Sadosky*, Interpolation of Operators and Singular Integrals (1979)
54. *J. Cronin*, Differential Equations (1980)
55. *C. W. Groetsch*, Elements of Applicable Functional Analysis (1980)
56. *I. Vaisman*, Foundations of Three-Dimensional Euclidean Geometry (1980)
57. *H. I. Freedan*, Deterministic Mathematical Models in Population Ecology (1980)
58. *S. B. Chae*, Lebesgue Integration (1980)
59. *C. S. Rees et al.*, Theory and Applications of Fourier Analysis (1981)
60. *L. Nachbin*, Introduction to Functional Analysis (R. M. Aron, trans.) (1981)
61. *G. Orzech and M. Orzech*, Plane Algebraic Curves (1981)
62. *R. Johnsonbaugh and W. E. Pfaffenberger*, Foundations of Mathematical Analysis (1981)
63. *W. L. Voxman and R. H. Goetschel*, Advanced Calculus (1981)
64. *L. J. Corwin and R. H. Szcarba*, Multivariable Calculus (1982)
65. *V. I. Istrățescu*, Introduction to Linear Operator Theory (1981)
66. *R. D. Järvinen*, Finite and Infinite Dimensional Linear Spaces (1981)
67. *J. K. Beem and P. E. Ehrlich*, Global Lorentzian Geometry (1981)
68. *D. L. Armacost*, The Structure of Locally Compact Abelian Groups (1981)
69. *J. W. Brewer and M. K. Smith, eds.*, Emily Noether: A Tribute (1981)
70. *K. H. Kim*, Boolean Matrix Theory and Applications (1982)
71. *T. W. Wieting*, The Mathematical Theory of Chromatic Plane Ornaments (1982)
72. *D. B. Gauld*, Differential Topology (1982)
73. *R. L. Faber*, Foundations of Euclidean and Non-Euclidean Geometry (1983)
74. *M. Carmeli*, Statistical Theory and Random Matrices (1983)
75. *J. H. Carruth et al.*, The Theory of Topological Semigroups (1983)
76. *R. L. Faber*, Differential Geometry and Relativity Theory (1983)
77. *S. Barnett*, Polynomials and Linear Control Systems (1983)
78. *G. Karpilovsky*, Commutative Group Algebras (1983)
79. *F. Van Oystaeyen and A. Verschoren*, Relative Invariants of Rings (1983)
80. *I. Vaisman*, A First Course in Differential Geometry (1984)
81. *G. W. Swan*, Applications of Optimal Control Theory in Biomedicine (1984)
82. *T. Petrie and J. D. Randall*, Transformation Groups on Manifolds (1984)
83. *K. Goebel and S. Reich*, Uniform Convexity, Hyperbolic Geometry, and Nonexpansive Mappings (1984)
84. *T. Albu and C. Năstăsescu*, Relative Finiteness in Module Theory (1984)
85. *K. Hrbacek and T. Jech*, Introduction to Set Theory: Second Edition (1984)
86. *F. Van Oystaeyen and A. Verschoren*, Relative Invariants of Rings (1984)
87. *B. R. McDonald*, Linear Algebra Over Commutative Rings (1984)
88. *M. Namba*, Geometry of Projective Algebraic Curves (1984)
89. *G. F. Webb*, Theory of Nonlinear Age-Dependent Population Dynamics (1985)
90. *M. R. Bremner et al.*, Tables of Dominant Weight Multiplicities for Representations of Simple Lie Algebras (1985)
91. *A. E. Fekete*, Real Linear Algebra (1985)
92. *S. B. Chae*, Holomorphy and Calculus in Normed Spaces (1985)
93. *A. J. Jerri*, Introduction to Integral Equations with Applications (1985)
94. *G. Karpilovsky*, Projective Representations of Finite Groups (1985)
95. *L. Narici and E. Beckenstein*, Topological Vector Spaces (1985)
96. *J. Weeks*, The Shape of Space (1985)
97. *P. R. Gribik and K. O. Kortanek*, Extremal Methods of Operations Research (1985)
98. *J.-A. Chao and W. A. Woyczynski, eds.*, Probability Theory and Harmonic Analysis (1986)
99. *G. D. Crown et al.*, Abstract Algebra (1986)
100. *J. H. Carruth et al.*, The Theory of Topological Semigroups, Volume 2 (1986)
101. *R. S. Doran and V. A. Belfi*, Characterizations of C*-Algebras (1986)
102. *M. W. Jeter*, Mathematical Programming (1986)
103. *M. Altman*, A Unified Theory of Nonlinear Operator and Evolution Equations with Applications (1986)
104. *A. Verschoren*, Relative Invariants of Sheaves (1987)
105. *R. A. Usmani*, Applied Linear Algebra (1987)
106. *P. Blass and J. Lang*, Zariski Surfaces and Differential Equations in Characteristic p > 0 (1987)
107. *J. A. Reneke et al.*, Structured Hereditary Systems (1987)

Additional Volumes in Preparation

SIMULATION AND CHAOTIC BEHAVIOR OF α-STABLE STOCHASTIC PROCESSES

Aleksander Janicki
Aleksander Weron
Technical University of Wrocław
Wrocław, Poland

MARCEL DEKKER, INC.   NEW YORK · BASEL

Library of Congress Cataloging-in-Publication Data

Janicki, Aleksander
    Simulation and chaotic behavior of α-stable stochastic processes /
Aleksander Janicki, Aleksander Weron.
        p.   cm. -- (Monographs and textbooks in pure and applied
mathematics; 178)
    Includes bibliographical references and index.
    ISBN 0-8247-8882-6
    1. Stochastic processes--Data processing.   I. Weron, A.   II. Title.
III. Series.
    QA274.J35   1994
    003'.76--dc20
                                                                        93-21359
                                                                            CIP

The publisher offers discounts on this book when ordered in bulk quantities. For more information, write to Special Sales/Professional Marketing at the address below.

This book is printed on acid-free paper.

MARCEL DEKKER, INC.
270 Madison Avenue, New York, New York 10016

Current printing (last digit):
10 9 8 7 6 5 4

PRINTED IN THE UNITED STATES OF AMERICA

Preface

Stochastic processes are recognized to play an important role in a wide range of problems encountered in mathematics, physics, chemistry, engineering, economics and finance. Recent developments show that in many practical applications leading to appropriate stochastic models a particular class of Lévy α–stable processes is involved. While the attempt at mathematical understanding of these processes leads to severe analytical difficulties, there exist very useful approximate numerical techniques.

This monograph is about Lévy α–stable processes, referred to as α–stable or stable processes. After preliminary remarks in Chapter 1, we demonstrate in Chapter 2 various properties of α–stable random variables and processes for $\alpha \in (0, 2]$, i.e., including the Gaussian case $\alpha = 2$. It turns out that with the use of suitable statistical estimation techniques, computer simulation procedures, and numerical discretization methods described in Chapter 3, it is possible to construct approximations of stochastic integrals with stable measures as integrators. Their updated mathematical theory is systematically presented in Chapters 4 and 5. As a consequence we obtain an effective general method allowing us to construct approximate solutions to a wide class of stochastic differential equations involving such integrals.

Applications of computer graphics, Chapters 6 and 7, provide useful quantitative and visual information on those features of stable variates that distinguish them from their commonly used Gaussian counterparts. It is possible to demonstrate time evolution of densities with heavy tails of various stochastic processes, visualize the effect of jumps of trajectories, etc. We try to demonstrate that stable variates can be very useful in stochastic modeling of various problems arising in science and engineering, and often provide better description of real life phenomena than their Gaussian counterparts.

The final part of the book, Chapters 9 and 10, contains a theoretical study of the hierarchy of chaos (ergodicity, weak mixing, p–mixing, Kolmogorov property and exactness) for α–stable stochastic processes. We are concerned with two basic questions: how are the ergodic and mixing properties of stable and more general infinitely divisible stationary processes related to the spectral representation introduced in Chapter 5, and how similar is the general symmetric α–stable situation to the Gaussian ?

The book can be followed without introductory studies by readers with basic knowledge of advanced probability and stochastic processes. It may be followed by less specialized readers as well. This is especially true of the computer simulations and graphic representations. Most of the results (including two original topics of this monograph: computer simulation of α–stable processes and their chaotic behavior) are presented with detailed proofs. However, in some places, basic concepts from stochastic integration, stochastic differential equations, convergence of approximate methods, statistical estimation, numerical discretization, etc., are introduced without proofs.

The only technique that touches every area dealt with is the use of computers for simulation and visualization. In the Appendix we present the computer program STOCH–Lm.c which was employed to produce many graphical examples contained in this book. Running this program one can solve approximately stochastic differential equations with respect to the α–stable Lévy motion.

The monograph should be of interest to mathematicians working in such fields as theory of stochastic processes, chaos, stochastic modeling, discrete and approximate methods in stochastic analysis, and applications of statistical methods. It can also be recommended as an auxiliary or basic reading for university graduate and postgraduate students interested in various mathematical and computer courses. It should be useful for those who are interested in numerical construction and computer simulation and visualization of solutions of various problems arising in stochastic analysis and stochastic modeling (also for physicists, chemists, economists, and many others).

The mathematical content, all figures and computer programs providing them, and the LaTeX source code of the whole text, i.e. everything in the book, were produced by the authors themselves. There is no one to blame for possible occasional mistakes but the authors. Nevertheless, we are happy to acknowledge our indebtedness to many colleagues who have offered their help and a variety of constructive criticisms. Among others these include Professors S. Cambanis, M. Maejima, S. T. Rachev, J. B. Robertson, J. Rosinski and W. A. Woyczynski. We also thank our former and present students, in particular, A. Gross, P. Kokoszka, Z. Michna, K. Podgórski and A. Rejman, for their assistance, lively reactions during the lectures, and a fruitful collaboration.

We express our gratitude to the Hugo Steinhaus Center for Stochastic Methods at Technical University of Wrocław for providing a rich intellectual environment and facilities indispensable for the writing of this text. We are grateful for the support provided by the KBN Grant No. 2 1153 9101 and the NSF U.S.–Eastern Europe Cooperative Research Project. Last but not least, we thank our families, who have been remarkably patient with our single–minded application of our time to this project.

Aleksander Janicki
Aleksander Weron

Contents

There is no applied mathematics in form of a ready doctrine. It originates in the contact of mathematical thought with the surrounding world, but only when both mathematical spirit and the matter are in a flexible state.

Hugo Steinhaus

SIMULATION AND CHAOTIC BEHAVIOR OF α-STABLE STOCHASTIC PROCESSES

Chapter 1

Preliminary Remarks

1.1 Historical Overview

We start with the historical overview of a development of investigations and applications of α–stable random variables and processes. The interested reader looking for documented sources on this topic is referred to Weron (1984), Zolotarev (1986), Janicki and Weron (1994).

The early problem considered by pioneering statisticians of the 18th and 19th centuries was to find the best fit of an equation to a set of observed data points. After some false starts they hit upon the method of least squares. Legendre's work seemed to be the most influential at the time. Laplace elaborated upon it, and finally, in a discussion of the distribution of errors, Gauss emphasized the importance of the normal or Gaussian distribution. Laplace was a great enthusiast of generating functions and solved many complicated probability problems exploiting them. As the theory of Fourier series and integrals emerged in the early 1800's, he and Poisson made the next natural step applying such representations of probability distributions as a new natural tool for analysis, thus introducing the powerful characteristic function method. Laplace seemed to be especially pleased noticing that the Gaussian density was its own Fourier transform. In the early 1850's Cauchy, Laplace's former student, became interested in the theory of errors and extended the analysis, generalizing the Gaussian formula to a new one

$$f_N(x) = \frac{1}{\pi} \int_0^\infty \exp(-ct^N) \cos(tx) \, dt,$$

expressed as a Fourier integral with t^N replacing t^2. He succeeded in evaluating the integral (in the non–Gaussian case) only for $N = 1$, thus obtaining the famous Cauchy law defined by the density

$$f_1(x) = \frac{c}{\pi(c^2 + x^2)}.$$

It was realized only much later (in 1919, thanks to Bernstein) that f_N is positive-definite, and hence is a probability density function only when $0 < N \leq 2$.

Replacing N by a real parameter α with values in $(0, 2]$, we find that the integral defining functions $f_\alpha = f_\alpha(x)$ is a source of remarkable surprises. After Cauchy there was a decline in mathematicians' interest in this subject until 1924, when the theory of stable distributions originated with Paul Lévy. When fashion sought the most general conditions for the validity of the Central Limit Theorem, Lévy found simple exceptions to it, namely the class of α–stable distributions with index of stability $\alpha < 2$. The ambiguous name *stable* has been assigned to these distributions because, if X_1 and X_2 are random variables having such distribution, then X defined by the linear combination $cX = c_1 X_1 + c_2 X_2$ has a similar distribution with the same index α for any positive real values of the constants c, c_1 and c_2 with $c^\alpha = c_1^\alpha + c_2^\alpha$. Lévy noted that the Gaussian case ($\alpha = 2$) is "singular" because for all $\alpha \in (0, 2)$ all nondegenerate densities $f_\alpha(x)$ have inverse power tails, i.e. $\int_{\{|x|>\lambda\}} f_\alpha(x)\, dx \approx C \cdot \lambda^{-\alpha}$ for large λ. Since these distributions have no second moments, the second moment existence condition for the CLT is violated, allowing for the possibility of unusual results.

Research concerning stable stochastic processes and models has been directed towards delineating the extent to which they share the features of the Gaussian models, and even more significantly, towards discovering their own distinguishing and often surprising features, e.g., Weron (1984). In the last ten years many important results characterizing different properties of these processes (and of other subclasses of processes with independent increments) have been obtained by several authors. Of particular importance are the results concerning representations involving stochastic integrals. A collection of papers edited by Cambanis, Samorodnitsky and Taqqu (1991) provides a review of the state of the art on the structure of stable processes as models for random phenomena.

Modern stochastic integration originated in the early work of Wiener. Stochastic integrals with respect to *Brownian motion* were defined by Itô (1944). Doob (1953) proposed a general integral with respect to L^2-martingales. On the basis of the Doob–Meyer decomposition theorem, Kunita and Watanabe (1967) further developed the theory of this integral. Meyer and Doleans–Dade (1970) extended the definition of the stochastic integral to all local martingales and subsequently to semimartingales. The natural role of semimartingales was made evident thanks to the contribution of Bichteler (1981) and Dellacherie (1980), who established that semimartingales are the most general class of integrators for which one can have a reasonable definition of stochastic integral against predictable integrands.

Our main tool of description of stochastic processes in which we are interested is a stochastic integral with respect to the α–stable Lévy motion or, in a more general setting, with respect to α–stable stochastic measures (see Samorodnitsky and Taqqu (1994)). The α–stable Lévy motion together with the *Poisson process* and Brownian motion are the most important examples of *Lévy processes*, which form the first class of stochastic processes being studied in the modern spirit.

They still provide the most important examples of *Markov processes* as well as of *semimartingales* (see Protter (1990)).

Thus, on the one hand, the class of stochastic processes in which we are interested is much broader than the class of Gaussian processes and, on the other, it is contained in the class of *infinitely divisible processes* (e.g., Lévy processes), which itself is contained in the class of semimartingales.

1.2   Stochastic α–Stable Modeling

In the past few years there has been an explosive growth in the study of physical and economic systems that can be successfully modeled with the use of stable distributions and processes. Especially infinite moments, elegant scaling properties and the inherent self–similarity property of stable distributions are appreciated by physicists. For a recent survey, see Janicki and Weron (1994). We believe that stable distributions and stable processes do provide useful models for many phenomena observed in diverse fields. The central–limit–type argument often used to justify the use of the Gaussian model in applications may also be applied to support the choice of the non–Gaussian stable model. That is, if the randomness observed is the result of summing many small effects, and those effects themselves follow a heavy–tailed distribution, then a non–Gaussian stable model may be appropriate. An important distinction between Gaussian and non–Gaussian stable distributions is that stable distributions are heavy–tailed, always with the infinite variance, and in some cases with the infinite first moment. Another distinction is that they admit asymmetry, or skewness, while a Gaussian distribution is necessarily symmetric about its mean. In certain applications then, where an asymmetric or heavy–tailed model is called for, a stable model may be a viable candidate. In any case, non–Gaussian stable distributions furnish tractable examples of non–Gaussian behavior and provide points of comparison with the Gaussian case, highlighting the special nature of Gaussian distributions and processes.

In order to appreciate the basic difference between a Gaussian distribution and a distribution with a long tail, Montroll and Shlesinger (1983b) proposed to compare the distribution of heights with the distribution of annual incomes for American adult males. An average individual who seeks a friend twice his height would fail. On the other hand, one who has an average income will have no trouble to discover a richer person, who, with a little diligence, may locate a third person with twice his income, etc. The income distribution in its upper range has a Pareto inverse power tail; however, most of the income distributions follow a log–normal curve, but the last few percent have a stable tail with exponent $\alpha = 1.6$ (cf., Badgar (1980)), i.e., the mean is finite but the variance of the corresponding 1.6–stable distribution diverges.

1.3 Statistical versus Stochastic Modeling

The notions of a statistical model and a stochastic model may be understood differently and may be ambiguous in some situations. In order to clear up what we mean by these terms we formulate a few remarks on this subject. One of a number possible descriptions of *statistical model* is the following (see Clogg (1992))

> What statistical methodology refers to in most areas today is virtually synonymous with statistical modeling. A statistical model can be thought of as an equation, or set of equations, that (a) link "inputs" to "outputs"..., (b) have both fixed and stochastic components, (c) include either a linear or a nonlinear decomposition between the two types of components, and (d) purport to explain, summarize or predict levels of or variability in the "outputs".

A very interesting discussion about how to model the progression of cancer, the AIDS epidemic, and other real life phenomena, is contained in the chapter "Model building: Speculative data analysis" of Thompson and Tapia (1990). The main idea is to derive a stochastic process that describes as closely as possible an investigated problem. Starting from an appropriate system of axioms one has to arrive at a formula (a *stochastic model*) defining this process, construct it explicitly in some way, and verify its correctness and usefulness. Appealing to one of the problems they are interested in, Thompson and Tapia (1990) say "If we wish to understand the mechanism of cancer progression, we need to conjecture a model and then test it against a data base."

Quite often a stochastic model is a synonym of a stochastic differential equation or system of stochastic differential equations. Thanks to Itô's theory of stochastic integration with respect to Brownian motion, it is commonly understood that any continuous diffusion process $\{X(t) : t \geq 0\}$ with given drift and diffusion coefficients can be obtained as a solution of the stochastic differential equation

$$X(t) = X_0 + \int_0^t a(s, X(s)) \, ds + \int_0^t c(s, X(s)) \, dB(s), \quad t \geq 0,$$

where $\{B(t) : t \in [0, \infty)\}$ stands for a Brownian motion process and X_0 is a given Gaussian random variable. The theory of such stochastic differential equations is well developed (see, e.g., Arnold (1974)) and they are widely applied in stochastic modeling.

However, it is not so commonly understood that a vast class of diffusion processes $\{X(t) : t \geq 0\}$ with given drift and diffusion coefficients can be described by the stochastic differential equation

$$X(t) = X_0 + \int_0^t a(s, X(s)) \, ds + \int_0^t c(s, X(s)) \, dL_\alpha(s), \quad t \geq 0,$$

where $\{L_\alpha(t) : t \in [0, \infty)\}$ stands for a stable Lévy motion process and X_0 is a given α-stable random variable. Note that, in general, diffusion processes $\{X(t) : t \geq 0\}$ defined above do not belong to the class of α-stable processes. On the other hand, they belong to the import class of diffusions with jumps.

As sources of information on modern aspects of stochastic analysis (e.g., on various properties of stochastic integrals and existence of solutions of stochastic differential equations driven by stochastic measures of different kinds) we recommend, among others, Protter (1990) and Kwapień and Woyczyński (1992).

An application of stochastic differential equations to statistical or stochastic model building is not an easy task. So it seems that the use of suitable statistical estimation techniques, computer simulation procedures, and numerical discretization methods should prove to be a powerful tool.

> In most of non–trivial cases ... the "closed form" solution is itself so complicated that it is good for little other than as a device for pointwise numerical evaluation. The simulation route should generally be the method of approach for non–trivial time–based modeling. ... Unfortunately, at the present time, the use of the modern digital computer for *simulation based modeling* and computation is an insignificant fraction of total computer usage.
> (Thompson and Tapia (1990), p. 232–233.)

We agree with this opinion and add that, unfortunately, as far as we know, practical approximate methods for solving stochastic differential equations involving stochastic integrals with stable integrands or integrators are only now beginning to be developed.

In our exposition we emphasize the methods exploiting computer graphics. Let us cite Thompson's opinion:

> I feel that the graphics–oriented density estimation enthusiasts fall fairly clearly into the exploratory data analysis camp, which tends to replace statistical theory by the empiricism of a human observer. Exploratory data analysis, including nonparametric density estimation, should be a first step down a road to understanding the mechanism of data generating systems. The computer is a mighty tool in assisting us with graphical displays, but it can help us even more fundamentally in revising the way we seek out the basic underlying mechanisms of real world phenomena via stochastic modeling.
> (Thompson and Tapia (1990), p. xiv.)

In our approach we feel strongly inspired by the work of S. M. Ulam, who was one of the first enthusiasts of application of computers not only to scientific calculations or to the construction of mathematical models of physical phenomena but even to the investigation of new universal laws of nature; consult e.g., A. R. Bednarek and F. Ulam (1990).

1.4   Hierarchy of Chaos

The past few years have witnessed an explosive growth in interest in physical, chemical, engineering and economic systems that could be studied using stochastic and chaotic methods, see Berliner (1992) and Chatterjee and Yilmaz (1992). "Stochastic" and "chaotic" refer to nature's two paths to unpredictability, or uncertainty. To scientists and engineers the surprise was that chaos (making a very small change in the universe can lead to a very large change at some later time) is unrelated to randomness. Things are unpredictable if you look at the individual events; however, one can say a lot about averaged–out quantities. This is where the stochastic stuff comes in.

Our aim in the second part of the monograph is the theoretical investigation and computer illustration of the *hierarchy of chaos* for stochastic processes (Gaussian and non–Gaussian stable) with applications to stochastic modeling.

The ergodic theory is one of the very few parts of mathematics that has undergone substantial changes in recent decades. Previously, the ergodic theory had been solving rather general and qualitative problems but now it has become a powerful tool for studying statistical and chaotic properties of dynamical systems. This, in turn, makes ergodic theory quite interesting not only for mathematicians but also for physicists, biologists, chemists and many others.

Indeed, the ergodic theory of stochastic processes is a part of general ergodic theory that is now intensively being developed. Until recently, characterizations of ergodic properties were known only for Markov or Gaussian processes. However, due to modern results on representations of different classes of stochastic processes in terms of stochastic integrals with respect to stochastic measures, it is now possible to characterize ergodic properties for a much wider spectrum of stochastic processes and establish the whole hierarchy of chaos (ergodicity, weak mixing, p–mixing, exactness and K–property). A crucial role in our theoretical study is played by the dynamical functional, which describes dynamical properties of stochastic processes and facilitates the investigation of ergodic properties of stochastic processes having a spectral representation.

One of our ideas is to study and characterize ergodic properties of different classes of stochastic processes by means of purely theoretical as well as computer methods. We attempt to demonstrate usefulness of constructive numerical methods of approximation and computer simulations of such processes in investigation of their ergodic properties, providing some new quantitative information on their dynamical properties.

1.5   Computer Simulations and Visualizations

Using our own software packages of computer programs written in languages Turbo C or Turbo Pascal for IBM PC, we want to show that proper use of computer graphics can provide useful and sometimes surprisingly interesting information enabling us to better understand phenomena that are of complicated, chaotic or stochastic nature.

To our knowledge, up to now the numerical analysis of stochastic differential systems driven by Brownian motion has essentially focused on such problems as mean–square approximation, pathwise approximation or approximation of expectations of the solution, etc. (see, e.g., Pardoux and Talay (1985), Talay (1983) and (1986), Yamada (1976)). There are some results on convergence of approximate solutions. The results on the rate of convergence of approximations of stochastic integrals driven by Brownian motion can be found in Rootzén (1980).

Our aim is to adapt some of these constructive computer techniques based on discretization of the time parameter t to the case of stochastic integrals and stochastic differential equations driven by stable Lévy motion. We describe some results on the convergence of approximate numerical solutions. There are still many open questions concerning this problem.

The research in the theory of convergence of constructive approximations of stochastic integrals and stochastic differential equations driven by infinitely divisible measures or by semimartingales is in progress. There is a growing literature on the *stability* of stochastic integrals and stochastic differential equations with jumps (we refer the reader to Kasahara, Yamada (1991), Jakubowski, Mémin and Pages (1989), Słomiński (1989), Kurtz and Protter (1991) and (1992)). We have to rely on some results in this more general setting, though they are not so easily applicable in practice.

Our idea is to represent the discrete time processes approximating stochastic processes with continuous time parameter t by appropriately constructed finite sets of random samples in order to obtain kernel estimators of densities of these processes on finite sets of values of t and to get some useful quantitative information on their behavior.

We recommend the following monographs: Bratley, Fox and Schrage (1987), Devroye (1986) and (1987), Newton (1988), Rice (1990).

1.6   Stochastic Processes

We restricted ourselves precisely to problems strictly connected with the theory of *α–stable stochastic processes* and with computer techniques of their approximate construction and simulation. We tried to make the book as much self–contained as possible. Our main goal was to convince the reader that our constructive approach provides powerful tools for modeling and solving approximately a wide variety of problems from the physical, biological and social sciences, so we provide detailed descriptions of necessary algorithms and a lot of examples of graphical presentation and visualization of stochastic processes of different kinds. Concentrating our attention on computer simulations and ergodic properties of α–stable processes, we tried to make our exposition as simple as possible, but the interested reader, on the basis of the very quickly growing literature concerning modern stochastic integration and its various applications, has an opportunity to go further in the investigation of the problems which are discussed here.

We assume that the reader is familiar with some of the well known text-books on probability theory, such as Breiman (1968), Feller (1966) and (1971), Shiryaev (1984) or many others. As the main source of information on α–stable processes we consider the book by Samorodnitsky and Taqqu (1993) (see also a survey paper of Weron (1984)). Basic facts concerning some aspects of stochastic analysis and especially the theory of stochastic integrals and stochastic differential equations of different types can be found, for example, in Arnold (1974), Elliot (1982), Ikeda and Watanabe (1981), Jacod (1979), Kallianpur (1980), Kwapień and Woyczyński (1992), Liptser and Shiryaev (1977) and (1978), McKean (1969), Métivier (1982), Protter (1990) or Revuz and Yor (1991).

Chapter 2

Brownian Motion, Poisson Process, α–Stable Lévy Motion

2.1 Introduction

In this chapter we recall briefly the main, most important properties of the Brownian motion and Poisson processes, list some properties of α–stable random variables, and introduce an α–stable Lévy motion, placing it somewhere in between Brownian motion and Poisson processes, among a vast class of infinitely divisible processes. These three classes of "elementary" processes will serve as principal tools for constructing stochastic measures, allowing us to describe vast classes of stochastic processes with the use of stochastic integrals of different kinds.

We emphasize the importance of constructive methods of description of stochastic processes. We believe that a few graphs of trajectories of the Brownian motion, the Poisson process and the α–stable Lévy motion will convince the reader that computer simulation methods are quite powerful and provide some useful information on behavior of stochastic processes. We apply such methods in the investigation of more complicated problems presented in the next chapters of the book.

2.2 Brownian Motion

We find it interesting to start with the presentation of some methods of construction of Brownian motion processes and computer graphs of their trajectories, which these methods provide (see Figures 2.2.1 - 2.2.3). We present briefly three different methods: the first is based on the Lévy–Ciesielski Representation, the second (from our point of view the most important) is based on the summation of independent increments, and the third exploits random walk process. All of them provide "essentially the same result", thus from the point of view of practical computations and applications, justifying their correctness.

Taking as a given a complete probability space $(\Omega, \mathcal{F}, P)$ (together with a filtration $\{\mathcal{F}_t\}$), let us recall the definition of the standard, one–dimensional Brownian motion.

Definition 2.2.1 *The standard, one–dimensional Brownian motion is a process $\{B(t) : t \geq 0\}$ (or in full notation: $\{B(t, \omega) : t \geq 0, \ \omega \in \Omega\}$), satisfying the following conditions*

1. $P\{\omega : B(0, \omega) = 0\} = 1$;

2. $\{B(t, \omega)\}$ *has independent increments, i.e. for any sequence $0 = t_0 < t_1 < ... < t_n$, the random variables $B(t_j) - B(t_{j-1})$, $j = 1, 2, ..., n$, are independent (in other words, for any $0 \leq s < t$, the increment $B(t) - B(s)$ is independent of $\mathcal{F}_s$);*

3. *for any $0 \leq s < t$, the random variable $B(t) - B(s)$ is normally distributed with mean 0 and variance $t - s$, i.e.*

$$P\{a \leq B(t) - B(s) \leq b\} = \frac{1}{\sqrt{2\pi(t - s)}} \int_a^b e^{-\frac{x^2}{2(t-s)}} \, dx;$$

4. $P\{\omega : B(\cdot, \omega) \text{ is a continuous function}\} = 1$.

Notice that

$$E\left[\exp\left(i \sum_{j=1}^n a_j \left(B(t_j) - B(t_{j-1})\right)\right)\right] = \prod_{j=1}^n \exp\left(-\frac{1}{2} a_j^2 (t_j - t_{j-1})\right), \quad (2.2.1)$$

for any sequence $0 = t_0 < t_1 < ... < t_n < \infty$ and any real numbers a_j, $j = 1, 2, ..., n$, $(i = \sqrt{-1})$.

It can be derived easily from the definition that the marginal distribution of $B(t, \omega)$ for Borel sets A in $\mathbb{R}$ can be given by the formula

$$P\left\{(B(t_1), ..., B(t_n)) \in A\right\} = \int_A \left(\prod_{j=1}^n \frac{1}{\sqrt{2\pi(t_j - t_{j-1})}} e^{-\frac{(x_j - x_{j-1})^2}{2(t_j - t_{j-1})}}\right) \, dx_1 ... dx_n,$$

for any sequence $0 = t_0 < t_1 < ... < t_n$ and $x_0 = 0$.

In other words, if we define the Gaussian kernel

$$p_t(x, y) = \frac{1}{\sqrt{2\pi t}} e^{-\frac{(x-y)^2}{2t}}, \quad t > 0, \quad x, y \in \mathbb{R},$$

then the cumulative distribution function for $(B(t_1), B(t_2), ..., B(t_n))$ has the form

$$F_{(t_1,t_2,...,t_n)}(x_1, x_2, ..., x_n)$$

$$= \int_{-\infty}^{x_1} \int_{-\infty}^{x_2} ... \int_{-\infty}^{x_n} p_{t_1}(0, y_1) p_{t_2-t_1}(y_1, y_2)...p_{t_n-t_{n-1}}(y_{n-1}, y_n) \, dy_n...dy_2 \, dy_1,$$

for $(x_1, x_2, ..., x_n) \in \mathbb{R}^n$.

A stochastic process is very often thought of as a consistent family of marginal distributions. Thus we can regard a Brownian motion as specified by the above marginal distributions. Of course, by the Kolmogorov Extension Theorem, such a family of marginal distributions specifies a stochastic process which has a version $\{B(t, \omega) : t \geq 0\}$ as above.

Transition probabilities. Let us also recall that transition probabilities are defined by

$$p_t(x, A) = \frac{1}{\sqrt{2\pi t}} \int_A e^{-\frac{(u-x)^2}{2t}} \, du$$

and satisfy the Chapman–Kolmogorov Equation, i.e.

$$p_{t+s}(x, A) = \int_{\mathbb{R}} p_t(x, dy) \, p_s(y, A).$$

Semigroup representation. The semigroup representation has the form

$$(P_t f)(x) = \int_{\mathbb{R}} f(y) \, p_t(x, dy), \quad \text{for } t > 0, \text{ with } P_0 = I.$$

Here $\{P_t : t \geq 0\}$ denotes a strongly continuous contraction semigroup on the Banach space of bounded uniformly continuous functions on $\mathbb{R}$. The infinitesimal generator of $\{P_t : t \geq 0\}$ is given by

$$\lim_{t \downarrow 0} \frac{(P_t f)(x) - f(x)}{t} = \frac{1}{2} f''(x).$$

Now we would like to present a few possible methods of constructing Brownian motion processes.

Interpolation technique of approximate construction of Brownian motion on $[0, 1]$. Observe that with fixed $0 \leq t_1 < t_2 < \infty$, the random variable $B(s)$ for $s = (t_1 + t_2)/2$ is normal with mean $w = (z_1 + z_2)/2$ and variance $\sigma^2 = \tau/2$ with $\tau = (t_2 - t_1)/2$, under conditions: $z_1 = B(t_1)$, $z_2 = B(t_2)$. Indeed, using the formulas

$$P[B(t_1) \in A_x, \ B(s) \in A_y, B(t_2) \in A_z] = \int_{A_x} \int_{A_y} \int_{A_z} p_{t_1}(0, x) p_\tau(x, y) p_\tau(y, z) \, dx \, dy \, dz$$

$$= \int_{A_x} \int_{A_y} \int_{A_z} p_{t_1}(0, x) p_{t_2-t_1}(x, z) \left(\frac{1}{\sigma\sqrt{2\pi}} \exp \left[-\frac{(y - w)^2}{2\sigma^2} \right] \right) \, dx \, dy \, dz$$

describing the joint density of $B(t_1)$, $B(s) - B(t_1)$, $B(t_2) - B(s)$, and

$$P[B(t_1) \in A_x, \ B(t_2) \in A_z] = \int_{A_x} \int_{A_z} p_{t_1}(0, x) p_{t_2 - t_1}(x, z) \, dx dz,$$

we obtain

$$P[B((t_1 + t_2)/2) \in A_y | \ B(t_1) = z_1, \ B(t_2) = z_2] = \frac{1}{\sigma \sqrt{2\pi}} \int_{A_y} e^{-(y - w)^2/(2\sigma^2)} \, dy.$$

This suggests that we can construct the Brownian motion on some finite interval of t, say, on the interval $[0, 1]$.

Assume that the probability space $(\Omega, \mathcal{F}, P)$ is rich enough to carry a countable collection $\{\xi_k : k = 0, 1, ...\}$ of independent, standard normal random variables (i.e. $\xi_k \sim \mathcal{N}(0, 1)$ for $k = 0, 1, ...$). Using linear interpolation and a recursion formula we define the sequence of processes $B^{(n)} = \{B^{(n)}(t) : t \in [0, 1]\}$, as follows.

Starting with the processes

$$B^{(0)}(t) = t\xi_0, \quad t \in [0, 1],$$

we can assume that for any fixed $n \geq 1$ we are given the real–valued stochastic processes $\{B^{(n-1)}(t) : t \in [0, 1]\}$ which is piecewise–linear on subintervals $[k/2^{n-1}, (k + 1)/2^{n-1}]$ of the interval $[0, 1]$ for $k = 0, 1, ..., 2^{n-1} - 1$.

In order to construct the process $\{B^{(n)}(t) : t \in [0, 1]\}$ we put

$$B^{(n)}(k/2^{n-1}) = B^{(n-1)}(k/2^{n-1}), \quad k = 0, 1, ..., 2^{n-1},$$

and, for $k = 0, 1, ..., 2^{n-1} - 1$,

$$\begin{aligned} B^{(n)}((2k + 1)/2^n) &= \frac{1}{2} \{ B^{(n-1)}(k/2^{n-1}) + B^{(n-1)}((k + 1)/2^{n-1}) \} \\ &+ \ 2^{-(n+1)/2} \ \xi_{2^{n-1} + k}. \end{aligned}$$

Linear interpolation on subintervals $[k/2^n, (k + 1)/2^n]$ of $[0, 1]$ completes the construction of $B^{(n)}$ on $[0, 1]$.

Lévy–Ciesielski construction.

The algorithm exploiting the Gaussian property of the Brownian motion described above is derived from the Lévy–Ciesielski series representation of this process. In order to obtain the theorem on the convergence of the sequence $\{B^{(0)}(t), \ B^{(1)}(t), ...\}$ to the Brownian motion on $[0, 1]$, we recall this theoretical construction giving another description of these processes. We define the system $\{H_k^{(n)}\}$ of *Haar functions*, putting

$$H_0^{(0)}(t) = 1 \qquad \text{for } t \in [0, 1];$$

$$H_k^{(n)}(t) = \begin{cases} 2^{(n-1)/2}, & \text{if } t \in [2k/2^n, (2k + 1)/2^n), \\ -2^{(n-1)/2}, & \text{if } t \in [(2k + 1)/2^n, (2k + 2)/2^n), \\ 0, & \text{elsewhere in } [0, 1], \end{cases}$$

for $n = 1, 2, ...,$ $k = 0, 1, ..., 2^{n-1} - 1$.

We define also the *Schauder functions* by

$$S_k^{(n)}(t) = \int_0^t H_k^{(n)}(u)\, du, \qquad \text{for } t \in [0,1],$$

where $n = 0, 1, ..., \ k = 0, 1, ..., 2^{n-1} - 1$.

It is clear, by induction with respect to n, that

$$B^{(n)}(t,\omega) = \sum_{m=0}^{n} \sum_{k=0}^{2^{n-1}-1} \xi_{2^{n-1}+k}(\omega) S_k^{(m)}(t), \qquad \text{for } t \in [0,1], \tag{2.2.2}$$

for $n = 0, 1, ...$.

Theorem 2.2.1 *As $n \to \infty$, the sequence $\{B^{(n)}(t,\omega)\}_{n=0}^{\infty}$ given by (2.2.2) converges uniformly with respect to t on $[0,1]$ to a continuous function $B(\cdot,\omega)$ for a.a. $\omega \in \Omega$ and the process $\{B(t,\omega) : t \in [0,1], \ \omega \in \Omega\}$ is the Brownian motion on $[0,1]$.*

PROOF. Define $b_n = \max_{0 \leq k \leq 2^{n-1}-1} |\xi_{2^{n-1}+k}|$. For $x > 0$ we have

$$
\begin{aligned}
P[|\xi_{2^{n-1}+k}| > x] &= P[|\xi_0| > x] = \sqrt{\frac{2}{\pi}} \int_x^{\infty} e^{-u^2/2}\, du \\
&\leq \sqrt{\frac{2}{\pi}} \int_x^{\infty} \frac{u}{x} e^{-u^2/2}\, du \\
&= \sqrt{\frac{2}{\pi}} \frac{e^{-x^2/2}}{x},
\end{aligned}
$$

which, for $n \geq 1$, gives

$$P[b_n > n] = P\left[\bigcup_{k=0}^{2^{n-1}-1} \{|\xi_{2^{n-1}+k}| > n\} \right] \leq 2^{n-1} P[|\xi_0| > n] \leq \sqrt{\frac{2}{\pi}} \frac{2^{n-1} e^{-n^2/2}}{n}.$$

Observe that $\sum_{n=1}^{\infty} 2^{n-1} e^{-n^2/2}/n < \infty$, thus the Borel–Cantelli lemma implies that there is a subset $\tilde{\Omega}$ of Ω such that $P(\tilde{\Omega}) = 1$ and for each $\omega \in \tilde{\Omega}$ there is an integer $n(\omega)$ satisfying $b_{n(\omega)} \leq n$ for all $n \geq n(\omega)$. Then

$$\sum_{n=n(\omega)}^{\infty} \sum_{k=0}^{2^{n-1}-1} |\xi_{2^{n-1}+k} S_k^{(n)}(t)| \leq \sum_{n=n(\omega)}^{\infty} n 2^{-(n+1)/2} < \infty,$$

so for $\omega \in \tilde{\Omega}$ the sequence $\{B^{(n)}(t,\omega)\}$ converges uniformly in t to a limit $B(t,\omega)$. The continuity of $\{B(t,\omega) : t \in [0,1]\}$ follows from the uniformity of the convergence.

The Haar system $\left\{ H_k^{(n)} \right\}$ forms a complete orthonormal basis in the Hilbert space $L^2[0,1]$, so applying the Parseval Equality to functions $f(u) = I_{[0,t]}(u)$ and $g(u) = I_{[0,s]}(u)$ gives

$$\sum_{n=0}^{\infty} \sum_{k=0}^{2^{n-1}-1} S_k^{(n)}(t)\, S_k^{(n)}(s) = s \wedge t, \qquad \text{for } s, \ t \in [0,1]. \tag{2.2.3}$$

In order to prove that the process $\{B(t, \omega) : t \in [0,1]\}$ is the Brownian motion process, i.e. that the increments $\{B(t_j) - B(t_{j-1})\}$, for $0 = t_0 < t_1 < ... < t_n \leq 1$, are independent, normally distributed, with mean zero and variance $t_j - t_{j-1}$, it suffices to prove (2.2.1).

Using the independence and standard normality of random variables $\{\xi_j\}$, we have from (2.2.2)

$$
\begin{aligned}
& E\left[\exp\left\{-i \sum_{j=1}^{n} (a_{j+1} - a_j) B^{(M)}(t_j)\right\}\right] \\
&= E\left[\exp\left\{-i \sum_{m=0}^{M} \sum_{k=0}^{2^{m-1}-1} \xi_{2^{m-1}+k} \sum_{j=1}^{n} (a_{j+1} - a_j) S_k^{(m)}(t_j)\right\}\right] \\
&= \prod_{m=0}^{M} \prod_{k=0}^{2^{m-1}-1} E\left[\exp\left\{-i \xi_{2^{m-1}+k} \sum_{j=1}^{n} (a_{j+1} - a_j) S_k^{(m)}(t_j)\right\}\right] \\
&= \prod_{m=0}^{M} \prod_{k=0}^{2^{m-1}-1} \exp\left[-\frac{1}{2}\left\{\sum_{j=1}^{n} (a_{j+1} - a_j) S_k^{(m)}(t_j)\right\}^2\right] \\
&= \exp\left[-\frac{1}{2} \sum_{j=1}^{n} \sum_{i=1}^{n} (a_{j+1} - a_j)(a_{i+1} - a_i) \sum_{m=0}^{M} \sum_{k=0}^{2^{m-1}-1} S_k^{(m)}(t_j) S_k^{(m)}(t_i)\right],
\end{aligned}
$$

for any $0 = t_0 < t_1 < ... < t_n \leq 1$ and any real parameters a_j, $j = 1, 2, ..., n+1$ with $a_{n+1} = 0$. Letting $M \to \infty$ and using (2.2.3), we obtain

$$
\begin{aligned}
& E\left[\exp\left\{i \sum_{j=1}^{n} a_j (B(t_j) - B(t_{j-1}))\right\}\right] \\
&= E\left[\exp\left\{-i \sum_{j=1}^{n} (a_{j+1} - a_j) B(t_j)\right\}\right] \\
&= \exp\left\{-\sum_{j=1}^{n-1} \sum_{i=j+1}^{n} (a_{j+1} - a_j)(a_{i+1} - a_i) t_j - \frac{1}{2} \sum_{j=1}^{n} (a_{j+1} - a_j)^2 t_j\right\} \\
&= \exp\left\{-\sum_{j=1}^{n-1} (a_{j+1} - a_j)(-a_{j+1}) t_j - \frac{1}{2} \sum_{j=1}^{n} (a_{j+1} - a_j)^2 t_j\right\} \\
&= \exp\left\{\frac{1}{2} \sum_{j=1}^{n-1} (a_{j+1}^2 - a_j^2) t_j - \frac{1}{2} a_n^2 t_n\right\} \\
&= \prod_{j=1}^{n} \exp\left\{-\frac{1}{2} \sum_{j=1}^{n-1} a_j^2 (t_j - t_{j-1})\right\}.
\end{aligned}
$$

This ends the proof.   □

The figure below (Fig. 2.2.1) contains the result of computer realization of this method.

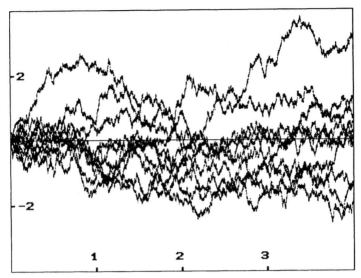

Figure 2.2.1. Trajectories of Brownian motion (Lévy–Ciesielski representation).

The technique used to produce all figures in this chapter will be elaborated and applied later on several times to visualize various types of stochastic processes and will be explained there in detail.

Construction of Brownian motion on $[0, \infty)$.

The construction of the Brownian motion on $[0, 1]$ described above and a simple patching–together technique provide a Brownian motion process defined on $[0, \infty)$.

Starting with a sequence $(\Omega_n, \mathcal{F}_n, P_n)$, $n = 1, 2, ...$, of probability spaces together with the Brownian motion $\{X^{(n)}(t) : t \in [0, 1]\}$ on each space, define $\Omega = \Omega_1 \times \Omega_2 \times ...$, $\mathcal{F} = \mathcal{F}_1 \otimes \mathcal{F}_2 \otimes ...$, $P = P_1 \times P_2 \times ...$, and finally

$$B(t, \omega) = \sum_{j=1}^{n-1} X^{(j)}(1, \omega) + X^{(n)}(t - n + 1, \omega), \quad n - 1 \leq t < n, \quad \text{with } n = 1, 2, ...,$$

The process $\{B(t) : t \in [0, \infty)\}$ is clearly continuous and has independent increments, which are symmetric Gaussian random variables with proper variances.

Approximation of Brownian motion by summation of increments.

In order to obtain an approximate construction of the Brownian motion on a given interval $[0, T]$ one can proceed as follows.

Introduce a mesh $\{t_i = i\tau : i = 0, 1, ..., I\}$ on $[0, T]$ with fixed natural number I and $\tau = T/I$. For a given finite sequence $\{\zeta_i\}$, $i = 1, 2, ..., I$, of independent

Gaussian variables with mean 0 and variance τ (i.e. $\zeta_i \sim \mathcal{N}(0,\tau)$), put

$$B^\tau(0) = 0 \quad a.e.,$$

for $i = 1, 2, ..., I$; compute

$$B^\tau(t) = B^\tau(t_{i-1}) + (t - t_{i-1})\zeta_i, \quad \text{for } t \in (t_{i-1}, t_i].$$

The process $\{B^\tau(t) : t \in [0,T]\}$ converges to the Brownian motion process on $[0,T]$, when $\tau \to 0$.

The result of application of this method is presented in Fig. 2.2.2 below, where $T = 4$ and $\tau = 0.002$ ($I = 2000$).

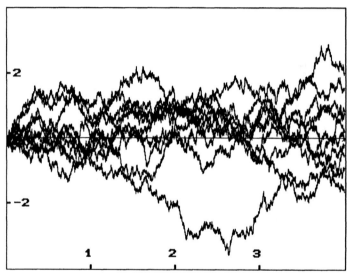

Figure 2.2.2. Trajectories of Brownian motion (summation of increments).

Random walk approximating Brownian motion.

Now we are going to state and prove a theorem on convergence of constructions described above, but in a slightly more general framework.

Let us consider a sequence $\{\xi_j\}_{j=1}^\infty$ of independent, identically distributed random variables with mean 0 and variance 1 and define the sequence of partial sums

$$S_0 = 0; \qquad S_k = \sum_{j=1}^k \xi_j, \quad \text{for } k = 1, 2, ...$$

By applying a linear interpolation technique we can define the continuous time process

$$Y(t) = S_{[t]} + (t - [t])\xi_{[t]+1}, \qquad \text{for } t \in [0, \infty),$$

where $[t]$ denotes the greatest integer not greater than t. Next, let us define the sequence of *random walk* processes

$$X^{(n)}(t) = \frac{1}{\sqrt{n}}Y(nt), \qquad \text{for } t \in [0, \infty). \qquad (2.2.4)$$

Note that $\{X^{(n)}(t)\}$ can be interpreted as a good approximation of the Brownian motion because, with $s = k/n$ and $t = (k+1)/n$, increments $X^{(n)}(t) - X^{(n)}(s) = (1/\sqrt{n})\xi_{k+1}$ are independent of $\mathcal{F}^{X^{(n)}} = \sigma(\xi_1, \xi_2, ..., \xi_k)$ with zero mean and variance $t - s$.

An example of realization of this method is visualized on Fig. 2.2.3 below. Random variables ξ_i are chosen to be independent discrete random variables with masses concentrated on points $0, \sqrt{3}, -\sqrt{3}$ with probabilities $\frac{2}{3}, \frac{1}{6}, \frac{1}{6}$, respectively. (Notice that five first moments of ξ_i are the same as those of the standard normal random variable.)

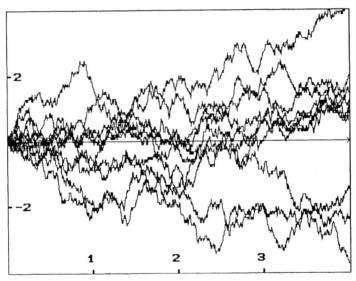

Figure 2.2.3. Trajectories of a random walk process approximating Brownian motion.

The result on convergence of this approximate construction is proved in the following theorem.

Theorem 2.2.2 *With $\{X^{(n)}(t) : t \geq 0\}$ defined by (2.2.4) and for any $0 \leq t_1 < t_2 < ... < t_d < \infty$, we have*

$$\left(X^{(n)}(t_1), X^{(n)}(t_2), ..., X^{(n)}(t_d)\right) \xrightarrow{D} \left(B(t_1), B(t_2), ..., B(t_d)\right), \qquad \text{as } n \to \infty,$$

where $\{B(t) : t \geq 0\}$ is the standard, one–dimensional Brownian motion.

PROOF. For the sake of simplicity we take the case $d = 2$. Set $s = t_1$, $t = t_2$. We have to show that

$$\left(X^{(n)}(s), X^{(n)}(t)\right) \xrightarrow{\mathcal{D}} (B(s), B(t)).$$

Since

$$\left|X^{(n)}(t) - \frac{1}{\sqrt{n}}S_{[tn]}\right| \le \frac{1}{\sqrt{n}}\left|\xi_{[tn]+1}\right|,$$

so, by the Čebyšev Inequality, we have

$$P\left[\left|X^{(n)}(t) - \frac{1}{\sqrt{n}}S_{[tn]}\right| > \epsilon\right] \le \frac{1}{\epsilon^2 n} \to 0,$$

as $n \to \infty$. It is clear then that

$$\left\|\left(X^{(n)}(s), X^{(n)}(t)\right) - \frac{1}{\sqrt{n}}\left(S_{[sn]}, S_{[tn]}\right)\right\| \to 0 \qquad \text{in probability,}$$

so it suffices to show that

$$\frac{1}{\sqrt{n}}\left(S_{[sn]}, S_{[tn]}\right) \xrightarrow{\mathcal{D}} (B(s), B(t)),$$

which is equivalent to

$$\frac{1}{\sqrt{n}}\left(\sum_{j=1}^{[sn]}\xi_j, \sum_{j=[sn]+1}^{[tn]}\xi_j\right) \xrightarrow{\mathcal{D}} (B(s), B(t) - B(s)).$$

The independence of the random variables $\{\xi_j\}_{j=1}^{\infty}$ implies

$$\lim_{n\to\infty} E\left[\exp\left\{\frac{iu}{\sqrt{n}}\sum_{j=1}^{[sn]}\xi_j + \frac{iv}{\sqrt{n}}\sum_{j=[sn]+1}^{[tn]}\xi_j\right\}\right]$$

$$= \lim_{n\to\infty} E\left[\exp\left\{\frac{iu}{\sqrt{n}}\sum_{j=1}^{[sn]}\xi_j\right\}\right] \cdot \lim_{n\to\infty} E\left[\exp\left\{\frac{iv}{\sqrt{n}}\sum_{j=[sn]+1}^{[tn]}\xi_j\right\}\right], \quad (2.2.5)$$

provided both limits on the right–hand side exist. We have

$$\lim_{n\to\infty} E\left[\exp\left\{\frac{iu}{\sqrt{n}}\sum_{j=1}^{[sn]}\xi_j\right\}\right] = e^{-u^2 s/2},$$

because

$$\left|\frac{1}{\sqrt{n}}\sum_{j=1}^{[sn]}\xi_j - \frac{\sqrt{s}}{\sqrt{[sn]}}\sum_{j=1}^{[sn]}\xi_j\right| \to 0, \qquad \text{in probability}$$

and, thanks to the Central Limit Theorem, the sequence $\sqrt{s/[sn]}\sum_{j=1}^{[sn]}\xi_j$ converges in distribution to the normal random variable with mean 0 and variance s. Similarly,

$$\lim_{n\to\infty} E\left[\exp\left\{\frac{iv}{\sqrt{n}}\sum_{j=[sn]+1}^{[tn]}\xi_j\right\}\right] = e^{-v^2(t-s)/2}.$$

Substitution in (2.2.5) completes the proof. □

Canonical probability space for Brownian motion.
Let us denote by $C[0,\infty)$ the space of all real–valued continuous functions $\omega = \omega(t)$ defined on $[0,\infty)$ with the metric

$$\rho(\omega_1,\omega_2) \overset{df}{=} \sum_{n=1}^{\infty}\frac{1}{2^n}\max_{0\le t\le n}\left(|\omega_1(t)-\omega_2(t)| \wedge 1\right).$$

Let $\mathcal{C}$ be the collection of finite–dimensional cylinder sets of the form

$$C = \{\omega \in C[0,\infty);\ (\omega(t_1),...,\omega(t_n)) \in D\},$$

for any natural $n \ge 1$, $0 < t_1 < ... < t_n < \infty$ and any $D \in \mathcal{B}(\mathbb{R}^n)$.

The smallest σ–field containing $\mathcal{C}$ is equal to the Borel σ–field generated by all open sets in $C[0,\infty)$. We denote it by $\mathcal{B}(C[0,\infty))$.

If μ, acting from $\mathcal{C}$ into $[0,1]$, is defined by

$$\mu(C) = \int_D \left(\prod_{j=1}^{n}\frac{1}{\sqrt{2\pi(t_j-t_{j-1})}}e^{-\frac{(x_j-x_{j-1})^2}{2(t_j-t_{j-1})}}\right)\,dx_1...dx_n,$$

for $C \in \mathcal{C}$ (with $t_0 = 0$ and $x_0 = 0$), then μ extends to a probability measure on the space $(C[0,\infty),\mathcal{B}(C[0,\infty))$. The measure μ is called the *Wiener measure* and the probability space $(C[0,\infty),\mathcal{B}(C([0,\infty)),\mu)$ is known as the *Wiener space*. Then $B(t,\omega) = \omega(t)$ is a Brownian motion and this probability space is called the *canonical probability space* for Brownian motion.

We want to end this section by recalling the theorem, which asserts convergence of random walks defined by (2.2.4).

Theorem 2.2.3 (The Invariance Principle). *Let $(\Omega,\mathcal{F},P)$ be a probability space on which is given a sequence $\{\xi_j\}_{j=1}^{\infty}$ of independent, identically distributed random variables with mean 0 and variance 1. Define the sequence of processes $X^{(n)} = \{X^{(n)}(t) : t \in [0,\infty)\}$ by (2.2.4). Let P_n be the measure induced by $X^{(n)}$ on $(C[0,\infty),\mathcal{B}(C[0,\infty)))$. Then the sequence of measures $\{P^{(n)}\}_{n=1}^{\infty}$ converges weakly to a measure P, under which the coordinate mapping process $W(t,\omega) \overset{df}{=} \omega(t)$ on $C[0,\infty)$ is the standard one–dimensional Brownian motion.*

For the proof we refer the interested reader to Karatzas and Shreve (1988).

Remark 2.2.1 *The standard one–dimensional Brownian motion defined on any probability space can be thought of as a random variable with values in $C[0,\infty)$. Regarded this way, the Brownian motion induces the Wiener measure P on the measure space $(C[0,\infty), \mathcal{B}(C[0,\infty)))$. This explains why the probability space of the form $(C[0,\infty), \mathcal{B}(C[0,\infty)), P)$ is called the canonical probability space for the Brownian motion.*

This characterization of the Brownian motion and more sophisticated constructions describing processes with jumps (and further – semimartingales) play an important role in the modern approach to the theory of stochastic processes, but are not so crucial in approximate constructions of stochastic processes, applicable in computer simulations.

Multidimensional Brownian motion.

In order to construct d–*dimensional Brownian motion* it is enough to take d independent copies of standard, one–dimensional Brownian motion $B^{(i)} = \{B^{(i)}(t) : t \in [0,\infty)\}$, $i = 1,...,d$, defined on probability spaces $(\Omega^{(i)}, \mathcal{F}^{(i)}, P^{(i)})$ (with filtrations $\mathcal{F}_t^{(i)}$) and on the product space

$$\left(\Omega^{(1)} \times ... \times \Omega^{(d)}, \mathcal{F}^{(1)} \otimes ... \otimes \mathcal{F}^{(d)}, P^{(1)} \times ... \times P^{(d)}\right)$$

to define

$$B(t,\omega) = \left(B^{(1)}(t,\omega_1), ..., B^{(d)}(t,\omega_d)\right),$$

with the filtration $\mathcal{F}_t = \mathcal{F}_t^B$.

2.3   The Poisson Process

Along with Brownian motion the Poisson process plays a fundamental role in the theory of continuous–time stochastic processes, so taking as given a complete probability space $(\Omega, \mathcal{F}, P)$ (together with a filtration $\{\mathcal{F}_t\}$), let us recall the definition of this process.

Suppose we are given a strictly increasing sequence $\{T_n\}_{n \geq 0}$ of positive random variables with $T_0 = 0$ a.s. and $\sup_n T_n = \infty$ a.s.

Definition 2.3.1 *The process* $\mathbf{N} = \{N(t) : t \geq 0\}$ *defined by*

$$N(t) = \sum_{n=1}^{\infty} I_{[T_n, \infty)}(t),$$

with values in $\mathbb{N} \cup \{0\}$ *is called* ***the counting process*** *(without explosion and associated to the sequence* $\{T_n\}_{n \geq 1}$*).*

$(I_A = I_A(x)$ denotes the characteristic function of the set A.)
It is not difficult to notice that

$$[T_n, \infty) = \{\mathbf{N} \geq n\} = \{(t,\omega) : N(t,\omega) \geq n\},$$

$$[T_n, T_{n+1}) = \{\mathbf{N} = n\}.$$

Note that for $0 \le s < t < \infty$ we have

$$N(t) - N(s) = \sum_{n=1}^{\infty} I_{[T_n, \infty)}(t) \cdot I_{(0, T_n)}(s).$$

The increment $N(t) - N(s)$ counts the number of random times T_n that occur between the fixed times s and t.

Remark 2.3.1 *In order to have this process adapted to the filtration $\{\mathcal{F}_t\}$ it is enough to assume that T_n are stopping times.*

Definition 2.3.2 *An adapted counting process $\mathbf{N}$ without explosion is a Poisson process if*

1. $P\{\omega : N(0, \omega) = 0\} = 1;$

2. $\{N(t, \omega)\}$ *has independent increments, i.e. for any sequence $0 = t_0 < t_1 < ... < t_n$, the random variables $N(t_j) - N(t_{j-1})$, $j = 1, 2, ..., n$, are independent (in other words, for any $0 \le s < t$, the increment $N(t) - N(s)$ is independent of $\mathcal{F}_s$);*

3. $\{N(t, \omega)\}$ *has stationary increments, i.e. for any $0 \le s < t < \infty$ and $0 \le u < v < \infty$ random variables $N(t) - N(s)$ and $N(v) - N(u)$ have the same distribution, whenever $t - s = v - u$.*

Theorem 2.3.1 *Let $\mathbf{N}$ be a Poisson Process. Then*

$$P(N(t) = k) = \frac{e^{-\lambda t}(\lambda t)^k}{k!}; \qquad k = 0, 1, ...,$$

for some $\lambda > 0$, which means that $N(t)$ has the Poisson distribution with parameter λt.

For the proof we refer the reader to Protter (1990).

In two figures below (Fig.2.3.1 and Fig.2.3.2) we present (exact, not approximated!) trajectories of the Poisson process with the compensator $\{N(t) - \lambda t\}_{t \ge 0}$, for two values of λ : 1.0 and 0.1, respectively. The random variables T_n in the sequence $\{T_n\}$ have been chosen as *arrival times*, i.e. T_n is a sum of n independent, exponentially distributed random variables with the distribution function $F(x) = (1 - e^{-\lambda x})I_{(0, \infty)}(x)).$

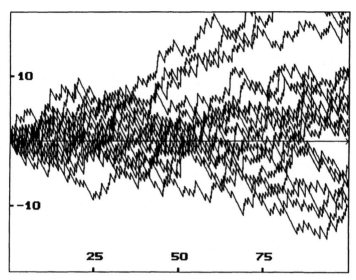

Figure 2.3.1. Trajectories of the Poisson process with compensator for $\lambda = 1.0$.

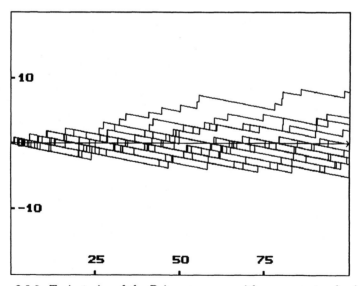

Figure 2.3.2. Trajectories of the Poisson process with compensator for $\lambda = 0.1$.

2.4 α–Stable Random Variables

As the title of this monograph indicates, we focus our sttention on the class of α–stable (or stable) random variables. From the literature on this topic let us mention among others: Feller (1966) and (1971), Lévy (1937) and (1948), Zolotarev (1986) or Samorodnitsky and Taqqu (1994).

Characteristic function. The most common and convenient way to introduce α–stable random variable is to define its characteristic function.

Definition 2.4.1 *The characteristic function of an α–stable random variable involves four parameters: α - index of stability, β - skewness parameter, σ - scale parameter and μ - shift. This function is given by*

$$\log \phi\left(\theta\right) = -\sigma^{\alpha}|\theta|^{\alpha}\left\{1 - i\beta\mathrm{sgn}\left(\theta\right)\tan\left(\alpha\pi/2\right)\right\} + i\mu\theta,$$

when $\alpha \in (0,1) \cup (1,2]$, $\beta \in [-1,1]$, $\sigma \in \mathbb{R}_+$, $\mu \in \mathbb{R}$; *and by*

$$\log \phi\left(\theta\right) = -\sigma|\theta| + i\mu\theta,$$

when $\alpha = 1$, *which gives a very well–known symmetric Cauchy distribution.*

It seems to us that this is the best way to describe the most important and suitable for practical use subfamily of α–stable laws, e.g. for construction of α–stable random measures and stochastic integrals.

For the random variable X distributed according to the rule described above we use the notation $X \sim S_\alpha(\sigma, \beta, \mu)$. When $\mu = \beta = 0$, i.e., X is a symmetric α–stable random variable, we will write $X \sim S\alpha S$. In order to shorten the notation we denote by S_α any random variable X such that $X \sim S_\alpha(1,0,0)$.

Domain of attraction of X**.** A random variable X has a stable distribution if and only if it has *a domain of attraction*, i.e., if there exist a sequence $Y_1, Y_2, ...$ of i.i.d. random variables and sequences $\{d_n\}$ and $\{a_n\}$ of positive real numbers such that

$$\frac{Y_1 + Y_2 + ... + Y_n}{d_n} + a_n \xrightarrow{\mathcal{D}} X.$$

According to Feller (1971), we have in general $d_n = n^{1/\alpha}h(n)$, where the function $h = h(x)$, $x \geq 0$ varies slowly at infinity; the sequence $\{Y_i\}$ is said to belong to the normal domain of attraction of X when $d_n = n^{1/\alpha}$. Observe that if $Y_i's$ are i.i.d. random variables with finite variance, then X is Gaussian and we obtain an ordinary version of the Central Limit Theorem.

Lévy measure. To justify what was said above one can recall the Lévy–Khintchine Representation Theorem (see Feller (1971)). Let us introduce the *Lévy measure*

$$d\nu_\alpha(x) = \frac{P}{x^{1+\alpha}}I_{(0,\infty)}(x)\,dx + \frac{Q}{|x|^{1+\alpha}}I_{(-\infty,0)}(x)\,dx$$

ive numbers P, Q, and a function

$$\psi(\theta, x) = e^{i\theta x} - 1 - \frac{i\theta x}{1 + x^2}.$$

Then for X we have the following representation

$$E \exp(i\theta X) = \begin{cases} exp\left\{ib\theta - c^2\theta^2\right\} & \text{if } \alpha = 2, \\ exp\left\{ib\theta + \int_{\mathbb{R}\backslash\{0\}} \psi(\theta, x)\, d\nu_\alpha(x)\right\} & \text{if } 0 < \alpha < 2, \end{cases}$$

where b i c are real numbers (looking for a more general form of the Lévy-Khintchine Formula, see Definition 4.4.2).

Some arithmetic properties. Now we are going to recall a few simple but important properties of random variables $S_\alpha(\sigma, \beta, \mu)$.

1. If we have $X_i \sim S_\alpha(\sigma_i, \beta_i, \mu_i)$ for $i = 1, 2$ and X_1, X_2 are independent random variables, then

$$X_1 + X_2 \sim S_\alpha(\sigma, \beta, \mu),$$

with

$$\sigma = (\sigma_1^\alpha + \sigma_2^\alpha)^{1/\alpha}, \qquad \beta = \frac{\beta_1\sigma_1^\alpha + \beta_2\sigma_2^\alpha}{\sigma_1^\alpha + \sigma_2^\alpha}, \qquad \mu = \mu_1 + \mu_2.$$

2. If we have $X_1, X_2 \sim S_\alpha(\sigma, \beta, \mu)$ and A, B are real positive constants and C is a real constant, then

$$AX_1 + BX_2 + C \sim S_\alpha\left(\sigma(A^\alpha + B^\alpha)^{1/\alpha}, \beta, \mu(A^\alpha + B^\alpha)^{1/\alpha} + C\right).$$

3. $X \sim S_\alpha(\sigma, \beta, \mu)$ is a symmetric random variable if and only if $\beta = 0$ and $\mu = 0$. It is symmetric about μ if and only if $\beta = 0$.

4. If we have $X \sim S_\alpha(\sigma, \beta, \mu)$ and $\alpha \in (0, 2)$ and $p \in (0, \alpha)$, then

$$E|X|^p < \infty$$

and if $p \in [\alpha, 2)$, then

$$E|X|^p = \infty.$$

5. If we have $X \sim S_\alpha(\sigma, 0, \mu)$ and $\alpha \in (1, 2]$, then

$$EX = \mu.$$

Covariation. Let (X_1, X_2) denote a jointly $S\alpha S$ random vector, where $\alpha \in (1, 2]$. Considering the $S\alpha S$ random variable $Y = \theta_1 X_1 + \theta_2 X_2$ for any real

θ_1, θ_2 we get

$$Y \sim S_\alpha(\sigma, 0, 0) \quad \text{with} \quad \sigma = \sigma(\theta_1, \theta_2).$$

Definition 2.4.2 *The* **covariation** $[X_1, X_2]_\alpha$ *of the jointly SαS random vector* (X_1, X_2) *is*

$$[X_1, X_2]_\alpha = \frac{1}{\alpha} \frac{\partial}{\partial \theta_1} \sigma^\alpha(\theta_1, \theta_2) \mid_{\theta_1 = 0, \ \theta_2 = 1}.$$

The covariation is designed to replace the covariance when $\alpha \in (1, 2)$. In the case of $\alpha = 2$ we have the following relation between these two expressions

$$[X_1, X_2]_2 = \frac{1}{2} \operatorname{Cov}(X_1, X_2).$$

Asymptotic behavior of tail probabilities. Using the Central–Limit–Theorem–type argument, one can prove that if $X \sim S_\alpha(\sigma, \beta, \mu)$ and $\alpha \in (0, 2)$, then

$$\lim_{y \to \infty} y^\alpha P\{X > y\} = C_\alpha \frac{1 + \beta}{2} \sigma^\alpha,$$

$$\lim_{y \to \infty} y^\alpha P\{X < -y\} = C_\alpha \frac{1 - \beta}{2} \sigma^\alpha,$$

where

$$C_\alpha = \left(\int_0^\infty x^{-\alpha} \sin(x) \, dx \right)^{-1}.$$

Density functions. When we start to work with α–stable distributions, the main problem is that except for a few values of four parameters describing the characteristic function, their density functions are not known explicitly. The most interesting exceptions are the following:
– the Gaussian distribution $S_2(\sigma, 0, \mu)$, whose density is

$$f(x) = \frac{1}{2\sigma\sqrt{\pi}} \exp\left\{ -(x - \mu)^2 / (4\sigma^2) \right\};$$

– the Cauchy distribution $S_1(\sigma, 0, \mu)$, whose density is

$$f(x) = \frac{2\sigma}{\pi((x - \mu)^2 + 4\sigma^2)};$$

– the Lévy distribution $S_{1/2}(\sigma, 1, \mu)$, whose density

$$f(x) = \left(\frac{\sigma}{2\pi} \right)^{1/2} (x - \mu)^{-3/2} \exp\left\{ -\frac{\sigma}{2(x - \mu)} \right\}$$

is concentrated on (μ, ∞), i.e., $f(x) = 0$ for $x \in (-\infty, \mu]$.

So, in order to obtain α–stable density functions, we have to take into account the definition describing characteristic functions of α–stable random variables and apply the Fourier transform, namely

$$f(x) = \frac{1}{2\pi} \int_{-\infty}^{\infty} e^{-itx} \phi(t) dt.$$

Using a numerical approximation of this formula we were able to construct such densities in the general case. The result of computer calculations is presented here in the form of a series of graphs of such densities for different values of parameters α, β, σ (the role of parameter μ is obvious), obtained with the use of IBM PC/386 graphics.

Figures 2.4.1 – 2.4.6 below demonstrate in a very precise way how α–stable density functions depend on parameters:

- α – the index of stability from the interval $(0, 2]$,

- β – the skewness parameter from the interval $[-1, 1]$,
  with the exception: $\alpha = 1 \Rightarrow \beta = 0$,

- σ – the scale parameter from $(0, \infty)$.

Figure 2.4.1 shows the dependence of densities on α and Figure 2.4.2 – the dependence on σ. Figures 2.4.3 and 2.4.4 demonstrate how densities fluctuate when parameter β goes throughout the set of different values from -1 to 1, for two different values of α, respectively. Each of these two figures contains one example of the so called *totally skewed* density (in Figure 2.4.3 we have the density of $X \sim S_{0.8}(1, -1, 0)$ and in Figure 2.4.4 – $X \sim S_{0.5}(1, 1, 0)$. Figures 2.4.5 and 2.4.6 show what is going on, when parameter α approaches 1 with fixed, different from 0 value of parameter β. This corresponds to the remark on the definition of the characteristic function of α–stable random variable (exceptional case of $\alpha = 1$) at the beginning of this section. We believe that it is an easy and instructive task to find out which curves correspond to which values of parameters in all figures presented here. Figures 2.4.1 – 2.4.5 preserve the same scaling of both axes; the next (Figure 2.4.6) has the vertical one changed.

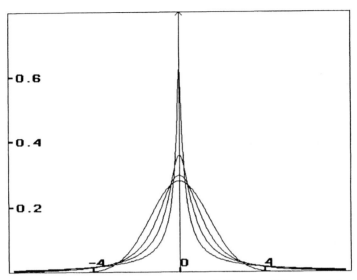

Figure 2.4.1. The case of $\alpha \in \{2.0,\ 1.2,\ 0.8,\ 0.5\}$, $\beta = 0.0$, $\sigma = 1.0$, $\mu = 0.0$.

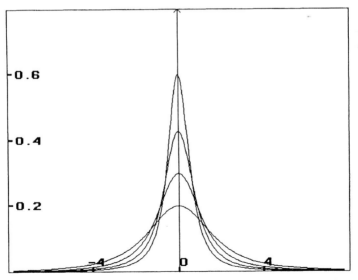

Figure 2.4.2. The case of $\alpha = 1.2$, $\beta = 0.0$, $\sigma \in \{0.5,\ 0.7,\ 1.0,\ 1.5\}$, $\mu = 0.0$.

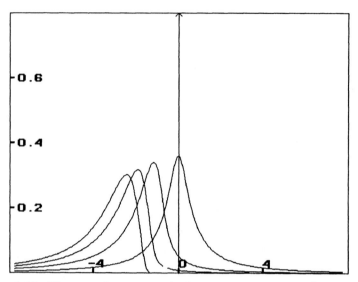

Figure 2.4.3. The case of $\alpha = 0.8$, $\beta \in \{-1.0, -0.8, -0.5, 0.0\}$, $\sigma = 1.0$ and $\mu = 0.0$.

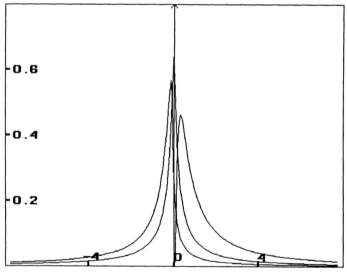

Figure 2.4.4. The case of $\alpha = 0.5$, $\beta \in \{-0.5, 0.0, 1.0\}$, $\sigma = 1.0$, $\mu = 0.0$.

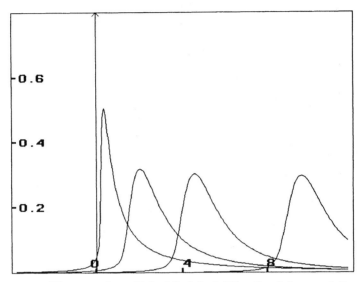

Figure 2.4.5. The case of $\alpha \in \{0.5, \ 0.8, \ 0.9, \ 0.95\}$, $\quad \beta = 0.8$, $\quad \sigma = 1.0$, $\quad \mu = 0.0$.

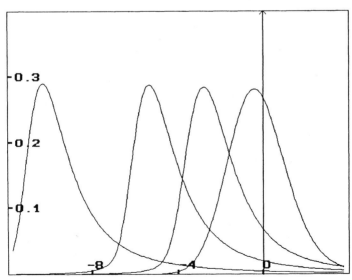

Figure 2.4.6. The case of $\alpha \in \{1.8, \ 1.2, \ 1.1, \ 1.05, \}$, $\quad \beta = 0.8$, $\quad \sigma = 1.0$, $\quad \mu = 0.0$.

2.5 α–Stable Lévy Motion

Roughly speaking, an α–stable process is a random element whose finite dimensional distribution is α–stable, where $0 < \alpha \leq 2$.

Definition 2.5.1 *A stochastic process $\{X(t) : t \in \mathbb{T}\}$, where $\mathbb{T}$ is an arbitrary set, is **stable** if all its finite dimensional distributions*

$$X(t_1), X(t_2), ..., X(t_n), \quad t_1, t_2, ..., t_n \in \mathbb{T}, \quad n \geq 1$$

*are stable. It is **symmetric stable** if all its finite–dimensional distributions are symmetric stable.*

If the finite dimensional distributions are stable or symmetric stable, then by consistency, they must all have the same index of stability α. It is known that $\{X(t) : t \in \mathbb{T}\}$ is symmetric stable if and only if all linear combinations

$$\sum_{k=1}^{n} a_k \, X(t_k), \quad n \geq 1, \quad t_1, t_2, ..., t_n \in \mathbb{T}, \quad a_1, a_2, ..., a_n \in \mathbb{R},$$

are symmetric stable. For non-symmetric stable processes this property holds only when $\alpha > 1$.

The best known example of an α–stable process is the α–stable Lévy motion. Let us recall the definition.

Definition 2.5.2 *A stochastic process $\{X(t) : t \geq 0\}$ is called the **(standard) α–stable Lévy motion** if*

1. $X(0) = 0$ a.s.;

2. $\{X(t) : t \geq 0\}$ has independent increments;

3. $X(t) - X(s) \sim S_\alpha((t-s)^{1/\alpha}, \beta, 0)$ for any $0 \leq s < t < \infty$.

Observe that the α–stable Lévy motion has stationary increments. It is the Brownian motion, when $\alpha = 2$. The α–stable Lévy motions are $S\alpha S$ when $\beta = 0$ and they are $1/\alpha$ self–similar. That is, for all $c > 0$ the processes $\{X(ct) : t \geq 0\}$ and $\{c^{1/\alpha} X(t) : t \geq 0\}$ have the same finite–dimensional distributions.

The first step toward a description of basic properties of stable processes would be the discussion of some properties of the α–stable Lévy motion. In order to illustrate the most important features of this stochastic process distinguishing it from the processes presented in Sections 2.2 and 2.3, we present some computer graphs of trajectories of standard α–stable Lévy motion for a few different values of the parameter α. At a first glance one can notice a remarkable qualitative and quantitative difference between graphical representations of the Brownian motion (see Section 2.2) and the α–stable Lévy motion presented here (with the parameter α taking on smaller values, the "jumps" of the trajectories become bigger, so in Figures 2.2.1 and 2.5.1 – 2.5.3 we had to change constantly the scaling of the vertical axis). We will come back to the discussion of this effect later on.

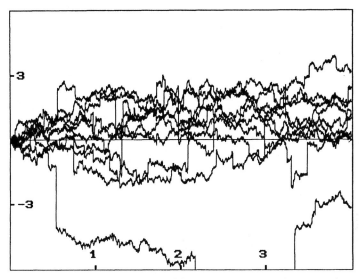

Figure 2.5.1. Trajectories of α–stable Lévy motion in the case of $\alpha = 1.7$.

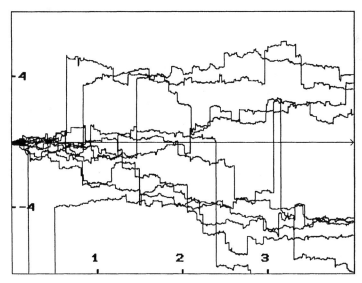

Figure 2.5.2. Trajectories of α–stable Lévy motion in the case of $\alpha = 1.2$.

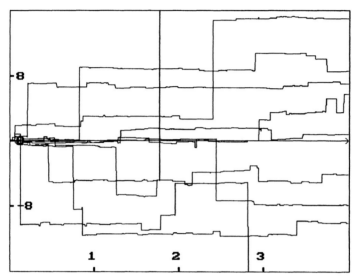

Figure 2.5.3. Trajectories of $S\alpha S$ Lévy motion in the case of $\alpha = 0.7$.

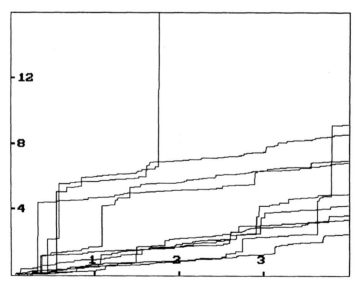

Figure 2.5.4. Trajectories of totally skewed Lévy motion in the case of $\alpha = 0.7$.

Construction of the $S\alpha S$ Lévy Motion from the Brownian Motion. It was Bochner (1955) who noted an interesting relationship between the processes plotted in figures presented in this chapter. It is possible to obtain the $S\alpha S$ Lévy motion process from the Brownian motion by a random time change defined by the totally skewed Lévy motion. Here we present this construction.

Observe that if the measure ν_α on $(0, \infty)$ is defined by the formula

$$d\nu_\alpha(x) = \frac{P}{x^{1+\alpha}}\, dx,$$

then for $0 < \alpha < 1$ the following integral

$$I(u) = \int_0^\infty (e^{iux} - 1)\, d\nu_\alpha(x)$$

is finite (see Sections 2.4 and 4.4 for the Lévy–Khintchine Formula in different settings).

Consider the stochastic process $\mathbf{X} = \{X(t)\}$ with stationary independent increments, and such that the characteristic function

$$\boldsymbol{E} e^{iuX(t)} = e^{t\psi(u)}$$

has the exponent function $\psi(u) = I(u)$. Then, the process $\mathbf{X}$ is the limit of the sequence of processes $\{\mathbf{X}_n\}$ with exponent functions

$$\psi_n(u) = \int_{1/n}^\infty (e^{iux} - 1)\, d\nu_\alpha(x).$$

These processes have only upward jumps. Hence, all paths of $\{X(t)\}$ are nondecreasing pure jump functions, (see Breiman (1968)).

Take on a given probability space the normalized Brownian motion $\{B(t)\}$ and an $\alpha/2$-stable totally skewed Lévy motion $\{X(t)\}$ with nondecreasing sample paths ($\alpha/2 \in (0,1)$, $\beta = 1$), and such that $\mathcal{F}\{X(t) : t \geq 0\}$, $\mathcal{F}\{B(t) : t \geq 0\}$ are independent. If we put

$$Y(t) = B(X(t)),$$

then the process $\{Y(t)\}$ is an $S\alpha S$ Lévy motion.

The idea becomes clear when one notice that the process defined by

$$Y(t + s) - Y(t) = B((X(t + s) - X(t)) + X(t)) - B(X(t))$$

looks just as if it were the $Y(s)$, independent of $Y(\tau)$, $\tau \leq t$.

For a formal proof we will follow Breiman (1968). Take as $X_n(t)$ the process of jumps of $X(t)$ larger than $[0, 1/n)$. According to the above considerations its characteristic function is described by the exponent function

$$\psi_n(u) = \int_{1/n}^\infty (e^{iux} - 1)\, d\nu_{\alpha/2}(x).$$

The jumps of $X_n(t)$ occur at the jump times of a Poisson process with intensity

$$\lambda_n = \int_{1/n}^{\infty} d\nu_{\alpha/2}(x).$$

and the jumps have magnitude $Y_1, Y_2, \ldots$ independent of one another and of the jump times, and are identical distributed. Thus $B(X_n(t))$ has jumps only at the jump times of the Poisson process. The size of the k–th jump is

$$U_k = \begin{cases} B(Y_k + \ldots + Y_1) - B(Y_{k-1} + \ldots + Y_1) & k > 1, \\ B(Y_1) & k = 1. \end{cases}$$

By an argument almost exactly the same as that used in the proof of the strong Markov property for the Brownian motion, U_k is independent of $U_{k-1}, \ldots, U_1$ and has the same distribution as U_1. Therefore $B(X(t))$ is a process with stationary, independent increments. Take $\{n'\}$ such that $X_{n'}(t) \to X(t)$ a.s. and uniformly on compact sets. Use continuity of the Brownian motion $B(t)$ to get

$$\lim_{n' \to \infty} B(X_{n'}(t)) = B(X(t)) \quad a.s. \quad \text{for every} \ t \in [0, \infty).$$

Thus, $B(X(t))$ is a process with stationary independent increments. To obtain its characteristic function write the conditional expectation

$$E\{\exp[iuB(X(1))] \mid Z(1) = z\} = E\exp[iuB(z)] = \exp[-zu^2/2].$$

Therefore,

$$E\exp[iuB(X(1))] = E\exp(-(u^2/2)X(1)) = \exp(-c_X|u|^{\alpha}),$$

so, $Y(t) = B(X(t))$ belongs to the class of $S\alpha S$ Lévy motion processes.

Chapter 3

Computer Simulation of α–Stable Random Variables

3.1 Introduction

One of our main goals in this book is to describe some computational methods of constructing and simulating the most common stochastic processes. These methods are based on computational algorithms providing random numbers or random variables with given distributions.

There is a vast literature concerning this topic. Among others, let us only mention Devroye (1986), Bratley, Fox and Schrage (1987), Rice (1990), Newton (1988), Deák (1990), James (1990).

Here we will recall only a few of the possible general methods consisting in the transformation of a given random variable, which is uniformly distributed in $(0, 1)$, into another random variable with the desired distribution. Theoretically, it is possible to use as a starting point a random variable obeying another probability law (e.g., Gaussian or gamma random variables), but the transformation of uniform random variables is the most suitable and the most common. They can be generated in a fast and easy way. For generating a random variable with a given distribution we prefer to apply "the most suitable" of a few possible methods, supporting our choice by computational experience, though we are aware that if highly accurate numerically data are required, then only a combination of two or more different methods should provide proper results.

For us, one of the most important classes of random variates is the class of α–stable (or stable) random variables. Thus we present and discuss here the results of our investigations concerning the properties of some methods of theoretical representation and computer simulation of such distributions. An especially important role in the theory of stable and infinitely divisible processes is played by their series representations (see Rajput and Rosinski (1989), Rosinski (1990), Tallagrand (1990) and others). In the simplest case we have the so called LePage series representations of stable random variables and processes (see LePage, Woodroofe and Zinn (1981), or LePage (1980) and (1989)). We show that, contrary to the common belief, they converge very slowly, so they are not the most suitable means of generating stable random variables. We discuss some methods that perform much better.

Accidentally, we have formulated a nice and simple lemma providing a solution of an important problem of calculating tail probabilities of stable random variables. The algorithm describing them involves only one deterministic function of two real variables. Thanks to some computer experiments we were able to provide a deeper quantitative insight into the structure of stable laws, varying with some parameters defining them.

Finally we recall basic facts concerning the problem of statistical estimation of densities of some classes of random variables (see e.g., Tapia and Thompson (1978)).

Computer methods of constructing stochastic processes involve at least two kinds of discretization techniques: discretization of the time parameter and approximate representation of random variates with the aid of artificially produced finite time series data sets or statistical samples so we are interested in statistical methods of data analysis such as constructions of empirical cumulative distribution functions or kernel probability density estimates, etc. Applying computer graphics, we attempt to explain to what extent they can provide results good enough to be applied to solve approximately quite complicated problems involving α–stable random variates that are discussed in the subsequent chapters of the book.

3.2   Computer Methods of Generation of Random Variables

In computer simulations of random variables and processes of primary importance is the problem of the proper construction of random quantities that follow a particular probability law. As we mentioned above, when saying *"computer"* we will always refer to IBM PC/AT/386 microcomputers, so it is natural to suppose that using computer programming languages Turbo C or Turbo Pascal, we have at our disposal a built-in function called *rand* that generates pseudo-random uniformly distributed positive integers (consult the Appendix, where this function is included in an exemplary program). It is not our objective to argue whether the method applied there is better or worse than other known methods of construction of pseudo-random uniformly distributed numbers. As a matter of fact, we are not interested in such questions as whether distributions of generated pseudo-random sequences would pass some statistical or other tests. Supported by the results of various numerical experiments we believe that the function *rand* is quite good for our purposes and we will rather focus our attention on such questions as weather given random quantities can be applied properly in the construction of solutions of given problems or not.

Linear congruential method.   Among devices generating large sets of random variates with "proper" statistical properties we would like to mention, currently the most common, the *linear congruential method*. It provides a sequence of integers $I_0, I_1, I_2, \ldots$, each between 0 and $m - 1$, by the recurrence relation

$$I_{n+1} = a\, I_n + c \ (\ mod\ m\ ),$$

where m, a, c are given positive integers. In calculations with IBM PC/386 the following exemplary triples (m, a, c) can serve our purposes:

$$(134456, 8121, 28411), \quad (243000, 4561, 51349),$$

$$(259200, 7141, 54733), \quad (714025, 4096, 150889).$$

Here, in all four cases, the period of the sequence $I_0, I_1, I_2, \ldots$ is of maximal length, i.e., of length m. Further application of the so–called shuffling procedure provides new efficient algorithms producing random sequences with still much longer periods, helps to break up possible sequential correlations or diminish the effects of any cycle or bias.

Add–with–carry and subtract–with–borrow generators.

A description of a new powerful class of methods for generating random numbers with very long periods can be found in Marsaglia and Zaman (1991).

Let us recall here only one example of add–with–carry generator useful for experiments. Fix base 6 to represent any integer I in the range from 0 up to $6^{21} - 1$ as a sequence of twenty one digits from the set $\{0, 1, 2, 3, 4, 5\}$. Now, start an iterating formula by choosing any sequence of twenty one seed digits and any initial carry bit $c \in \{0, 1\}$ (with two exceptions: twenty one 0's for $c = 0$ and twenty one 5's for $c = 1$), i.e., let

$$I_0 = x_0 x_1 ... x_{19} x_{20}, \quad c_0 \in \{0, 1\}.$$

With

$$I_{n-1} = x_{n-1} x_n ... x_{n+18} x_{n+19}, \quad c_{n-1} \in \{0, 1\},$$

already constructed, one has to calculate

$$x_{n+20} = x_{n-1} + x_{n+18} + c_{n-1} \ (\ mod\ 6\ ),$$

$$c_n = \begin{cases} 0, & \text{if} \quad x_{n-1} + x_{n+18} + c_{n-1} < 6, \\ 1, & \text{if} \quad x_{n-1} + x_{n+18} + c_{n-1} \geq 6 \end{cases}$$

and to output finally

$$I_n = x_n x_{n+1} ... x_{n+19} x_{n+20},$$

where positive n can increase up to $6^{21} + 6^2 - 1$, because this number is a prime for which 6 is a primitive root and thus this generator will always have period $6^{21} + 6^2 - 2 = 21\ 936\ 950\ 640\ 377\ 890$. (Every possible set of twenty one successive "throws" from the set $\{0, 1, 2, 3, 4, 5\}$ will appear in the sequence, with frequencies for shorter strings consistent with uniformity for the full period!)

From now on we can assume that, at any time needed, we have at our disposal a random variable γ which is uniformly distributed on the interval (0,1). This allows us to construct random variates that follow different probability laws.

Inversion method. This method provides a random number with the desired distribution function by making use of the inverse of the distribution function. Consider a random variable ξ with density function $f(x)$, $x \in [a, b]$. Its distribution function is

$$F(x) = \int_a^x f(y)dy.$$

If its inverse F^{-1} exists and γ is uniformly distributed on (0,1), then the distribution function of the random variable $\xi = F^{-1}(\gamma)$ is $F(x)$.

Exponential distribution. Consider the random variable τ exponentially distributed with parameter λ and apply the inversion method to get

$$f(x) = \lambda e^{-\lambda x} \text{ and } F(x) = 1 - e^{-\lambda x} \text{ for } x \geq 0.$$

So, if γ is distributed uniformly on (0,1), then

$$\tau = \frac{-1}{\lambda} \log(1 - \gamma).$$

Normal distribution. One of the variations of the Central Limit Theorem states that if $\xi_1, \xi_2, \ldots, \xi_n, \ldots$ are independent, identically distributed random variables with finite variance and $E(\xi_n) = \mu$, $D^2(\xi_n) = \sigma^2$, for $n = 1, 2, \ldots$, then

$$\lim_{n \to \infty} P\left\{\frac{\xi_1 + \xi_2 + \ldots + \xi_n - n\mu}{\sigma\sqrt{n}} < x\right\} = \frac{1}{\sqrt{2\pi}} \int_{-\infty}^x e^{-u^2/2} du.$$

Taking into account the sequence $\gamma_1, \gamma_2, \ldots, \gamma_{12}$ of independent random variables uniformly distributed on (0,1) (here $\mu = 1/2$, $\sigma^2 = 1/12$), we have

$$P\{\gamma_1 + \gamma_2 + \ldots + \gamma_{12} - 6 < x\} \approx \int_{-\infty}^x e^{-u^2/2} du,$$

so we obtain an algorithm generating approximately the standard normal distribution $\mathcal{N}(0, 1)$.

Another exact method of computer construction of standard normal distributions, the so-called *Box–Muller method*, is based on the observation that if γ_1, γ_2 are two independent random variables uniformly distributed on (0, 1), then

$$\eta_1 = \sqrt{-2\ln(\gamma_1)} \cos(2\pi\gamma_2), \quad \eta_2 = \sqrt{-2\ln(\gamma_1)} \sin(2\pi\gamma_2)$$

are independent, standard normal.

The random iteration algorithm for computing fractals.

It is a very nice random algorithm producing approximate attractors of IFS (iterated function systems; see Barnsley (1988)). Here we present an example of such an algorithm which produces the famous "fern" fractal.

Let us fix $(x_0, y_0) = (0,0)$.

According to the law of a discrete random variable k, that takes values $1, 2, 3, 4$ with probabilities 0.01, 0.85, 0.07, 0.07, respectively, one has to compute

$$x_{n+1} = a[k]x_n + b[k]y_n + e[k],$$

$$y_{n+1} = c[k]x_n + d[k]y_n + f[k],$$

for $n = 0, 1, \dots$, where

a[1] = 0.00;	a[2] = 0.85;	a[3] = 0.20;	a[4] = -0.15;
b[1] = 0.00;	b[2] = 0.04;	b[3] = -0.26;	b[4] = 0.28;
c[1] = 0.00;	c[2] = -0.04;	c[3] = 0.23;	c[4] = 0.26;
d[1] = 0.16;	d[2] = 0.85;	d[3] = 0.22;	d[4] = 0.24;
e[1] = 0.00;	e[2] = 0.00;	e[3] = 0.00;	e[4] = 0.00;
f[1] = 0.00;	f[2] = 1.60;	f[3] = 1.60;	f[4] = 0.44.

The results of the computer experiment are contained in two figures presented below. We believe that they demonstrate that, to some extent, the computer generator of discrete random variables and computer graphics (see Remark 3.5.2) provide acceptable results.

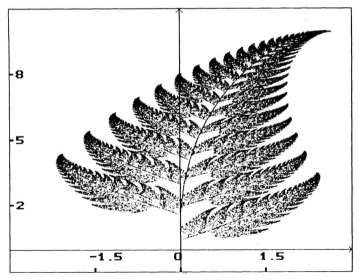

Figure 3.2.1. The fern fractal obtained by IFS method for 100000 iterations.

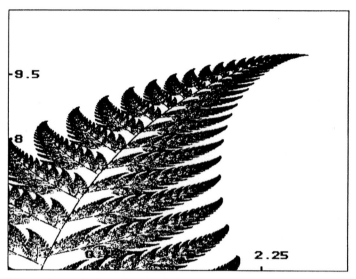

Figure 3.2.2. Enlarged part of the fern fractal (400000 iterations).

3.3   Series Representations of Stable Random Variables

From our point of view the most important class of random variates is the class of α-stable random variables and processes. An explicit description provides the so-called *LePage series representation* formula. It plays a very important role in theoretical investigations of stable and infinitely divisible variables and processes.

It seems to be an important problem to answer the question whether this formula would be useful in computer simulations. So, in this section we discuss the problem of a speed of convergence of such series representations.

In the sequel we will restrict ourselves to the case of $\alpha \in (1,2)$ and $\beta \in (-1,1)$ and we will consider only "normalized" stable variables $X \sim S_\alpha(1, \beta, 0)$. (The other cases can be treated in a very similar way.) So, let $\{\tau_1, \tau_2, \ldots\}$ be a sequence of arrival times (jump points) of the right continuous Poisson process with unit rate. Thus the random variable τ_j has the density

$$f_j(x) = x^{j-1} e^{-x} I_{[0,\infty)}(x)/\Gamma(j).$$

(Here Γ stands for the gamma function and $I_A = I_A(x)$ denotes the characteristic function of the set A.) Let us also define the sequence $\{\xi_1, \xi_2, \ldots\}$ of i.i.d. random variables as follows:

$$P\{\xi_j = t_1\} = p_1, \qquad P\{\xi_j = t_2\} = p_2, \qquad j = 1, 2, \ldots,$$

where

$$t_1 = (1 + \beta)^{1/(\alpha-1)}, \qquad t_2 = -(1 - \beta)^{1/(\alpha-1)},$$

$$p_1 = (1 - \beta)^{1/(\alpha-1)} / \left\{ (1 + \beta)^{1/(\alpha-1)} + (1 - \beta)^{1/(\alpha-1)} \right\}, \qquad p_2 = 1 - p_1.$$

We will assume the sequence $\{\xi_1, \xi_2, \ldots\}$ to be independent of the sequence $\{\tau_1, \tau_2, \ldots\}$.

Let us define

$$X_n = L(\alpha, \beta)^{-1/\alpha} \sum_{j=1}^{n} \xi_j \tau_j^{-1/\alpha},$$

$$X = L(\alpha, \beta)^{-1/\alpha} \sum_{j=1}^{\infty} \xi_j \tau_j^{-1/\alpha},$$

where

$$L(\alpha, \beta) = -2\alpha \Gamma(-\alpha) \cos(\alpha\pi/2) p_1 t_1.$$

Our aim is to give a quantitative information on the rate of convergence of LePage–type sums X_n to X with respect to different values of parameters α, β.

Before stating the main theorem we need two lemmas.

Lemma 3.3.1 *If $s > 0$, then*

$$s \sum_{j=1}^{\infty} \xi_j \tau_j^{-1/\alpha} \sim S_\alpha(\sigma, \beta, 0),$$

i.e. the series on the left–hand side converges a.s. to a stable random variable $X \sim S_\alpha(\sigma, \beta, 0)$ for

$$\sigma^\alpha = s^\alpha L(\alpha, \beta).$$

Lemma 3.3.2 *For X_n defined above we have*

$$E|X_{n+m} - X_n|^2 = C(\alpha, \beta) \sum_{j=n+1}^{n+m} E\tau_j^{-2/\alpha},$$

where

$$C(\alpha, \beta) = L(\alpha, \beta)^{-2/\alpha} (1 - \beta^2)^{1/(\alpha-1)}.$$

Applying both lemmas and the formula

$$E\tau_j^{-2/\alpha} = \frac{\Gamma(j - 2/\alpha)}{\Gamma(j)}, \qquad for \quad j > 2/\alpha,$$

we obtain the main result.

Theorem 3.3.1 *If $\alpha \in (1,2)$, $\beta \in (-1,1)$, then for $n \geq \alpha/2$ we have*

$$E|X - X_n|^2 = W_n(\alpha, \beta),$$

where

$$W_n(\alpha, \beta) = C(\alpha, \beta)V_n(\alpha),$$

$$V_n(\alpha) = \sum_{j=n+1}^{\infty} \Gamma(j - 2/\alpha)/\Gamma(j).$$

It is not easy to handle the above formula in computer calculations, so we propose an asymptotic estimate.

Theorem 3.3.2 *If $\alpha \in (1,2)$, $\beta \in (-1,1)$, then for $n \geq \alpha/2$ we have*

$$E|X - X_n|^2 \approx R_n(\alpha, \beta),$$

where

$$R_n(\alpha, \beta) = C(\alpha, \beta)S_n(\alpha),$$

$$S_n(\alpha) = \sum_{j=n+1}^{\infty} (j - 2/\alpha)^{-2/\alpha}.$$

Here the relation "$\approx$" means that with respect to n, functions $W_n(\alpha, \beta)$ and $R_n(\alpha, \beta)$ are asymptotically the same, i.e.

$$e^{-2}S_n(\alpha) \leq V_n(\alpha) \leq S_n(\alpha),$$

$$\lim_{j \to \infty} (j - 2/\alpha)^{2/\alpha} \Gamma(j - 2/\alpha)/\Gamma(j) = 1.$$

It is also possible to obtain some other estimates.

Theorem 3.3.3 *If $n > \alpha/2$, $\eta > 0$ and $p \in (1, \alpha)$, then*

$$P\{|X - X_n| > \eta\} \leq \eta^{-2}R_n(\alpha, \beta),$$

$$E\{|X - X_n|^p\}^{1/p} \leq \{R_n(\alpha, \beta)\}^{1/2}.$$

Now we are going to present some results of computer experiments that allowed us to obtain quantitative information on the rate of convergence of the LePage series to α–stable random variables and to check their usefulness in the construction of computer simulation algorithms.

Hypothetical algorithm. Stable random variable $X \sim S_\alpha(1, \beta, 0)$ can be approximated by the sums $X_n = \sum_{j=1}^{n} \xi_j \tau_j^{-1/\alpha}$.

In order to obtain a statistical (approximate) sample $\left\{\overline{X}^{(m)}\right\}_{m=1}^{M}$ of X with fixed natural M, it is enough to do the following:

- fix an appropriate (large enough) natural number n;

- according to the definition of discrete random variables ξ_j, construct their statistical samples $\left\{\xi_j^{(m)}\right\}_{m=1}^{M}$ for $j = 1, 2, \ldots, n$;

- notice that each τ_j can be represented as a sum of j exponentially distributed independent random variables with mean parameter $\lambda = 1$, so the method of computation of samples $\left\{\tau_j^{(m)}\right\}_{m=1}^{M}$ for $j = 1, 2, \ldots, n$ is obvious;

- construct as a final result the corresponding sample for the sum defining X_n.

Unfortunately, as we shall in the next section, this algorithm is very costly as far as the time of calculations is concerned, even for small values of n.

3.4 Convergence of LePage Random Series

Let us present the results of computer experiments concerning the speed of convergence of LePage series. Precise quantitative information on the rate of convergence of LePage sums X_n to X give functions $C(\alpha, \beta)$, $S_n(\alpha)$ and $R_n(\alpha, \beta)$ defined below.

The number series S_n converges awfully slowly. Thus, applying some techniques that accelerate the convergence of number series, we propose the following approximate algorithm

$$
\begin{aligned}
S_n(\alpha) \quad &\approx \quad \frac{1}{2}(n + \nu + 1 - 2/\alpha)^{-2/\alpha} + \frac{\alpha}{2 - \alpha}(n + \nu + 1 - 2/\alpha)^{(\alpha-2)/\alpha} \\
&+ \quad \sum_{j=n+1}^{n+\nu} (j - 2/\alpha)^{-2/\alpha},
\end{aligned}
$$

where $n \geq 2/\alpha$ and ν is an appropriately chosen positive integer.

The best thing to do is to make use of IBM PC/386 graphics to demonstrate how these functions depend on α, β and n.

To obtain nicer technical effects we decided to present in Fig. 3.4.1 the function $\bar{C}(\alpha, \beta) = min\{C(\alpha, \beta),\ 0.5\}$ instead of $C(\alpha, \beta)$.

It is not difficult to notice that the function $C(\alpha, \beta)$ depends only on the sequence of discrete random variables ξ_1, ξ_2, ... The behavior of this function reflects perfectly what we should expect looking closely at the definition of the characteristic function of a stable random variable (see Chapter 2, Section 2.4). We are able to see the subset of the rectangle $[1, 2] \times [-1, 1]$ of "good" (α, β) and two subsets of "troublesome" (α, β). In particular, if $\alpha = 1$, then the only acceptable value of β is $\beta = 0$.

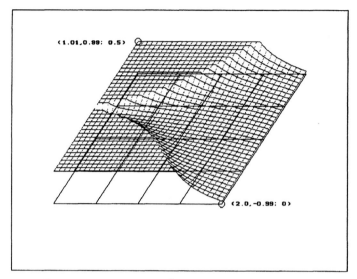

Figure 3.4.1. Function $\bar{C}(\alpha, \beta) = min \{C(\alpha, \beta), 0.5\}$.

Remark 3.4.1 *It is not difficult to notice that $S_n(\alpha)$ depends only on the sequence $\{\tau_{n+1}{}^{-2/\alpha}, \tau_{n+2}{}^{-2/\alpha}, \ldots \}$ and the only cause of very bad behavior of $E|X - X_n|^2$ (especially for α close to 2) is the presence of arrival times $\tau_1, \tau_2, \ldots$ of the Poisson process in the LePage representation of X.*

Figures 3.4.2 and 3.4.3 contain the graphs of $\bar{R}_n(\alpha, \beta) = min\{R_n(\alpha, \beta), 0.1\}$ for $n = 100$ and $n = 10000$. They allow us to see how slowly $\{X_n\}$ converges to X and how this convergence depends on parameters α, β. By comparing them one can see that, roughly speaking, the approximation X_n of X is acceptable in practice only for (α, β) such that $(\alpha - 1 - \delta)^2 + \beta^2/\rho \leq \delta$, with small $\delta > 0$ and $\rho \approx 4$.

Computer experiments indicate that methods based on the LePage series representations are not the most suitable means of generating stable random variables. Figures 3.4.4 and 3.4.5 illustrate this situation. They present histograms of X_n, which should approximate the densities of the stable random variable $X = X(\alpha, \beta)$. For $\alpha = 1.2$, $\beta = 0.3$ and $\alpha = 1.2$, $\beta = 0.45$, respectively, and $n = 100$, we obtained rather poor results.

According to the situation presented in Fig.3.4.2, one can check that for values of α, β slightly different from those chosen above the histograms might be even worse. Nevertheless, the mathematical fact of very slow convergence of the LePage series for a large part of the possible values of parameters α, β (as suggested by Figs. 3.4.2 and 3.4.3) forces us to look for other, less wasteful as far as computer time is concerned, methods.

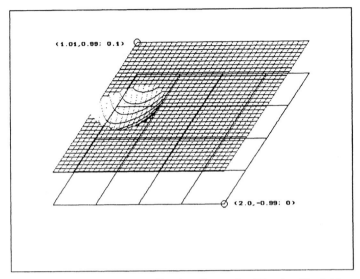

Figure 3.4.2. Function $\bar{R}_{100}(\alpha, \beta) = min \{R_{100}(\alpha, \beta), 0.1\}$.

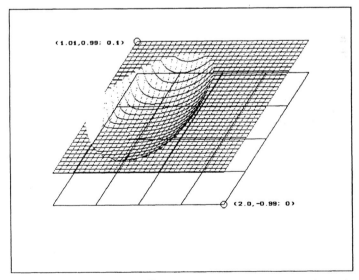

Figure 3.4.3. Function $\bar{R}_{10000}(\alpha, \beta) = min \{R_{10000}(\alpha, \beta), 0.1\}$.

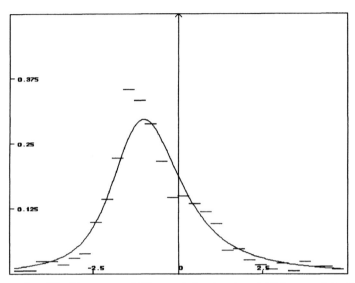

Figure 3.4.4. The histogram of X_{100} and the density of X for $\alpha = 1.2$, $\beta = 0.30$.

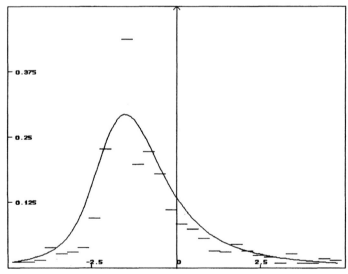

Figure 3.4.5. The histogram of X_{100} and the density of X for $\alpha = 1.2$, $\beta = 0.45$.

3.5 Computer Generation of α–Stable Distributions

When we start working with α–stable distributions, the main problem is that, except for a few values of four parameters describing the characteristic function, their density functions are not known explicitly. (The most interesting exceptions are presented in Chapter 2.)

As we know, the LePage series representations are not the most suited for an application in computer calculations, so our aim is to describe now a few of the possible general techniques of constructing and computer simulating stable random variables.

Application of the fast Fourier transform. Taking into account the definition describing characteristic functions of α–stable random variables, we can apply the Fourier transform to obtain densities, namely

$$f(x) = \frac{1}{2\pi} \int_{-\infty}^{\infty} e^{-itx} \phi(t) dt.$$

In order to construct a computer representation of f, we have to calculate a finite number of values $f(x_k)$ of f. Omitting the detailed description of a numerical method applied to solve this problem, let us only mention that it is necessary to define a proper set of points x_k, find a criterion to decide how to approximate the infinite integral defining Fourier transform by a sum of integrals on intervals $[2k\pi, 2(k+1)\pi]$ and to adjust to our needs one of the methods of numerical integration in such a way that the application of the fast Fourier transform (FFT method) would be possible.
Now we are in a position to propose an effective algorithm to generate a stable random variable X:

- starting from a density function construct numerically the inverse $F^{-1}(y)$ of the probability distribution,

- generate a random variable γ uniformly distributed on $(0,1)$ and finally compute $X = F^{-1}(\gamma)$.

Application of gamma distribution. Let us recall that in LePage representation

$$X = L(\alpha, \beta)^{-1/\alpha} \sum_{j=1}^{\infty} \xi_j \tau_j^{-1/\alpha}$$

of stable variable $X \sim S_\alpha(1, \beta, 0)$, where ξ_j denote discrete random variables and τ_j – the random variables with densities

$$f_j(x) = x^{j-1} e^{-x} I_{[0,\infty)}(x)/(j-1)!.$$

An obvious procedure generating the sample $\left\{\tau_j^{(m)}\right\}_{m=1}^{M}$ for $j = 1, 2, ..., n$ consists in calculating

$$\tau_j^{(m)} = -\log\left(\prod_{i=1}^{j} U_i^{(m)}\right),$$

where $\{U_1^{(m)}, U_2^{(m)}, ..., U_j^{(m)}\}$ denotes the statistical sample of the random variable U distributed uniformly on $(0, 1)$.

Remark 3.5.1 *In Bratley, Fox and Schrage (1987) one can find the suggestion that variance reduction techniques (see e.g., Marsaglia (1977) or Schmeiser, Lal (1980)) perform much better.*

Taking this into account we still insist that computer methods of construction of α–stable random variables based on the LePage series representations, even with the best possible generators of gamma distribution, are not well suited to computer simulations of stable variates. The mathematical fact of very slow convergence of LePage sums definitively kills such methods.

Series description of a distribution function. Bergström (see Bergström (1952)) obtained the following formulas:

$$F_\alpha(x) = \frac{1}{2} + \frac{1}{\pi\alpha}\sum_{k=1}^{\infty}(-1)^{k-1}\frac{\Gamma((2k-1)/\alpha)}{(2k-1)!}x^{2k-1},$$

or

$$F_\alpha(x) = 1 + \left\{\frac{1}{\pi}\sum_{k=1}^{\infty}(-1)^k\frac{\Gamma(\alpha k)}{k!}x^{-\alpha k}\sin\left(\frac{k\pi\alpha}{2}\right)\right\}$$

for the distribution function of the random variable $S_\alpha(1, 0, 0)$. Partial sums of these series can probably be applied with care in constructing approximate values of the distribution function $F_\alpha(x)$.

Direct method. The best method of computer simulation of a very important class of symmetric α–stable random variable $X \sim S_\alpha(1, 0, 0)$ for $\alpha \in (0, 2]$ consists in the following:

- generate a random variable V uniformly distributed on $(-\pi/2, \pi/2)$ and an exponential random variable W with mean 1;

- compute

$$X = \frac{\sin(\alpha V)}{\{\cos(V)\}^{1/\alpha}} \times \left\{\frac{\cos(V - \alpha V)}{W}\right\}^{(1-\alpha)/\alpha}. \tag{3.5.1}$$

The formula (3.5.1) is generalized below by (3.5.2).

Tail probabilities from samples. In practical simulations an important question is what size of statistical sample is good enough to expect acceptable accuracy in Monte–Carlo–type calculations of some parameters. In order to give an idea of what one should expect when working with α–stable random variables, we present here two tables containing values of tail probabilities calculated on the basis of statistical samples S_α^- representing random variables $S_\alpha(1, 0, 0)$ and generated by the direct method described above for a few different values of the parameter α. The first table contains the result obtained for two different samples of the size 2000, the second – of the size 4000. Comparing the numbers included in Tables 3.5.1, 3.5.2 and 3.6.1 one can conclude, that when working with the IBM PC built–in function *rand*(), a reasonable choice of a sample is somewhere between 2000 and 4000 (the result for 1000 "trials" is poorer). Unfortunately, the exactness of computer calculations is not impressive (see Remark 3.5.2). The only way to improve them is to work with generators with much larger periods (see some propositions above in this chapter) and samples bigger in size, but it would consume much more of computer program execution time, which is a very significant factor in complicated calculations.

Table 3.5.1. Tail Probabilities from the sample of 2000 "trials".

α	$P\{S_\alpha^- > 0.3\}$		$P\{S_\alpha^- > 3.0\}$		$P\{S_\alpha^- > 30.0\}$	
1.7	0.403	0.412	0.038	0.034	0.000	0.000
1.2	0.413	0.417	0.076	0.079	0.005	0.006
0.8	0.402	0.393	0.138	0.129	0.023	0.022
0.5	0.355	0.373	0.164	0.175	0.057	0.061
0.3	0.364	0.353	0.225	0.231	0.139	0.128

Table 3.5.2. Tail Probabilities from the sample of 4000 "trials".

α	$P\{S_\alpha^- > 0.3\}$		$P\{S_\alpha^- > 3.0\}$		$P\{S_\alpha^- > 30.0\}$	
1.7	0.409	0.416	0.034	0.037	0.000	0.000
1.2	0.408	0.413	0.087	0.082	0.005	0.005
0.8	0.397	0.409	0.129	0.134	0.022	0.025
0.5	0.380	0.363	0.194	0.181	0.071	0.073
0.3	0.345	0.346	0.223	0.223	0.132	0.141

Computer generator of skewed α–stable variables. In the chapters that follow we develop mainly the methods and algorithms involving symmetric stable variables ($\beta = 0$) (e.g., symmetric stable random measures, stochastic integrals with respect to such measures, symmetric stable stationary processes, etc.). The algorithm providing skewed stable random variables $Y \sim S_\alpha(1, \beta, 0)$ with $\alpha \in (0, 1) \cup (1, 2]$ and $\beta \in [-1, 1]$ consists in the following:

- generate a random variable V uniformly distributed on $(-\pi/2, \pi/2)$ and an exponential random variable W with mean 1;

- compute

$$Y = D_{\alpha,\beta} \times \frac{\sin(\alpha(V + C_{\alpha\beta}))}{\{\cos(V)\}^{1/\alpha}} \times \left\{ \frac{\cos(V - \alpha(V + C_{\alpha,\beta}))}{W} \right\}^{(1-\alpha)/\alpha} \quad (3.5.2)$$

and

$$C_{\alpha,\beta} = \frac{\text{arc tan}(\beta \tan(\pi\alpha/2))}{\alpha},$$

$$D_{\alpha,\beta} = [\cos(\text{arc tan}(\beta \tan(\pi\alpha/2)))]^{-1/\alpha}.$$

Generalizing the result of Kanter (1975) or slightly modifying the algorithm of Chambers, Mallows and Stuck (1976), one can see that Y belongs to the class of $S_\alpha(1, \beta, 0)$ random variables.

We regard the method defined by (3.5.1) and (3.5.2) as a good technique of computer simulation of α-stable random variables, stochastic measures and processes of different kinds. Of course, it has its own limitations in applicability as any computer technique has. Here we provide some information on this subject.

Remark 3.5.2 *Let us mention the following sources of possible computer instability or errors:*

1. *computer arithmetic works on the finite set of rational numbers (not on the real line* R*);*

2. *computer generators provide pseudo–random numbers which are not really random (appropriate methods of statistical testing provide useful information on the quality of obtained results);*

3. *computer graphics provides a finite number of "pixels" on a computer screen (not a continuum in the form of a rectangle in* R × R*);*

4. *mathematical and numerical instability of all calculations connected with α-stable random variables, when parameter α approaches 0 or when α converges to 1, with $\beta \neq 0$.*

3.6   Exact Formula for Tail Probabilities

According to Feller (1971), the asymptotic behavior of tail probabilities of the symmetric random variable $X \sim S_\alpha(\sigma, 0, 0)$ is described by

$$\lim_{\lambda \to \infty} \lambda^\alpha P\{X > \lambda\} = D_\alpha \sigma^\alpha,$$

where

$$D_\alpha = \frac{1}{2} \left(\int_0^\infty x^{-\alpha} \sin(x) \, dx \right)^{-1}.$$

Without loss of generality we can restrict ourselves in the sequel to the case of $X = S_\alpha$ (which means that $X \sim S_\alpha(1, 0, 0)$).

The following result gives an exact explicit formula, which serves as a basis for two algorithms: an applicable algorithm of computation of tail probabilities and a method of computer simulation of α–stable variables and random measures.

Theorem 3.6.1 *Let us define in the square of* $(x, y) \in (0, 1) \times (0, 1)$ *the function*

$$f_\alpha(x, y) = \frac{\cdot \, \sin(\alpha \pi (x - \frac{1}{2}))}{\{\cos(\pi(x - \frac{1}{2}))\}^{1/\alpha}} \times \left\{ \frac{\cos((1 - \alpha)\pi(x - \frac{1}{2}))}{-\log(y)} \right\}^{(1-\alpha)/\alpha}. \qquad (3.6.1)$$

Then we have

$$P\{S_\alpha > \lambda\} = |\{(x, y); \ f_\alpha(x, y) > \lambda\}|, \qquad (3.6.2)$$

where $|A|$ *denotes the volume of a set A in* $\mathbb{R}^2$.

A formula more general then (3.6.1) concerning skewed stable variables can be easily derived from (3.5.2). It was Kanter (1975) who noticed that a certain integral formula dealing with totally skewed to the right stable random variables ($\beta = 1$ and $\alpha \in (0, 1)$) derived by Ibragimov and Chernin (1959) implies a simulation method given in (3.5.2), for this special case. Then Chambers, Mallows and Stuck (1976) noticed that a formula of Zolotarev (1966) can be used similarly to give a simulation method for the general case.

In the five following figures we present graphs of the function

$$f_\alpha^\lambda(x, y) = \begin{cases} \min \ \{f_\alpha(x, y), \ \lambda\}, & \text{if} \quad f_\alpha(x, y) \geq 0, \\ \max \ \{f_\alpha(x, y), -\lambda\}, & \text{if} \quad f_\alpha(x, y) < 0, \end{cases}$$

for fixed value $\lambda = 3$ and five different values of the parameter α.

We hope that the series of data presented below (Figures 3.6.1 – 3.6.4 and Table 3.6.1) illustrates exactly how tail probabilities change with α. In each figure this probability is equal to the 2–dimensional volume of a flat upper part of the function f_α^λ.

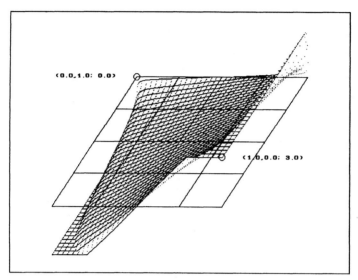

Figure 3.6.1. Function $f_{1.7}^{\lambda} = f_{1.7}^{\lambda}(x,y)$; $P\{S_{1.7} > 3\} = 0.0362$.

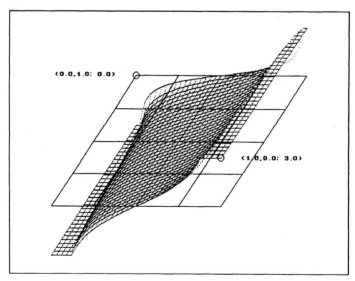

Figure 3.6.2. Function $f_{1.2}^{\lambda} = f_{1.2}^{\lambda}(x,y)$; $P\{S_{1.2} > 3\} = 0.0795$.

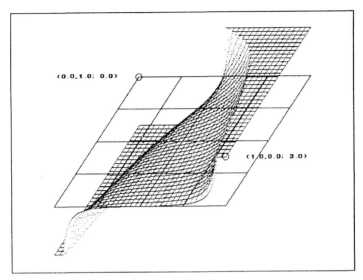

Figure 3.6.3. Function $f_{0.5}^{\lambda} = f_{0.5}^{\lambda}(x, y)$; $P\{S_{0.5} > 3\} = 0.1835$.

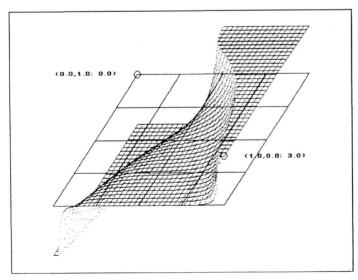

Figure 3.6.4. Function $f_{0.3}^{\lambda} = f_{0.3}^{\lambda}(x, y)$; $P\{S_{0.3} > 3\} = 0.2297$.

Table 3.6.1. Table of Stable Tail Probabilities

α	$P\{S_\alpha > 0.3\}$	$P\{S_\alpha > 3.0\}$	$P\{S_\alpha > 30.0\}$
2.0	0.4160	0.0169	0.0000
1.9	0.4160	0.0228	0.0000
1.8	0.4158	0.0293	0.0000
1.7	0.4156	0.0362	0.0000
1.6	0.4152	0.0436	0.0002
1.5	0.4147	0.0516	0.0009
1.4	0.4140	0.0601	0.0017
1.3	0.4131	0.0695	0.0030
1.2	0.4118	0.0795	0.0048
1.1	0.4099	0.0904	0.0071
1.0	0.4075	0.1027	0.0106
0.9	0.4036	0.1156	0.0156
0.8	0.3986	0.1300	0.0228
0.7	0.3918	0.1461	0.0330
0.6	0.3832	0.1638	0.0475
0.5	0.3733	0.1835	0.0677
0.4	0.3622	0.2054	0.0956
0.3	0.3507	0.2297	0.1334
0.2	0.3388	0.2563	0.1828
0.1	0.3267	0.2849	0.2449
0.01	0.3163	0.3130	0.3088

Remark 3.6.1 *The numbers included in the table presented here (Table 3.6.1) give an idea of an extension of the well–known "rule of 3 σ's" for normal standard distributions to the case of stable variables. (Remember that $S_2 \sim \mathcal{N}(0,2)$; if $X \sim \mathcal{N}(0,1)$, then $P\{X > \mu\} = P\{S_2 > \sqrt{2}\mu\}$.)*

Without developing any necessary mathematical formalism let us mention that

$$\lim_{\alpha \downarrow 0} S_\alpha = Z,$$

where Z is described by the following conditions: $P\{Z = 0\} = p$, $P\{Z = +\infty\} = q/2$, $P\{Z = -\infty\} = q/2$, with $p > 0$, $q > 0$ and $p + q = 1$.

When looking at the data presented in this section one can imagine what is going on when we simulate stable variables on a computer for changing values of the parameter α.

3.7   Density Estimators

Computer algorithms for constructing various random variates often involve approximate methods of representation of random variables with the aid of artificially produced finite statistical samples, so we are interested in statistical methods of data analysis such as constructions of empirical cumulative distribution functions or kernel probability density estimates and so on.

Here we are not interested in statistical methods of estimation of parameters of α–stable random variables from a given statistical sample. The reader interested in these kinds of problems can find some information on the subject in Feuerverger, McDunnough (1981). The results of computer experiments with maximum likelihood estimations can provide additional important information on constructed variates.

In this section we will discuss the important and very interesting problem of approximation of univariate density functions $f = f(x)$. There are some very well known monographs concerning this subject. Let us mention Noether (1967), Tapia and Thompson (1978), Csörgö and Révész (1981), Devroye and Györfi (1985), etc. Our aim is to recall some simple, useful methods describing kernel density estimators.

Let us suppose that on a probability space $(\Omega, \mathcal{F}, P)$ we are given a sequence $\{\xi_1, \xi_2, ..., \xi_n, ...\}$ of i.i.d. random variables distributed according to the law described by the unknown density function $f = f(x)$. To abbreviate the notation let us define $\Xi_n = \{\xi_1, \xi_2, ..., \xi_n\}$. The observed sequence of values (or realizations) of Ξ_n, i.e. its statistical sample we denote by $\Xi^{(n)} = \{\xi^{(1)}, ..., \xi^{(n)}\}$ or, in full notation, $\Xi^{(n)}(\omega) = \{\xi^{(1)}(\omega), ..., \xi^{(n)}(\omega)\}$. A value of the density f estimator for any fixed $x \in \mathbb{R}$ is denoted by $f_n(x; \omega)$, when we view it as a random variable on $(\Omega, \mathcal{F})$ depending on $\Xi_n(\omega)$, or simply by $f_n(x)$ when we regard it as an approximate value of the density f, derived from a fixed sample $\Xi^{(n)}$.

In most cases in which we are interested in $\Xi^{(n)}$ will be obtained as a result of computer simulation of α–stable variates with the use of the direct method, described in Section 3.5.

We will emphasize general methods and theoretical facts suitable in this context.

The simplest particular example of a nonparametric density estimator belonging to the class of kernel density estimators represents commonly used histograms.

Histograms.   Suppose we are given a statistical sample $\Xi^{(n)}$. Let us fix appropriately a finite interval $[a, b]$ contained in the support of an unknown density f. Let us fix $N \in \mathbb{N}$ and introduce a set of mesh points $x_i = a + ih$, $i = 0, 1, ..., N$, dividing the interval $[a, b]$ into N subintervals of equal length $h = (b-a)/N$. A *histogram* is a function $f_n = f_n(x)$, constant on each of subintervals $[x_i, x_{i+1})$ and defined as follows.

For $i = 0, 1, ..., N - 1$ and for $x \in [x_i, x_{i+1})$ define

$$f_n(x) = \frac{K_n(x_i, x_{i+1})}{nh},$$

where function $K_n(x_i, x_{i+1})$ counts how many of the data values $\xi^{(k)}$ are in each of the intervals, i.e.

$$K_n(x_i, x_{i+1}) = \#\{k; \ \xi^{(k)} \in [x_i, x_{i+1}) \ \}.$$

A bar graph of f_n gives an approximation of the graph of the unknown density function f. Joining middles of bars over intervals $[x_i, x_{i+1})$, one can obtain continuous, piecewise linear approximation of the density f.

Now we are going to present a few methods providing algorithms for calculating kernel density estimators. We will assume that we are given a continuous kernel function $K = K(u)$ on $\mathbb{R}$, satisfying conditions

$$\int_{-\infty}^{\infty} K(u) \, du = 1; \quad K(u) \geq 0 \quad \text{for} \quad u \in \mathbb{R}.$$

Appropriate and commonly used kernel functions are the following:

1. rectangular:
$$K(u) = \begin{cases} \frac{1}{2}, & \text{if} \quad |u| \leq 1, \\ 0, & \text{if} \quad |u| > 1; \end{cases}$$

2. triangular:
$$K(u) = \begin{cases} \frac{1}{\sqrt{6}} - \frac{|u|}{6}, & \text{if} \quad |u| \leq \sqrt{6}, \\ 0, & \text{if} \quad |u| > \sqrt{6}; \end{cases}$$

3. Gaussian:
$$K(u) = (2\pi)^{-1/2} e^{-u^2/2};$$

4. "optimal":
$$K(u) = \begin{cases} \frac{3}{4\sqrt{5}} \left(1 - \frac{u^2}{5}\right), & \text{if} \quad |u| \leq \sqrt{5}, \\ 0, & \text{if} \quad |u| > \sqrt{5}; \end{cases}$$

5. biweight:
$$K(u) = \begin{cases} \frac{15}{16\sqrt{7}} \left(1 - \frac{u^2}{7}\right)^2, & \text{if} \quad |u| \leq \sqrt{7}, \\ 0, & \text{if} \quad |u| > \sqrt{7}. \end{cases}$$

We will also assume that we are given a sequence of positive real numbers $\{b_n\}_{n=1}^{\infty}$ such that

$$\lim_{n \to \infty} b_n = 0, \quad \lim_{n \to \infty} n b_n = \infty.$$

From the well known results on the asymptotic behavior of such density estimators (see, e.g. Devroye (1987)) it follows that in practical computer calculations it is reasonable to choose

$$b_n = c n^{-1/5},$$

where c stands for an appropriately chosen constant.

Rosenblatt–Parzen kernel density estimator. Starting from a given statistical sample $\Xi^{(n)}$, the method provides for any fixed $x \in \mathbb{R}$ an approximate value $f_n(x) = f_n(x; \omega)$ of $f(x)$ in the following form

$$f_n(x) = \frac{1}{n} \sum_{i=1}^{n} \frac{1}{b_n} K\left(\frac{x - \xi^{(i)}}{b_n}\right). \tag{3.7.1}$$

Overlooking a method of constructing a bar graph of any histogram, we can think of the height of the bar for a particular interval $[x_i, x_{i+1})$ as a quantity proportional to the average number of total data that are in that interval. In practical realization of a more general kernel density estimator like (3.7.1) the range should also be divided into intervals, but these intervals are allowed to overlap, and the aim is to estimate the density at the center point x of each interval of the form $[x - b_n, x + b_n]$. Instead of just counting the number of points in the interval, the method assigns a non–negative real number to each point $\xi^{(i)}$ in this interval: "throws" $\xi^{(i)}$ close to the center of the interval get highest possible values; other, farther and farther from the center, gain less and less of the weight coefficient, according to the rule defined by kernel function.

Wolverton–Wagner–Yamato recursive kernel density estimator. The method is just a recursive version of the previous one and consists in the following

$$f_n(x) = \frac{1}{n} \sum_{i=1}^{n} \frac{1}{b_i} K\left(\frac{x - \xi^{(i)}}{b_i}\right). \tag{3.7.2}$$

The recursive formula is obvious

$$f_n(x) = \frac{n-1}{n} f_{n-1}(x) + \frac{1}{n b_n} K\left(\frac{x - \xi^{(n)}}{b_n}\right).$$

Now we are going to justify these methods recalling only a few facts from the vast literature related to the problems of convergence of nonparametric density estimators. We do not pretend to present them in the most general possible form. Our aim is to recall some theorems on convergence of methods described above, applicable when α–stable laws are discussed.

Theorem 3.7.1 *Let us suppose that an unknown density function $f = f(x)$ and a kernel $K = K(u)$ are continuous and bounded on $\mathbb{R}$. Then for the Rosenblatt–Parzen method (3.7.1) we have*

1. for any $x \in \mathbb{R}$

$$\lim_{n \to \infty} \boldsymbol{E} f_n(x) = f(x);$$

2. for any $x \in \mathbb{R}$

$$\lim_{n \to \infty} f_n(x) = f(x), \quad \text{in probability;}$$

3. *if for any positive δ the series $\sum_{n=1}^{\infty} \exp(-\delta n b_n)$ converges, then for any $x \in \mathbb{R}$*

$$P\{\lim_{n \to \infty} f_n(x) = f(x)\} = 1;$$

4. *if for any positive δ the series $\sum_{n=1}^{\infty} \exp(-\delta n b_n^2)$ converges, then*

$$P\left\{\lim_{n \to \infty} \sup\{|f_n(x) - f(x)|;\ x \in \mathbb{R}\} = 0\right\} = 1;$$

5. *if the density f has continuous and bounded second derivative f'' and $K = K(u)$ is an even function satisfying the condition $\int_{\mathbb{R}} u^2 K(u)\, du < \infty$, then for any $x \in \mathbb{R}$*

$$\boldsymbol{E} f_n(x) - f(x) = b_n^2 \frac{f''(x)}{2} + o(b_n^2).$$

Theorem 3.7.2 *Let an unknown density function $f = f(x)$ and a kernel $K = K(u)$ be continuous and bounded on $\mathbb{R}$. Then for the Wolverton–Wagner–Yamato method (3.7.2) we have*

1. *for any $x \in \mathbb{R}$*

$$\lim_{n \to \infty} \boldsymbol{E} f_n(x) = f(x);$$

2. *for any $x \in \mathbb{R}$*

$$\lim_{n \to \infty} f_n(x) = f(x), \quad in\ probability;$$

3. *if $\lim_{n \to \infty} \frac{n b_n}{\log \log n} = \infty$, then*

$$P\{\lim_{n \to \infty} f_n(x) = f(x)\} = 1;$$

4. *if the density f has continuous and bounded second derivative f'' and $K = K(u)$ is an even function satisfying the condition $\int_{\mathbb{R}} u^2 K(u)\, du < \infty$, and $\sum_{n=1}^{\infty} b_n^2 = \infty$, then for any $x \in \mathbb{R}$*

$$\boldsymbol{E} f_n(x) - f(x) = \frac{\sum_{i=1}^{n} b_i^2}{n} \frac{f''(x)}{2} + o(\sum_{i=1}^{n} b_i^2/n).$$

Now we present some graphical results of computer experiments that allow us to demonstrate what we should expect doing some statistics out of samples of α-stable random variables. Figures 3.7.1–3.7.3 show, in particular, that one has to work with at least a few thousands of approximate paths of stable (and Gaussian too) stochastic processes in order to obtain a minimum of appropriate accuracy required.

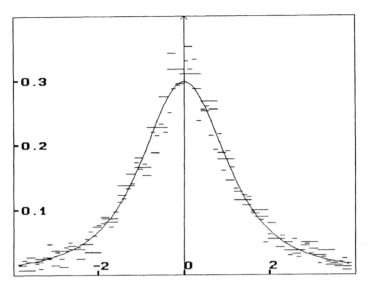

Figure 3.7.1. Density of $S_{1.2}(1,0,0)$ and histograms for three sample sizes: 1000, 2000, 4000.

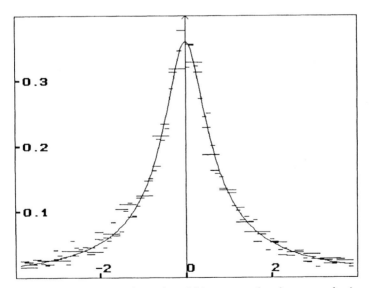

Figure 3.7.2. Density of $S_{0.8}(1,0,0)$ and histograms for three sample sizes: 2000, 4000, 8000.

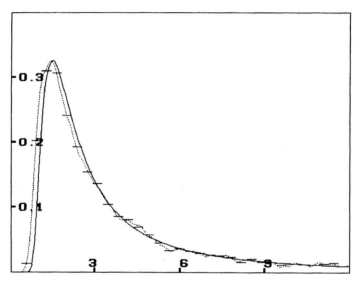

Figure 3.7.3. Density of $S_{0.7}(1,1,0)$ and kernel density estimator and histogram for the sample of 8000 "trials".

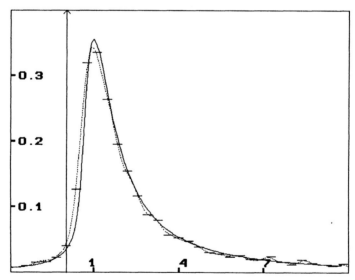

Figure 3.7.4. Density of $S_{0.7}(1,0.7,0)$ and kernel density estimator and histogram for the sample of 8000 "trials".

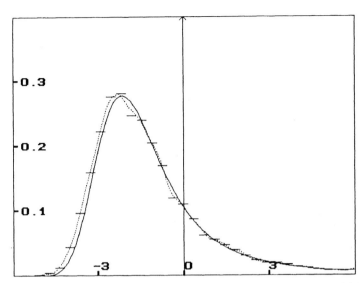

Figure 3.7.5. Density of $S_{1.3}(1,1,0)$ and kernel density estimator and histogram for the sample of 8000 "trials".

Figure 3.7.1 represents the density of the random variable $S_{1.2}(1,0,0)$ and three histograms obtained from 1000, 2000, 4000 "trials" and 20, 40, 80 subintervals of the interval $[-4,4]$, respectively (with tails properly cut off). Figure 3.7.2 presents the density and histograms of $S_{0.8}(1,0,0)$, produced with 2000, 4000, 8000 "trials" and 20, 40, 80 subintervals, so comparing it with Figure 3.7.1 one can see what happens when the size of a statistical sample is doubled.

Figures 3.7.3 – 3.7.5 include graphs of three totally skewed and skewed α-stable random variables, their density estimators obtained by the application of the Rosenblatt–Partzen method with "optimal" kernel and $n = 8000$, and histograms as well.

Figures 3.7.1 – 3.7.5 show that computer methods of simulation of α-stable random variates demand statistical samples of large sizes, especially when high accuracy is needed.

The quality of the kernel estimator heavily depends on the chosen value of the parameter b_n. Acceptable graphs were obtained for values of b_{8000} close to 0.2.

Remark 3.7.1 *The reason of representing Figures 3.7.6–3.7.12 is at least threefold:*

- *they indicate that the world of bivariate α–stable vectors is surprisingly reach;*

- *they confirm that smoothing techniques of statistical data are efficient (the densities were obtained by application of the kernel density estimator acting on the 2–dimensional sample produced by the computer generator of α–stable random variables);*

- *they suggest the possibility of application of α–stable random vectors in modeling of, non–Gaussian in nature, real life problems.*

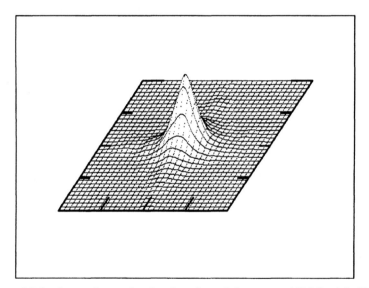

Figure 3.7.6. Approximate density function of the vector (X, Y) with X, $Y \sim S_{1.3}(1.0, 0.0, 0.0)$ on the rectangle $[-10, 10] \times [-10, 10]$.

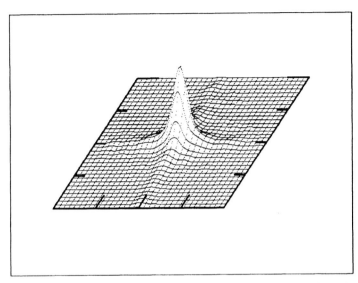

Figure 3.7.7. Approximate density function of the vector (X, Y) with X, $Y \sim$ $S_{0.7}(1.0, 0.0, 0.0)$ on the rectangle $[-10, 10] \times [-10, 10]$.

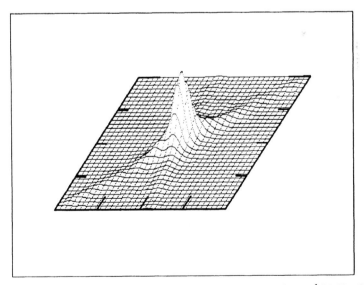

Figure 3.7.8. Approximate density function of the vector $(X - \frac{1}{10}Y, X + Y)$ with X, $Y \sim S_{0.7}(1.0, 0.0, 0.0)$ on the rectangle $[-10, 10] \times [-10, 10]$.

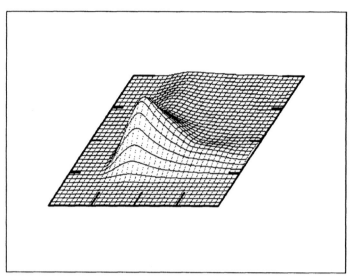

Figure 3.7.9. Approximate density function of the vector (X, Y) with X, $Y \sim$ $S_{0.7}(1.0, 1.0, 0.0)$ on the rectangle $[-2, 8] \times [-2, 8]$.

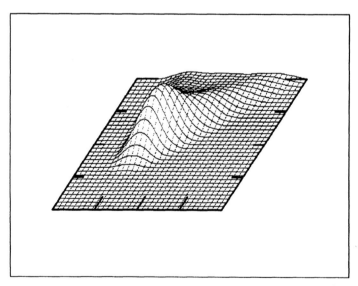

Figure 3.7.10. Approximate density function of the vector $(X - \frac{1}{10}Y, X + Y)$ with X, $Y \sim S_{0.7}(1.0, 1.0, 0.0)$ on the rectangle $[-2, 8] \times [-2, 8]$.

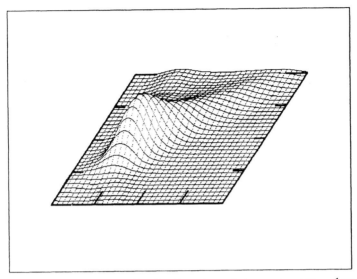

Figure 3.7.11. Approximate density function of the vector $(X - \frac{1}{10}Y, X + Y)$ with X, $Y \sim S_{0.7}(1.0, 0.7, 0.0)$ on the rectangle $[-6, 4] \times [-6, 4]$.

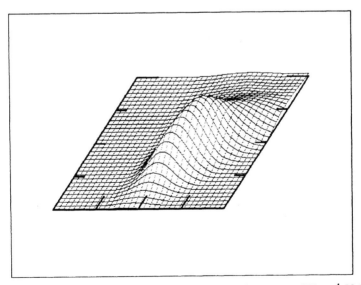

Figure 3.7.12. Approximate density function of the vector $(X - \frac{1}{10}Y, X + Y)$ with X, $Y \sim S_{1.3}(1.0, 1.0, 0.0)$ on the rectangle $[-8, 2] \times [-8, 2]$.

Chapter 4

Stochastic Integration

4.1 Introduction

Now our aim is to introduce briefly a few basic notions of stochastic analysis and to make a few remarks on their usefulness in the book. We consider only real–valued stochastic processes with continuous time, i.e. our main object of study will be random functions $\mathbf{X} = \{X(t) : t \geq 0\}$ defined on the set $\mathbb{R}_+ = [0, \infty)$, with values in appropriate spaces of random variables or finite dimensional random vectors.

Many processes, including d–dimensional Brownian motion and d–dimensional α–stable Lévy motion, can be defined in terms of their finite–dimensional distributions irrespective of their probability spaces, avoiding modern axiomatic probability theory setting. There are many textbooks, especially oriented towards engineering applications or statistical methods, written in this style. Trying to avoid unnecessary complications, however, we have to rely sometimes on facts and theorems concerning stochastic processes involving evaluation of conditional expectations not only with respect to finite families of random variables but with respect to appropriate σ–fields of events. (We recommend an inexperienced reader to check Karlin and Taylor (1975), where this subject is carefully introduced into investigation of some classes of stochastic processes.)

So, let us briefly recall the main notions of stochastic analysis which we find useful in the sequel, using the simplest possible formulation. We take as given a usual *complete probability space* $(\Omega, \mathcal{F}, P)$, together with a *filtration* $\{\mathcal{F}_t\}$, i.e., a nondecreasing family of sub–σ–fields of $\mathcal{F}$ ($\mathcal{F}_s \subseteq \mathcal{F}_t \subseteq \mathcal{F}$ for $0 \leq s < t < \infty$), which is assumed to satisfy the usual conditions of right–continuity and inclusion of all P-null sets. A stochastic process $\mathbf{X} = \{X(t, \omega) : t \geq 0, \omega \in \Omega\}$ (or in shortened notation: $\{X(t) : t \geq 0\}$) with values in $\mathbb{R}^d$ is a function $\mathbf{X} : \mathbb{R}_+ \times \Omega \to \mathbb{R}^d$ such that for each $t \geq 0$, $X(t) = X(t, \cdot)$ is a measurable function from $(\Omega, \mathcal{F})$ into $\mathbb{R}^d$ equipped with the σ–field of Borel sets $\mathcal{B}(\mathbb{R}^d)$, i.e. the smallest σ–field containing all open sets of $\mathbb{R}^d$. $\mathbf{X}$ is said to be *adapted* (to $\{\mathcal{F}_t\}$) if for each $t \geq 0$, $X(t)$ is a measurable function from $(\Omega, \mathcal{F}_t)$ into $\mathbb{R}^d$; $\mathbf{X}$ is said to be *cadlag* (after the French: continu à droite, limité à gauche) if its sample paths $\{t \to X(t, \omega) : \omega \in \Omega\}$ are right–continuous with finite left limits. All *martingales* will be assumed to have cadlag paths. A *semimartingale* is a cadlag adapted stochastic process that can be decomposed as the sum of a cadlag local martingale and a cadlag adapted process whose paths are locally of finite variation.

Given a stochastic process $\mathbf{X}$, the simplest choice of a filtration is that generated by the process itself, i.e. the family $\mathcal{F}_t = \sigma(X_s : 0 \le s \le t)$, the smallest σ-field with respect to which X_s is measurable for every $s \in [0, t]$. This will be a typical situation in the sequel.

As we have mentioned already, our main tool of investigation of α-stable processes is a stochastic integral with respect to an α-stable stochastic measure. A vast class of stochastic processes, especially important from the point of view of ergodic theory, consists of α-stable processes that can be represented by stochastic integrals in the following form

$$X(t) = \int f(t, s) \, dL_\alpha(s),$$

where $f = f(t, s)$ denotes a deterministic function satisfying appropriate conditions and $\{L_\alpha(t) : t \ge 0\}$ stands for α-stable Lévy motion.

Concentrating our attention on real-valued diffusion processes that solve stochastic differential equations of different kinds and leaving aside the discussion of maybe sometimes more important and relevant stochastic integrals, we would like to explain how the class of α-stable processes is situated among the vast family of processes studied in textbooks on stochastic analysis.

It is not so commonly understood that a vast class of $\mathbb{R}^n$-valued diffusion processes $\{X(t) : t \ge 0\}$ driven by α-stable stochastic measure, with given drift and dispersion coefficients, can be described by the following stochastic differential equation

$$d\, X(t) = a(X(t-)) \, dt + c(t) \, dL_\alpha(t); \quad t > 0, \quad X(0) = X_0,$$

that can be rewritten in the following integral form

$$X(t) = X_0 + \int_0^t a(X(s-)) \, ds + \int_0^t c(s) \, dL_\alpha(s);$$

one can consider still more general equation of the form

$$X(t) = X_0 + \int_0^t a(X(s-)) \, ds + \int_0^t c(X(s-)) \, dL_\alpha(s).$$

To the contrary, thanks to Itô's theory, it is commonly understood that any $\mathbb{R}^n$-valued diffusion process $\{X(t) : t \ge 0\}$ driven by the Wiener process (Brownian motion), with given drift and dispersion coefficients, can be obtained as a solution of the following stochastic differential equation

$$X(t) = X_0 + \int_0^t a(X(s)) \, ds + \int_0^t c(X(s)) \, dB(s),$$

where $\{B(t)\}$ stands for a Brownian motion process. The theory of such stochastic differential equations (see Arnold (1974) or Kallianpur (1980)) has been developed for a long time.

It is clear that stochastic differential equations involving stochastic integrals with a Lévy motion as an integrator includes the above equation as a special case.

Still more general are the so–called stochastic differential equations with jumps. In the case of real–valued process the problem is to find the solution $\{X(t) : t \in [0, \infty)\}$ of the equation

$$X(t) = X_0 + \int_0^t \int_{\mathrm{IR}\backslash\{0\}} f(X(s-), x)\, N(ds, dx),$$

where $N(ds, dx)$ is a Poisson measure of a suitable point process with a given intensity measure $ds\, d\nu(x)$.

All types of stochastic equations described above are special cases of general stochastic differential equations driven by semimartingales, i.e. equations of the form

$$X(t) = X_0 + \int_0^t f(X(s-))\, dY(s),$$

where $\{Y(t)\}$ stands for a given semimartingale process.

There is a vast literature concerning this topic (see, for example, Protter (1990) and the bibliography therein).

4.2   Itô Stochastic Integral

For a given complete probability space $(\Omega, \mathcal{F}, P)$ (with a filtration $\{\mathcal{F}_t : t \geq 0\}$), let us shortly list some properties of martingales.

Definition 4.2.1 *Let $\{X(t) : t \in [0, \infty)\}$ be a real–valued process, adapted to the filtration $\{\mathcal{F}_t\}$ and such that $E|X(t)| < \infty$ for all $t \geq 0$. We say that $\{X(t)\}$ is a **martingale** if*

$$\forall_{t \geq s} \quad E\ (X(t)|\mathcal{F}_s) = X(s). \tag{4.2.1}$$

It is known that if a process $\{X(t)\}$ is a martingale such that $E\,|X(t)|^p < \infty$ for $p \geq 1$, then

$$P\left\{\sup_{0 \leq s \leq t} |X(s)| \geq \lambda\right\} \leq \frac{E|X(t)|^p}{\lambda^p}. \tag{4.2.2}$$

For $p > 1$, estimation (4.2.2) can be sharpened to

$$E \sup_{0 \leq s \leq t} |X(t)|^p \leq \left(\frac{p}{p-1}\right)^p E\,|X(t)|^p. \tag{4.2.3}$$

Wiener processes.   Let us recall the definition of the standard real–valued Wiener process, equivalent to the definition 2.2.1 of a Brownian motion process.

Definition 4.2.2 *The **standard one–dimensional Wiener process** is a process $\{W(t) : t \geq 0\}$ (or in full notation: $\{W(t,\omega) : t \geq 0,\ \omega \in \Omega\}$), satisfying the following conditions*

(i) $\forall_{t_1,\dots,t_n}$ $W(t_1),\dots,W(t_n)$ is a Gaussian vector with mean 0;

(ii) $\forall_{t,\,s \geq 0}$ $E\ W(t)W(s) = \min\{t,s\}$;

(iii) $P\{\omega : W(\cdot,\omega)$ is a continuous function$\} = 1$.

Let us notice that

$$W(0) = 0, \qquad E\ W(t)^2 = t, \tag{4.2.4}$$

$$\forall_{0 \leq t_1 \leq t_2 \leq t_3 \leq t_4}\ \ W(t_2) - W(t_1),\ \ W(t_4) - W(t_3)\ \text{ are independent.} \tag{4.2.5}$$

The last statement follows from the equality

$$E\ \langle W(t_2) - W(t_1), W(t_4) - W(t_3)\rangle$$
$$= \min\{t_2,t_4\} - \min\{t_1,t_4\} - \min\{t_2,t_3\} + \min\{t_1,t_3\} = 0,$$

which shows that random variables $W(t_2) - W(t_1)$ and $W(t_4) - W(t_3)$ are uncorrelated.

Let $\mathcal{G}_t$ denote the σ–algebra generated by $\{W(s) : s \in [0,t]\}$. We have

$$\forall_{t \geq s}\ \ E\ [W(t)|\mathcal{G}_s] = W(s). \tag{4.2.6}$$

Indeed, $W(t) - W(s)$ is independent of all $W(\tau)$ for $\tau \in [0,s]$, and thus $\{W(t)\}$ is a martingale with respect to $\{\mathcal{G}_t\}$, which means that we have

$$\forall_{t \geq s}\ \ E\ [(W(t) - W(s))^2|\mathcal{G}_s] = t - s. \tag{4.2.7}$$

This equation characterizes martingales with mean 0 which satisfy the definition of the Wiener process. Moreover, we have the following theorem due to Lévy.

Theorem 4.2.1 *Let $\{W(t) : t \geq 0\}$ denote a real valued continuous process such that $W(0) = 0$. Let this process be a martingale with respect to an increasing family of σ–algebras $\{\mathcal{F}_t : t \geq 0\}$ and such that*

$$\forall_{t \geq s \geq 0}\ \ E\ [(W(t) - W(s))^2|\mathcal{F}_s] = t - s. \tag{4.2.8}$$

Then $\{W(t)\}$ is a Wiener process.

For the proof the interested reader is referred to Doob (1953).

Definition 4.2.3 *The **standard, n–dimensional Wiener process** is the vector-valued process $\{W(t) : t \geq 0\}$ composed of n scalar Wiener processes $\{W_i(t)\}_{i=1}^n\}$ such that*

$$E\ W_i(t)W_j(s) = \delta_{ij}\ \min\{t,s\}, \quad \text{for}\quad i,j = 1,2,\dots,n. \tag{4.2.9}$$

In a more general setting, the $\mathbb{R}^n$-valued Wiener process can be defined as the continuous $\{\mathcal{F}_t\}$ martingale which starts from 0 at 0 a.s. and such that

$$E\ [\langle W(t) - W(s), W(t) - W(s)\rangle | \mathcal{F}_s] = (t - s)\ I, \qquad (4.2.10)$$

where I denotes the identity matrix.

As usual, we can take for the filtration $\{\mathcal{F}_t\}$ the family of σ-algebras generated by $\{W(s) : 0 \le s \le t\}$.

Stochastic integral with respect to Wiener process.

With a given increasing family $\{\mathcal{F}_t : t \in [0, T]\}$ we fix a continuous stochastic process $\{W(t) : t \in [0, T]\}$ which is an $\{\mathcal{F}_t\}$–martingale, vanishes at 0 with probability 1, and satisfies condition (4.2.10).

Let us introduce for $p \ge 1$ the space of real valued and measurable adapted (equivalence classes of) processes $\{\Phi(t) = \Phi(t, \omega)\}$:

$$L^p{}_W(0, T) = \left\{ \Phi : \int_0^T |\Phi(t)|^p\ dt < \infty \ \text{a.s.} \right\}, \qquad (4.2.11)$$

and the space

$$L^p{}_{\mathcal{F}}(0, T) = \left\{ \Phi \in L^p{}_W(0, T) : \ E\ \left(\int_0^T |\Phi(t)|^p\ dt \right) < \infty \right\}. \qquad (4.2.12)$$

In the obvious way one can define the spaces $L^p{}_W(0, T; \mathbb{R}^n)$ and $L^p{}_{\mathcal{F}}(0, T; \mathbb{R}^n)$. Let us notice that $L^p{}_{\mathcal{F}}(0, T; \mathbb{R}^n)$ is contained in $L^p((0, T) \times \Omega, dt \otimes dP; \mathbb{R}^n)$.

When $p = 2$ we have an important special case of the Hilbert space

$$\mathfrak{F} \stackrel{\text{df}}{=} L^2{}_{\mathcal{F}}(0, T; \mathbb{R}^n).$$

Let N denote an integer tending to ∞, and let $\tau = T/N$. With any fixed process $\{\Phi(t)\}$ we can associate a process with "finite number of values" $\{\Phi_\tau(t)\}$ defined as follows

$$\Phi_\tau(t) = \frac{1}{\tau} \int_{(n-1)\tau}^{n\tau} \Phi(s)\ ds \ \text{ for } \ t \in [(n-1)\tau, n\tau) \ \text{ and } \ 1 \le n < N. \qquad (4.2.13)$$

The sequence $\{\Phi_\tau\}$ converges to Φ in the space $L^2((0, T); L^2(\Omega, \mathcal{F}, P; \mathbb{R}^n))$. It is also clear that $\Phi_\tau \in \mathfrak{F}$, so we have $\Phi_\tau \to \Phi$ in the space $\mathfrak{F}$. This means that all "simple processes" Φ_τ define a dense subspace in the Hilbert space $\mathfrak{F}$.

Definition 4.2.4 *Let* $\Phi \in \mathfrak{F}$ *be a **simple process**, i.e. a process such that*

$$\Phi(t) = \Phi_n \quad \text{for} \quad t \in [t_n, t_{n+1}) \quad \text{and} \quad n = 0, 1, ..., N - 1, \qquad (4.2.14)$$

for a given partition $0 = t_0 < t_1 < ... < t_N = T$ *of* $[0, T]$ *and where* Φ_n *are* $\mathcal{F}_{t_n}$*-measurable. It is natural to define the **Itô stochastic integral** for such processes as finite sums of the following form*

$$I = I(\Phi) \stackrel{\text{df}}{=} \sum_{n=0}^{N-1} \langle \Phi_n, W(t_{n+1}) - W(t_n) \rangle. \qquad (4.2.15)$$

We have

$$E\ I(\Phi)^2 = \sum_{n=0}^{N-1}(t_{n+1} - t_n)\ E|\Phi_n|^2 = E\int_0^T |\Phi(t)|^2\ dt. \qquad (4.2.16)$$

Indeed, for $t_{n+1} \leq t_m$ we have

$$E\langle\Phi_n, W(t_{n+1}) - W(t_n)\rangle\langle\Phi_m, W(t_{m+1})\rangle$$

$$= E\{\langle\Phi_n W(t_{n+1}) - W(t_n)\rangle\langle\Phi_m, E[W(t_{m+1}) - W(t_m)|\mathcal{F}_{t_m}]\rangle\} = 0,$$

because $E[W(t_{m+1})|\mathcal{F}_{t_m}] = W(t_m)$.

We also have

$$E\ |\langle\Phi_n, W(t_{n+1}) - W(t_n)\rangle|^2$$

$$= E\ \{\langle(\ E[\langle W(t_{n+1}) - W(t_n), W(t_{n+1}) - W(t_n)\rangle|\mathcal{F}_{t_n}]\ )\Phi_n, \Phi_n\rangle\}$$
$$= (t_{n+1} - t_n)\ E\ |\Phi_n|^2,$$

according to (4.2.10).

Now it is clear that the operator $I = I(\Phi)$ acts as a linear operator from a dense subspace of $\mathfrak{F}$ into $L^2(\Omega, \mathcal{F}, P)$. It is also continuous with respect to the norm of $\mathfrak{F}$. We can thus formulate the following definition.

Definition 4.2.5 *The operator $I = I(\Phi)$ defined in Definition 4.2.4 can be linearly extended onto all space $\mathfrak{F}$, so passing to the limit with an appropriately constructed sequence of simple processes converging to Φ, we obtain*

$$I(\Phi) \overset{\text{df}}{=} \int_0^T \langle\Phi(s), dW(s)\rangle. \qquad (4.2.17)$$

*The random variable $I(\Phi)$ is by definition the **Itô stochastic integral** of Φ.*

Incidentally, we also obtained the following properties of this integral.

Proposition 4.2.1 *The stochastic integral described by (4.2.17) defines a linear operator acting from the space $\mathfrak{F}$ into $L^2(\Omega, \mathcal{F}, P)$, which has the following properties*

$$\forall_{\Phi\in\mathfrak{F}}\ \ E\ I(\Phi) = 0; \qquad (4.2.18)$$

$$\forall_{\Phi\in\mathfrak{F}}\ \ E\ I(\Phi)^2 = \|\Phi\|_{\mathfrak{F}}^2; \qquad (4.2.19)$$

$$\forall_{\Phi,\Psi\in\mathfrak{F}}\ \ E\ I(\Phi)\ I(\Psi) = E\int_0^T \langle\Phi(s), \Psi(s)\rangle\ ds. \qquad (4.2.20)$$

Stochastic integral as a stochastic process. We want to discuss briefly some properties of the process

$$I(\Phi)(t) = \int_0^t \langle \Phi(s), dW(s) \rangle, \quad t \in [0, T], \tag{4.2.21}$$

defined for $\Phi \in \mathfrak{F} = L^2_{\mathcal{F}}(0, T; \mathbb{R}^n)$.

We can assume that we have chosen a separable version of the process defined above. This process is also an $\mathcal{F}_t$–martingale because

$$E\left[I(\Psi)(t)|\mathcal{F}_s\right] = I(\Phi)(s) + E\left[\int_s^t \langle \Phi(u), dW(u) \rangle | \mathcal{F}_s\right] = I(\Phi)(s).$$

We have also

$$E I(\Phi)(t)^2 = E \int_0^t |\Phi(s)|^2 \, ds,$$

and, thanks to properties (4.2.2) and (4.2.3) of martingales, we get

$$P\left\{ \sup_{0 \le t \le T} \left| \int_0^t \langle \Phi(s), dW(s) \rangle \right| > a \right\} \le \frac{1}{a^2} E \int_0^T |\Phi(s)|^2 \, ds, \tag{4.2.22}$$

$$E \sup_{0 \le t \le T} \left| \int_0^t \langle \Phi(s), dW(s) \rangle \right|^2 \le 4 E \int_0^T |\Phi(s)|^2 \, ds. \tag{4.2.23}$$

4.3 α–Stable Stochastic Integrals of Deterministic Functions

Denote by $(\Omega, \mathcal{F}, P)$ the underlying probability space and by $L^o(\Omega, \mathcal{F}, P)$ the set of all real random variables defined on it. Let $(E, \mathcal{E}, m)$ be a measure space, let

$$\beta : E \to [-1, 1]$$

be a measurable function, and let

$$\mathcal{E}_0 = \{A \in \mathcal{E} : m(A) < \infty\}$$

be the subset of $\mathcal{E}$ that contains sets of finite m–measure.

Definition 4.3.1 *An independently scattered σ–additive set function*

$$M : \mathcal{E}_0 \ni A \to M(A) \in L^o(\Omega, \mathcal{F}, P)$$

such that for each $A \in \mathcal{E}_0$

$$M(A) \sim S_\alpha \left((m(A))^{1/\alpha}, \frac{1}{m(A)} \int_A \beta(x) \, dm(x), 0 \right)$$

is called an α–stable random measure on $(E, \mathcal{E})$ with control measure m and skewness intensity β. Measure M is called an $S\alpha S$ random measure if the skewness intensity β is zero.

Now we want to define an α–stable stochastic integral

$$I(f) = \int_E f(x) \, dM(x)$$

with respect to an α–stable random measure on $(E, \mathcal{E})$ for all measurable functions $f : E \to \mathbb{R}$ satisfying the condition

$$\int_E |f(x)|^\alpha \, dm(x) < \infty, \tag{4.3.1}$$

i.e., for all $f \in L^\alpha(E, \mathcal{E}, m)$.

The first step. For a simple function $f(x) = \sum_j c_j I_{A_j}(x)$ with disjoint sets A_j belonging to $\mathcal{E}_0$, we put

$$I(f) = \int_E f(x) \, dM(x) = \sum_{j=1}^n a_j M(A_j). \tag{4.3.2}$$

The map $I(f)$ is obviously linear on the linear space of simple functions f. Since the random measure M is independently scattered, the α–stable random variables: $M(A_1), ..., M(A_n)$ are independent and, consequently, $I(f)$ is α–stable, say $I(f) \sim S_\alpha(\sigma_f, \beta_f, \mu_f)$.

The second step. Consider now any function f from the space $L^\alpha(E, \mathcal{E}, m)$, i.e., from the linear space of all functions satisfying (4.3.1). It is easy to notice that there exists a sequence of simple functions $\{f^{(n)}\}_{n=1}^\infty$ such that

$$f^{(n)}(x) \to f(x) \quad \text{for} \quad a.a. \quad x \in E;$$

$$|f^{(n)}(x)| \le g(x), \text{ for any } n, \ x \text{ and some } g \in L^\alpha(E, \mathcal{E}, m).$$

The sequence of integrals $\{I(f^{(n)})\}_{n=1,2,...}$ is well defined by (4.3.2). It is a Cauchy sequence in the complete space of α–stable random variables with the metric induced by converges in probability. So, there exists a random variable $I(f)$ which is the limit of $\{I(f^{(n)})\}$ in this space.

Definition 4.3.2 *Therefore an α–stable stochastic integral of any function $f \in L^\alpha(E, \mathcal{E}, m)$ is defined by the following formula*

$$I(f) \stackrel{\mathrm{df}}{=} \lim_{n \to \infty} I(f^{(n)}) \quad \text{in probability.} \tag{4.3.3}$$

This definition does not depend on the choice of the approximating sequence $\{f^{(n)}\}$ and $I(f)$ is linear in f. Thus, for any $f_1, ..., f_d$ from $L^\alpha(E, \mathcal{E}, m)$, the random vector $(I(f_1), I(f_2), ..., I(f_d))$ is α–stable with the characteristic function

$$\phi_{f_1,...,f_d}(\theta_1, ..., \theta_d) =$$

$$\exp\left\{-\int_E |\sum_{j=1}^d \theta_j f_j(x)|^\alpha \left(1 - i\beta(x)\text{sign}\left(\sum_{j=1}^d \theta_j f_j(x)\right) \tan(\pi\alpha/2)\right) dm(x)\right\}.$$

We refer the reader to Samorodnitsky and Taqqu (1993) for more details.

Now we recall some basic properties of an α–stable integral. (In the sequel the stochastic integral is discussed in more general setting.)

Theorem 4.3.1 *Let the sequence of integrals* $X_j = \int_E f_j(x) \, dM(x)$, *for* $j = 1, 2, ...,$ *and* $X = \int_E f(x) \, dM(x)$ *be defined by* f_j *and* f *from the space* $L^\alpha(E, \mathcal{E}, m)$, *respectively, where* $\mathbf{M}$ *denotes an* α–*stable random measure with control measure* m. *Then* $X = \lim_{j\to\infty} X_j$ *in probability if and only if* $f = \lim_{j\to\infty} f_j$ *in the corresponding space* $L^\alpha(E, \mathcal{E}, m)$.

Theorem 4.3.2 *Let be given two stochastic integrals* $X_j = \int_E f_j(x) \, dM(x)$, *for* $j = 1, 2$, *where* $\mathbf{M}$ *denotes an* α–*stable random measure with control measure* m. *Then* X_1 *and* X_2 *are independent if and only if* $f_1 f_2 \equiv 0$ m–*a.e. on* E *for* $0 < \alpha < 2$, *or* $\text{Cov}(X_1, X_2) = 0$ *for* $\alpha = 2$.

Theorem 4.3.3 *Let be given two stochastic integrals* $X_j = \int_E f_j(x) \, dM(x)$, *for* $j = 1, 2$, *where* $\mathbf{M}$ *denotes an* $S\alpha S$ *random measure with control measure* m. *If* $1 < \alpha \leq 2$, *then the covariation* $[X_1, X_2]_\alpha = \int_E f_1 f_2^{<\alpha-1>} dm(x)$ *is well defined and* X_2 *is James orthogonal to* X_1 *if and only if* $[X_1, X_2]_\alpha = 0$.

In the next few sections we present constructions of more general stochastic integrals.

Stochastic α–stable integrals defined in this section play, however, such an important role in our investigations, that in what follows we present another approach to the construction of this integral based on the Poisson representation of underlying α–stable random measures.

4.4 Infinitely Divisible Processes

Let be given a filtered probability space $(\Omega, \mathcal{F}, \{\mathcal{F}_t\}_{0\leq t\leq\infty}, P)$.

Definition 4.4.1 *An adapted process* $\mathbf{X} = \{X(t)\}_{0\leq t<\infty}$ *with* $X(0) = 0$ *a.s. is a Lévy process if*

1. $\mathbf{X}$ *has increments independent of the past; that is,* $X(t) - X(s)$ *is independent of* $\mathcal{F}_s$, $0 \leq s < t < \infty$;

2. $\mathbf{X}$ *has stationary increments; that is,* $X(t) - X(s)$ *has the same distribution as* $X(t - s)$, $0 \leq s < t < \infty$;

3. $\{X(t)\}$ *is continuous in probability; that is,* $\lim_{t\to s} X(t) = X(s)$ *and the limit is taken in probability.*

If we take the Fourier transform of each $X(t)$, we get a function $f(t, u) = f_t(u)$ defined by

$$f_t(\theta) = E\{e^{i\theta X(t)}\},$$

where $f_0(u) = 1$, and $f_{t+s}(u) = f_t(u)f_s(u)$, and $f_t(u) \neq 0$ for every (t, u). Using the (right) continuity in probability we conclude that $f_t(u) = e^{-t\psi(u)}$ for some continuous function $\psi(u)$ with $\psi(0) = 0$. (The Bochner Theorem can be used to show the converse: if ψ is continuous, $\psi(0) = 0$, and if for all $t \geq 0$, $f_t(u) = e^{-t\psi(u)}$ satisfies $\sum_{i,j} \alpha_i \overline{\alpha}_j f_t(u_i - u_j) \geq 0$, for all finite $u_1, ..., u_n; \alpha_1, .., \alpha_n$ then there exists a Lévy process corresponding to the function f.) In particular, it follows that if X is a Lévy process then for each $t > 0$, $X(t)$ has an infinitely divisible distribution. Conversely, it can be shown that for each infinitely divisible distribution there exists a Lévy process X such that μ is the distribution of $X(1)$.

Theorem 4.4.1 *Let X be a Lévy process. There exists a unique modification Y of X which is cadlag and also a Lévy process.*

PROOF. Let $M_t^u = \frac{e^{iuX(t)}}{f_t(u)}$. For each fixed $u \in \mathbb{Q}$, the rational in $\mathbb{R}$, the process $(M_t^u)_{0 \leq t < \infty}$ is a complex-valued martingale (relative to $\{\mathcal{F}_t\}$). It can be checked that M^u has a version with cadlag paths for each u. Let Θ^u be the null set such that for $\omega \in \Theta_u$ the map $t \to M_t^u(\omega)$ is right continuous with left limits. Let $\Theta = \bigcup_{u \in \mathbb{Q}} \Theta_u$. Then $P(\Theta) = 0$.

We first show that the paths of X cannot explode a.s. The process $\{M_t^u\}_{t \geq 0}$ is a (complex-valued) martingale for any real u and thus for a.a. ω the functions $t \to M_t^u(\omega)$ and $t \to e^{iuX(t,\omega)}$, with $t \in \mathbb{Q}$ are the restrictions to $\mathbb{Q}_+$ of cadlag functions. Let

$$\Theta = \{(\omega, u) \in \Omega \times \mathbb{R} : e^{iuX(t,\omega)}, t \in \mathbb{Q}\}.$$

One can check that Θ is a measurable set and we see that $\int I_\Theta(\omega, u)P(d\omega) = 0$ for each $u \in \mathbb{R}$. By the Fubini Theorem

$$\int \int_{-\infty}^{\infty} I_\Theta(\omega, u) du P(d\omega) = \int_{-\infty}^{\infty} \int I_\Theta(\omega, u)P(d\omega)du = 0,$$

hence we conclude that for a.a. ω the function $t \to e^{iuX(t,\omega)}, t \in \mathbb{Q}_+$ is a restriction of a cadlag function for almost all $u \in \mathbb{R}$. We can conclude that the function $t \to X(t,\omega), t \in \mathbb{Q}_+$, is the restriction of a cadlag function for every such ω. It is enough to make use of Lemma 4.4.1, which is proven below.

Next, choose $\omega \notin \Theta$ and suppose that $s \to X(s,\omega)$ has two or more distinct accumulation points, say x and y, when s increases (respectively, decreases) to t. Choose an $u \in \mathbb{Q}$ such that $u(y - x) \neq 2\pi m$ for all integers m. This then contradicts the assumption that $\omega \notin \Omega$, whence $s \to X(s,\omega)$ can have only unique limit points.

Set $Y_t = \lim_{s \in \mathbb{Q}_+, s \downarrow t} X(s)$ on $\Omega \backslash \Theta$ and $Y_t = 0$ on Θ, all t. Since $\mathcal{F}_t$ contains all P-null sets of $\mathcal{F}$ and $\{\mathcal{F}_t\}_{0 \leq t < \infty}$ is right continuous then $Y(t) \in \mathcal{F}_t$. Since X is continuous in probability, $P\{Y(t) \neq X(t)\} = 0$, hence Y is a modification of X. It is clear that Y is a Lévy process as well. $\square$

Now we have to prove the lemma which was used in the above theorem, making use of a purely probabilistic method.

Lemma 4.4.1 *Let* $\{x_n\}$ *be a sequence of real numbers such that* e^{iux_n} *converges as* $n \to \infty$ *for almost all* $u \in \mathbb{R}$. *Then* $\{x_n\}$ *converges to a finite limit.*

PROOF. We will verify the following Cauchy criterion: x_n converges if for any increasing sequences n_k and m_k $\lim_{k\to\infty}(x_{n_k} - x_{m_k}) = 0$. Let U be a random variable which has the uniform distribution on $[0,1]$. For any real t, by hypothesis $e^{itUx_{n_k}}$ and $e^{itUx_{m_k}}$ converge a.s. to the same limit. Thus we have

$$\lim_{k\to\infty} e^{itU(x_{n_k}-x_{m_k})} = 1 \quad \text{a.s.}$$

and therefore the characteristic functions converge, i.e.

$$\lim_{k\to\infty} E\{e^{it(x_{n_k}-x_{m_k})U}\} = 1,$$

for all $t \in \mathbb{R}$. Consequently $(x_{n_k} - x_{m_k})U$ converges to zero in probability, whence $\lim_{k\to\infty}(x_{n_k} - x_{m_k}) = 0$, as claimed. $\qquad\square$

We can henceforth always assume that we are using *the (unique) cadlag version of any given Lévy process*. Lévy processes provide us with examples of filtrations that satisfy the "usual hypotheses", as the next theorem shows.

Theorem 4.4.2 *Let* **X** *be a Lévy process and let* $\mathcal{G}_t = \mathcal{F}_t^0 \vee \mathcal{N}$, *where* $\{\mathcal{F}_t^0\}_{0 \leq t < \infty}$ *is the natural filtration of* **X** *and* $\mathcal{N}$ *are the P–null sets of* $\mathcal{F}$. *Then* $\{\mathcal{G}_t\}_{0 \leq t < \infty}$ *is right continuous.*

PROOF. We must show $\mathcal{G}_{t+} = \mathcal{G}_t$, where $\mathcal{G}_{t+} = \bigcap_{u>t}\mathcal{G}_u$. Note that since the filtration $\mathcal{G}$ is increasing, it suffices to show that $\mathcal{G}_t = \bigcap_{n\geq1}\mathcal{G}_{t+\frac{1}{n}}$. Thus we can take countable limits and it follows that if we take $s_1, \ldots s_n \leq t$ then for $(u_1, \ldots u_n)$ we have

$$E\{e^{i(u_1X(s_1)+\ldots+u_nX(s_n))} \mid \mathcal{G}_t\} = E\{e^{i(u_1X(s_1)+\ldots+u_nX(s_n))} \mid \mathcal{G}_{t+}\}$$
$$= e^{i(u_1X(s_1)+\ldots+u_nX(s_n))}$$

For $v_1, \ldots, v_n > t$ and $(u_1, \ldots, u_n)$ we present the proof only for $n = 2$ for notational convenience. Therefore, let $z > v > t$, and suppose that u_1, u_2 are given. We have

$$E\{e^{i(u_1X(v)+u_2X(z))} \mid \mathcal{G}_{t+}\} = \lim_{w\downarrow t} E\{e^{i(u_1X(v)+u_2X(z))} \mid \mathcal{G}_w\}$$

$$= \lim_{w\downarrow t} E\left\{e^{iu_1X(v)}\frac{e^{iu_2X(z)}}{f_z(u_2)} \mid \mathcal{G}_w\right\}$$

$$= \lim_{w\downarrow t} E\left\{e^{iu_1X(v)}\frac{e^{iu_2X(z)}}{f_z(u_2)}f_z(u_2) \mid \mathcal{G}_w\right\}$$

$$= \lim_{w\downarrow t} E\left\{e^{iu_1X(v)}\frac{e^{iu_2X(v)}}{f_v(u_2)}f_z(u_2) \mid \mathcal{G}_w\right\},$$

noticing that $M_v^{u_2} = \frac{e^{iu_2 X(v)}}{f_v(u_2)}$ is a martingale. Combining terms, the above becomes

$$E\{e^{i(u_1 X(v)+u_2 X(z))} \mid \mathcal{G}_{t+}\} = \lim_{w \downarrow t} E\left\{e^{i(u_1+u_2)X(v)} f_{z-v}(u_2) \mid \mathcal{G}_w\right\}$$

and the same martingale argument yields

$$E\left\{ e^{i(u_1 X(v)+u_2 X(z))} \mid \mathcal{G}_{t+}\right\}$$
$$= \lim_{w \downarrow t} e^{i(u_1+u_2)X(w)} f_{v-w}(u_1+u_2) f_{z-v}(u_2) = e^{i(u_1+u_2)X(t)},$$

which ends the proof.   □

Let us recall two important definitions.

Definition 4.4.2 *A probability measure μ on $\mathbb{R}$, or a real–valued random variable Y with law μ, is said to be **infinitely divisible** if for any $n \geq 1$ there is a probability measure μ_n such that $\mu = \mu_n^{*n}$ or equivalently if Y has the law of the sum of n independent identically distributed random variables.*

It is easy to see that Gaussian, Poisson or Cauchy variables are infinitely divisible.

Definition 4.4.3 *We say that a measure μ satisfies the **Lévy–Kchintchine** formula with the **Lévy measure** ν if its Fourier transform $\widehat{\mu}$ is of the form*

$$\widehat{\mu}(\theta) = \exp\{\psi(\theta)\},$$

with

$$\psi(\theta) = ib\theta - \frac{c\theta^2}{2} + \int \left(e^{i\theta x} - 1 - \frac{i\theta x}{1+x^2}\right) \nu(dx)$$

where $b \in \mathbb{R}$, $c \geq 0$ and ν is a Radon measure on $\mathbb{R}\setminus\{0\}$ such that

$$\int \frac{x^2}{1+x^2}\nu(dx) < \infty.$$

It seems worthwhile to state a few elementary facts about Lévy processes which will be used in the sequel. Lévy processes have been widely studied in their own right because their properties often hint at properties of general Markov processes. In what follows below we deal only with real–valued Lévy processes.

Obviously, if $\mathbf{X}$ is a Lévy process then any random variable $X(t)$ is infinitely divisible. Conversely, it was proved by Lévy that any infinitely divisible r.v. Y may be imbedded in a unique convolution semi–group, in other words, there exists a Lévy process $\mathbf{X}$ such that $Y \overset{d}{=} X(1)$. This can be proved as follows. By analytical methods, one can show that μ is infinitely divisible if and only if it satisfies the Lévy–Kchintchine Formula. For every $t \in \mathbb{R}_+$ the function $\exp(t\psi(\theta))$ is now clearly the Fourier transform of a probability measure μ_t and plainly $\mu_t * \mu_s = \mu_{t+s}$, and $\lim_{t \downarrow 0} \mu_t = \varepsilon_0$ which proves Lévy's theorem.

The different terms which appear in the Lévy–Kchintchine Formula have a probabilistic significance which will be further emphasized in the sequel. If $\mathfrak{c} = 0$ and $\nu = 0$ then $\mu = \varepsilon_{\mathfrak{b}}$ and the corresponding semi–group is that of transformation at speed $\mathfrak{b}$; if $\mathfrak{b} = 0$ and $\nu = 0$, the semi–group is that of a multiple of the Brownian motion and the corresponding Lévy process has continuous paths; if $\mathfrak{b} = 0, \mathfrak{c} = 0$, we get a "pure jump" process.

Every Lévy process is obtained as a sum of independent process of the three types above. Thus, the Lévy measure accounts for the jumps of $\mathbf{X}$ and the knowledge of ν permits to give a probabilistic construction of $\mathbf{X}$.

Remark 4.4.1 *Notice that Lévy processes include Brownian motion, the Poisson process and α–stable Lévy motion processes.*

4.5   Stochastic Integrals with ID Integrators

Before introducing integrals of deterministic functions with respect to stochastic measures it is convenient to consider the idea of an integral with respect to a general vector–valued measure.

Assume that E is an arbitrary set and let $\mathcal{E}_0$ be an *algebra of subsets* of E i.e. for any $A, B \in \mathcal{E}_0$ we have $A \cup B \in \mathcal{E}_0$ and $E \setminus A \in \mathcal{E}_0$. Now, let $(\mathfrak{F}, \|\cdot\|)$ be a complete linear metric space with distance $\|x - y\|$ for $x, y \in \mathfrak{F}$, and let

$$\Lambda : \mathcal{E}_0 \ni A \to \Lambda(A) \in \mathfrak{F}$$

be an *additive set function* which means that $\Lambda(A \cup B) = \Lambda(A) + \Lambda(B)$ whenever $A, B \in \mathcal{E}_0$, and $A \cap B = \emptyset$.

If one wants to construct a rich enough theory of integration with respect to Λ, the immediate natural question is: under what conditions can Λ be extended to a σ–*additive set function* or, in other words, $\mathfrak{F}$–*valued measure*, on the σ–field $\mathcal{E} = \sigma(\mathcal{E}_0)$? The measure, whenever it exists, will be also denoted by Λ. One can prove that Λ can be extended to an $\mathfrak{F}$–valued measure if and only if the following condition is satisfied:

$$\text{if} \quad A_n \in \mathcal{E}_0 \quad \text{and} \quad \limsup_{n \to \infty} A_n = \emptyset, \quad \text{then} \quad \lim_{n \to \infty} \Lambda(A_n) = 0. \qquad (4.5.1)$$

In the set of all real–valued functions f defined on the measure space $(E, \mathcal{E})$ let us distinguish the class $\Sigma = \Sigma(\mathcal{E}_0)$ of step functions on E which are $\mathcal{E}_0$–measurable, i.e.

$$f(s) = \sum_{i=1}^{n} \phi_i \boldsymbol{I}_{A_i}(s),$$

where $\phi_i \in \mathbb{R}$, and $A_i \in \mathcal{E}_0$. For such f the obvious definition of the integral is

$$\int_E f \, d\Lambda = \sum_{i=1}^{n} \phi_i \Lambda(A_i).$$

In order to extend this construction of the integral to wider classes of real-valued functions defined on the measure space $(E, \mathcal{E})$ in a general case one can make use of the following functional

$$\rho_\Lambda(f) = \sup \left\{ \left\| \int_E \nu f \, d\Lambda \right\| : \nu \in \Sigma(\mathcal{E}_0), \ |\nu| \le 1 \right\} \qquad (4.5.2)$$

defined for $f \in \Sigma(\mathcal{E}_0)$.

In what follows we restrict ourselves to the case of Λ taking values in subspaces of $L^o(\Omega, \mathcal{F}, P)$. We will refer to it as an *additive set function* or a *stochastic measure*.

The following property of ρ_Λ and the integral $\int f \, d\Lambda$ with $f \in \Sigma(\mathcal{E}_0)$ is directly related to the question of extendability of Λ to such a measure.

Theorem 4.5.1 *Let $\mathfrak{F} = L^p(\Omega, \mathcal{F}, P)$, $0 \le p < \infty$, and $\Lambda : \mathcal{E}_0 \to \mathfrak{F}$ be an additive stochastic set function. Then Λ can be extended to an $\mathfrak{F}$-valued stochastic measure $\Lambda : \mathcal{E} = \sigma(\mathcal{E}_0) \to \mathfrak{F}$ if and only if for each sequence of functions $f_n \in \Sigma(\mathcal{E}_0)$ with $|f_n(s)| \le 1$, and such that $\lim_{n \to \infty} f_n(s) = 0$ for all $s \in E$, we have $\lim_{n \to \infty} \rho_\Lambda(f_n) = 0$.*

For the proof we refer the reader to Kwapień and Woyczyński (1992).

In view of the above results, throughout the remaining part of this section Λ will be assumed to satisfy the condition formulated in the theorem. This assumption permits us to introduce the following definition of Λ-integrability.

Definition 4.5.1 *Function $f : E \to \mathbb{R}$ is said to be Λ-integrable if there exists a sequence $f_n \in \Sigma(\mathcal{E}_0)$ such that:*

(i) for any $s \in E$, $f_n(s) \to f(s)$ as $n \to \infty$;

(ii) $\rho_\Lambda(f_n - f_m) \to 0$, as $n, m \to \infty$.

For an Λ-integrable function f, condition (ii) implies, that there exists an element $I_f \in \mathfrak{F}$ such that $\| I_f - \int_E f_n \, d\Lambda \| \to 0$ as $n \to \infty$ and we can define the integral of f with respect to Λ by the formula

$$\int_E f \, d\Lambda = I_f.$$

Since not every bounded measurable function is a pointwise limit of a sequence of step functions, the above definition has to be augmented in the following *non-essential* way. First define a mapping $\Lambda^\sharp$ from the family of subsets $A \subset E$ into $\mathbb{R}^+$ by the formula

$$\Lambda^\sharp(B) = \inf \sum_{i=1}^{\infty} \rho_\Lambda(I_{A_i}),$$

where the infimum is taken over all the sequences $\{A_i\} \subset \mathcal{E}_0$ such that $A \subset \bigcup_i A_i$. Next, an $\mathcal{E}$-measurable function f is said to be Λ-integrable if there exists a function $\overline{f}$ satisfying Definition 4.5.1 and such that

$$\Lambda^\sharp(\{s \in E : f(s) \ne \overline{f}(s)\}) = 0.$$

In this case we define
$$\int_A f \, d\Lambda = \int_A \overline{f} \, d\Lambda.$$

It is easy to prove that the augmentation of Definition 4.5.1 is correct, i.e. that the value of $\int_E f d\Lambda$ does not depend on the choice of $\overline{f}$. The class of all Λ–integrable functions will be denoted by $L(\Lambda)$.

It follows from Kalton, Peck and Roberts (1982) that each stochastic measure Λ has a *control measure* λ, i.e. a positive measure λ on $(E, \mathcal{E})$ such that $\lambda(A) = 0$ if and only if $\Lambda^\sharp(A) = 0$. If λ is a control measure for Λ, then Definition 4.5.1 can be easily modified to include all Λ–integrable functions. Namely, (i) has to be replaced by the condition

$$\text{for } \lambda - \text{almost every } s \in E \quad f_n(s) \to f(s) \quad \text{as } n \to \infty. \tag{4.5.3}$$

A more important comment is that the above definition of $\int f \, d\Lambda$ does not depend on the choice of a sequence $\{f_n\}$.

Basic properties of the integral defined in this way are gathered in the following.

Proposition 4.5.1   *(i) If f, $g \in L(\Lambda)$ and a, $b \in \mathbb{R}$, then*

$$\int_E (af + bg) d\Lambda = a \int_E f d\Lambda + b \int_E g d\Lambda.$$

(ii) If $f \in L(\Lambda)$ and $|g(s)| \leq |f(s)|$, $s \in E$, then $g \in L(\Lambda)$.

(iii) If $f_n, g \in L(\Lambda)$ and $|f_n(s)| \leq g(s)$ and $\lim_{n\to\infty} f_n(s) = f(s)$ for all $s \in E$, then

$$\lim_{n\to\infty} \int_E f_n d\Lambda = \int_E f d\Lambda.$$

(iv) In order to extend formula (4.5.2) onto $\mathcal{E}$–measurable functions f one can write

$$\rho_\Lambda(f) = \sup \{\|\textstyle\int_E g d\Lambda\| : g \in L(\Lambda), \ |g| \leq |f|\}.$$

Indeed, if $f \in L(\Lambda)$, then

$$\rho_\Lambda(f) = \sup \{\|\textstyle\int_E \nu f d\Lambda\| : \nu \in \Sigma(\mathcal{E}_0), \ |\nu| \leq 1\} = \lim_{n\to\infty} \rho_\Lambda(f_n),$$

where $\{f_n\} \subset \Sigma(\mathcal{E}_0)$ is any sequence approximating f, satisfying condition (4.5.3). Moreover,

$$f \in L(\Lambda) \ \text{ if and only if } \ \lim_{\varepsilon\to 0} \rho_\Lambda(\varepsilon f) = 0.$$

(v) If $\{f_n\} \subset L(\Lambda)$ and

$$\lim_{n,m\to\infty} \rho_\Lambda(f_n - f_m) = 0,$$

then there exists an $f \in L(\Lambda)$ such that

$$\lim_{n\to\infty} \rho_\Lambda(f_n - f) = 0.$$

For the proof we refer the reader to Kwapień and Woyczyński (1992).

Now we present a series of examples of stochastic measures which are considered in this book.

Example 4.5.1 *Purely discrete stochastic measure.*

First let us give a very simple, but illuminating, example of a purely discrete random measure. Let $E = \mathbf{T} = \mathbf{N}$, $\mathcal{N}_0$ be the algebra of subsets of $\mathbf{N}$ generated by single points $\{n\}, n \in \mathbf{N}$, and let $\xi_1, \xi_2, ...$, be a sequence of random variables in $L^o(\Omega, \mathcal{F}, P)$ such that the series $\sum \xi_n$ converges in probability. If $A \in \mathcal{N}_0$, then we define

$$\Lambda(A) = \sum_{n \in A} \xi_n.$$

It is easy to see that Λ extends to a σ–additive stochastic measure with values in $\mathfrak{F} = L^o(\Omega, \mathcal{F}, P)$ if and only if the series $\sum \xi_n$ converges unconditionally, i.e., for each subsequence $\{n_k\} \subset \mathbf{N}$ the series $\sum \xi_{n_k}$ converges in $L^o(\Omega, \mathcal{F}, P)$. By the Kashin–Maurey–Pisier Theorem this is the case if and only if for each bounded sequence $\{\phi_n\} \subset \mathbf{R}$ the series $\sum \phi_n \xi_n$ converges. Thus it follows that the sequence $\{\phi_n\}$ is an Λ–integrable function on $\mathbf{N}$ if and only if the series $\sum \phi_n \xi_n$ converges unconditionally.

In the case when $\{\xi_n\}$ is a sequence of independent random variables the Three Series Theorem implies that the sequence $\{\phi_n\} \subset \mathbf{R}$ is Λ–integrable if and only if

$$\sum_{n=1}^{\infty} \chi_n(f_n) < \infty,$$

where

$$\chi_n(r) = \big| E[\![r\xi_n]\!] \big| + E[\![r\xi_n]\!]^2, \quad r \in \mathbf{R}.$$

We recall that above and in what follows we use

$$[\![X]\!] = X, \quad \text{if } \|X\| \leq 1, \quad \text{and} \quad [\![X]\!] = X/\|X\|, \quad \text{if } \|X\| > 1. \quad (4.5.4)$$

as a centering function easier to work with rather than any other possible centering function that is needed in the Lévy representation of appropriate measures.

Example 4.5.2 *Independently scattered stochastic measure.*

Let $E = \mathbf{T} = [0, T]$, $\mathcal{E}_0$ be the algebra generated by intervals $(s, t] \cap \mathbf{T}$, where $s, t \in \mathbf{R}$, and let $\{X(t) : t \in \mathbf{T}\}$ be a process with independent increments and $X(0) = 0$ a.s. Then the formula

$$\Lambda((s, t]) = X(t) - X(s)$$

defines an additive set function Λ on $\mathcal{E}_0$ with values in $L^o(\Omega, \mathcal{F}, P)$. One can prove that Λ extends to a stochastic measure if and only if

$$\sup_{0 \leq t_0 < t_1 < ... < t_k \leq T} \sum_{i=1}^{k} \big| E[\![X_{t_i} - X_{t_{i-1}}]\!] \big| < \infty,$$

or, equivalently, if for each $u \in \mathbb{R}$ the function $g(t, u) = \boldsymbol{E} e^{iuX(t)}$, where $t \in \mathbb{T}$, is of bounded variation on $\mathbb{T}$. Such a measure is called an *independently scattered stochastic measure* since for any disjoint $A, B \in \mathcal{E}$, random variables $\Lambda(A)$ and $\Lambda(B)$ are independent.

Example 4.5.3 *Predictable processes.*

Let $(\Omega, \mathcal{F}, P)$ be a probability space with filtration $\{\mathcal{F}_t\}_{t \in \mathbb{T}}$, where $\mathbb{T} = [0, T]$. Let us define $E = \mathbb{T} \times \Omega$, and let $\mathcal{E}_0$ be the (predictable) algebra generated by sets $(s, t] \cap \mathbb{T} \times A$, where $s, t \in \mathbb{T}$ and $A \in \mathcal{F}_s$. If $\{X(t) : t \in \mathbb{T}\}$ is an $\{\mathcal{F}_t\}$–adapted process, then the formula

$$\Lambda((s, t] \times A) = (X(t) - X(s)) \boldsymbol{I}_A$$

defines an additive stochastic set function on $\mathcal{E}_0$ with values in $L^0(\Omega, \mathcal{F}, P)$. It can be shown that $\boldsymbol{\Lambda}$ extends to a stochastic measure if and only if for each $\varepsilon > 0$ there exists an $s > 0$ such that

$$\sup_{0 \leq t_0 < t_1 < \ldots < t_k \leq T} P \left(\sum_{i=1}^{k} | \boldsymbol{E} \left([\![X_{t_i} - X_{t_{i-1}}]\!] \mid \mathcal{F}_{t_{i-1}} \right) | > s \right) < \varepsilon.$$

As it turns out, this condition characterizes **X** as a semimartingale.

Let us notice that in the case when $\mathcal{E}$ is a σ–algebra of subsets of E and $\boldsymbol{\Lambda}$ is a stochastic measure taking values in $L^p(\Omega, \mathcal{F}, P)$, $0 \leq p < \infty$, Definition 4.5.1 is equivalent to the following

Definition 4.5.2 *A function $f : E \to \mathbb{R}$ is said to be **integrable with respect to $\boldsymbol{\Lambda}$** if there exists a sequence $\{f_n\}$ of step functions from Σ such that*

(i) for each $t \in \mathbb{T}$, the sequence $f_n(t) \to f(t)$ as $n \to \infty$;

(ii) for each $A \in \mathcal{E}$, the sequence $\{\int_A f_n d\boldsymbol{\Lambda}\}$ converges in $L^p(\Omega, \mathcal{F}, P)$ as $n \to \infty$.

If f is integrable with respect to $\boldsymbol{\Lambda}$, then we put

$$\int_A f d\boldsymbol{\Lambda} = \lim_{n \to \infty} \int_A f_n d\boldsymbol{\Lambda}.$$

For more details we refer the reader to Kwapień and Woyczyński (1992).

4.6   Lévy Characteristics

Let $\mathbf{X} = \{X(t) : t \in \mathbb{T}\}$, where $\mathbb{T} = [0, T]$ with $T < \infty$, be a real stochastic process with independent increments, i.e. such that for each $n \in \mathbb{N}$ and any $0 = t_0 \leq t_1 \leq \ldots \leq t_n \leq T$, random variables

$$d_i = X(t_i) - X(t_{i-1}), \quad i = 1, 2, \ldots, n,$$

are independent.

We will always assume that the process X is cadlag. Denote

$$X^* = \sup_{t \in \mathbb{T}} |X(t)|.$$

In this case the set of points of stochastic discontinuity of X is at most countable. Recall that a process X is said to be *stochastically continuous* at $t \in \mathbb{T}$ if $\lim_{s \to t} \|X(t) - X(s)\|_0 = 0$.

Let

$$\pi^n = \{\{t_k^n\} : 0 = t_0^n < ... < t_{k_n}^n = T\}, \quad n = 1, 2, ...,$$

be a normal sequence of partitions of $\mathbb{T}$, i.e. $\max_k |t_k^n - t_{k-1}^n| \to 0$ as $n \to \infty$, which is also assumed to be nested, i.e. $\pi^n \subset \pi^{n+1}$, $n = 1, 2, ...$. For each $n = 1, 2, ...$ let us consider the sequence of increments

$$d_k^n = X(t_k^n) - X(t_{k-1}^n), \quad k = 1, ..., k_n.$$

Then it is not difficult to verify that for each $s > 0$ we have

$$P(X^* > s) < 3 \sup_{t \in \mathbb{T}} P(|X(t)| > s/3). \tag{4.6.1}$$

We will write that $f \in R_0$ if the function $f : \mathbb{R} \to \mathbb{R}$ satisfies the conditions:

$$|f(x)| \leq c \text{ for all } x, \quad f(x) = 0 \text{ for } |x| \leq r, \tag{4.6.2}$$

for some $r, c > 0$.

Now, for fixed f from R_0, we define

$$\mathfrak{b}_n(t) = \sum_{k : t_k^n \leq t} E[\![d_k^n]\!], \tag{4.6.3}$$

$$\mathfrak{v}_n(t) = \sum_{k : t_k^n \leq t} E[\![d_k^n]\!]^2 - (E[\![d_k^n]\!])^2, \tag{4.6.4}$$

$$\mathfrak{p}_n(t) = \sum_{k : t_k^n \leq t} E f(d_k^n). \tag{4.6.5}$$

Proposition 4.6.1 *Assume that $\{\pi^n\}$ is a normal nested sequence of partitions of $\mathbb{T}$ such that all points of stochastic discontinuity of X are contained in $\bigcup_{n=1}^{\infty} \pi^n$. Then*

(i) for each $t \in \mathbb{T}$, limits

$$\mathfrak{b}(t) = \lim_{n \to \infty} \mathfrak{b}_n(t), \tag{4.6.6}$$

$$\mathfrak{v}(t) = \lim_{n \to \infty} \mathfrak{v}_n(t), \tag{4.6.7}$$

$$\mathfrak{p}(t) = \lim_{n \to \infty} \mathfrak{p}_n(t), \tag{4.6.8}$$

exist and the convergence is uniform on $\mathbb{T}$;

(ii) $\sup_{n \in \mathbb{N}} \sum_{k=1}^{k_n} E|f(d_k^n)| \leq \frac{cP(X^* > r/2)}{1/3 - P(X^* > r/2)}$;

(iii) $\sup_{n \in \mathbb{N}} \sum_{k=1}^{k_n} |E[d_k^n] - (\mathfrak{b}(t_k^n) - \mathfrak{b}(t_{k-1}^n))| = 0$.

It follows from the above proposition that the function $\mathfrak{b}$ is of bounded variation if and only if

$$\sup_{0 \leq t_0 < t_1 < \ldots < t_k \leq T} \sum_{i=1}^{k} |E[X_{t_i} - X_{t_{i-1}}]| < \infty. \tag{4.6.9}$$

Definition 4.6.1 *(i) The function*

$$\mathfrak{b}(t) = \lim_{n \to \infty} \sum_{k:t_k^n \leq t} E[d_k^n],$$

*is called the **first Lévy characteristic** of the process X.*

*(ii) The **second Lévy characteristic** of the process X is a measure μ on $(\mathbb{R} \setminus \{0\}) \times \mathbb{T}$ determined by the following condition: for any $t \in \mathbb{T}$ and any continuous function $f \in R_0$ one has*

$$\int_{\mathbb{R}} \int_0^t f(x) \mu(ds, dx) = \lim_{n \to \infty} \sum_{k:t_k^n \leq t} E f(d_k^n).$$

*(iii) For the process X, for which the first Lévy characteristic $\mathfrak{b}$ is a function of bounded variation, the **third Lévy characteristic** $\mathfrak{c}$ is a function defined by the formula*

$$\mathfrak{c}(t) = \mathfrak{v}(t) - \int_{\mathbb{R} \setminus \{0\}} \int_0^t [x]^2 \, \mu(ds, dx). \tag{4.6.10}$$

Let us introduce the following set of functions

$$\mathcal{R} = \{ \ f : \mathbb{R} \to \mathbb{R}; \ f \text{ is bounded, continuous, } f(0) = 0,$$
$$\text{there exist real numbers } \dot{f}(0), \ddot{f}(0) \text{ such that}$$
$$\lim_{x \to 0} (f(x) - \dot{f}(0)x)/x^2 = \frac{1}{2} \ddot{f}(0) \}.$$

Proposition 4.6.2 *Let X and $\{\pi^n\}$ be as in Proposition 4.6.1. Then for each $f \in \mathcal{R}$*

$$\lim_{n \to \infty} \sum_{k:t_k^n \leq t} E f(d_k^n) = \dot{f}(0) \mathfrak{b}(t) + \frac{1}{2} \ddot{f}(0) \mathfrak{c}(t) + \int_{\mathbb{R} \setminus \{0\}} \int_0^t f(x) \, \mu(ds, dx),$$

uniformly in $t \in \mathbb{T}$.

The three Lévy characteristics $\mathfrak{b}$, μ and $\mathfrak{c}$ fully determine the process $\mathbf{X}$. Moreover, if $\mathbf{X}$ is stochastically continuous then Lévy characteristics determine its characteristic function via the Lévy–Khintchine Formula

$$E e^{i\theta X(t)} = \exp\left\{ i\mathfrak{b}(t)\theta - \frac{\mathfrak{c}(t)\theta^2}{2} + \int_{\mathbb{R}\setminus\{0\}} \int_0^t \left(e^{ix\theta} - 1 - i\theta[\![x]\!] \right) \, \mu(ds, dx) \right\}.$$

This fact follows from Proposition 4.6.2, applied to the function $f(x) = e^{i\theta x} - 1$. Moreover, $\mathfrak{c}$ is of bounded variation on $\mathbb{T}$ if and only if for each fixed $\theta \in \mathbb{R}$ the function $E e^{i\theta X(t)}$ is of bounded variation on $\mathbb{T}$.

4.7 Stochastic Processes as Integrators

Now we discuss the integration with respect to general stochastic independently scattered measures defined by increments of given stochastic processes.

So, let us fix $\mathbb{T} = [0, T]$ and $\mathcal{E}_0$ as the algebra generated by intervals $(s, t] \cap \mathbb{T}$, where s, $t \in \mathbb{R}$. It means that we have the measurable space $(E, \mathcal{E})$, where $E = \mathbb{T}$ and $\mathcal{E} = \sigma(\mathcal{E}_0)$. Let $\{X(t) : t \in \mathbb{T}\}$ be a process with independent increments and $X(0) = 0$ and, finally, let

$$\Lambda((s, t]) = X(t) - X(s),$$

define the associated additive stochastic set function on $\mathcal{E}_0$ with values in the space $L^o(\Omega, \mathcal{F}, P)$.

The goal here is to give necessary and sufficient conditions for Λ to be extendable to a σ–additive stochastic independently scattered measure, and to describe deterministic functions integrable with respect to Λ. To simplify the notation we will suppress dependence on Λ in favor of $\mathbf{X}$, unless mentioning the underlying stochastic measure is absolutely necessary. So, we shall write about $\mathbf{X}$–integrable functions instead of Λ–integrable, $L^{det}(\mathbf{X})$ instead of $L^{det}(\Lambda)$, $\rho_{\mathbf{X}}$ instead of ρ_Λ, etc.

Now we introduce on $(E, \mathcal{E})$ the *control measure* λ of Λ by the formula

$$\lambda(ds) = |d\mathfrak{b}(s)| + d\mathfrak{c}(s) + \int_{\mathbb{R}\setminus\{0\}} [\![x^2]\!] \, \mu(ds, dx), \qquad (4.7.1)$$

where $\mathfrak{b}$, μ and $\mathfrak{c}$ are Lévy characteristics of $\mathbf{X}$ and $\mathfrak{b}$ is assumed to be of bounded variation. In this way we obtain the measure space $(E, \mathcal{E}, \lambda)$.

Applying the Radon–Nikodym Theorem and disintegrating μ, we obtain functions $b(s)$, $c(s)$ on E and a measure ρ such that

$$d\mathfrak{b}(s) = b(s)\lambda(ds), \quad d\mathfrak{c}(s) = c(s)\lambda(ds), \quad \mu(ds, dx) = \rho(s, dx)\lambda(ds).$$

Furthermore, for $s \in E$ and $x \in \mathbb{R}$, let

$$k(s, x) = \int_{\mathbb{R}\setminus\{0\}} [\![xu]\!]^2 \, \rho(s, du) + c(s)x^2, \qquad (4.7.2)$$

$$l(s,x) = \left\{ \int_{\mathbb{R}\setminus\{0\}} (\llbracket xu \rrbracket - x\llbracket u \rrbracket)\ \rho(s,du) + b(s)x \right\} \qquad (4.7.3)$$

and define

$$\psi(s,x) = k(s,x) + \sup_{|y|\leq|x|} l(s,y). \qquad (4.7.4)$$

Then ψ satisfies the assumptions required in the theory of Musielak–Orlicz spaces (we refer the reader to Musielak (1983)), so we can consider the following Musielak–Orlicz space

$$L^\psi(E,\mathcal{E},\lambda) = \{f : \Psi_{\mathbf{X}}(f) = \int_E \psi(s,f(s))\lambda(ds) < \infty\}$$

The functional $\Psi_{\mathbf{X}}$, called *modular* induces on $L^\psi(E,\mathcal{E},\lambda)$ a topology of a complete linear metric space in which step functions are dense.

Now we are ready to formulate the theorem which gives a complete characterization of processes $\mathbf{X}$, for which the measure Λ extends to a σ–additive stochastic measure and gives a complete description of deterministic Λ–integrable functions.

Theorem 4.7.1 *Let* $\mathbf{X}$ *be a stochastic process with independent increments.*

(i) *The additive stochastic set function* Λ *generated by* $\mathbf{X}$ *extends to a stochastic measure if and only if the Lévy characteristic* b *is a function of bounded variation.*

(ii) *If* Λ *is a stochastic measure generated by* $\mathbf{X}$, *then* λ *is a control measure for* Λ, *and a function* $f : E \to \mathbb{R}$ *is* Λ–*integrable if and only if* $f \in L^\psi(E,\mathcal{E},\lambda)$.

If the condition mentioned above is satisfied one can define the stochastic integral of f with respect to $\mathbf{X}$ as a process

$$I(t) = \int_0^t f\ d\mathbf{X}, \quad t \in \mathbb{T},$$

with cadlag sample paths. If f is a step function then the definition of this process is obvious. If $f \in L(\mathbf{X})$, and $\{f_n\}$ is a sequence of simple functions as in Definition 4.5.1, then the defining condition $\rho_{\mathbf{X}}(f_n - f_m) \to 0$ and the inequality (4.6.1) imply than $\sup_{t \in \mathbb{T}} |\int_0^t (f_n - f_m)\ d\mathbf{X}| \to 0$, which gives the uniform convergence of processes $I_n(t) = \int_0^t f_n d\mathbf{X}$ to the desired process $I(t)$. It is also clear that for two different sequences defining the integral process, the limiting processes have a.s. identical sample paths.

Example 4.7.1 *Processes with stationary independent increments.*

The theory of integration simplifies when **X** has *stationary increments* i.e. if the distribution law of $X(t) - X(s)$ depends only on $t - s$. In this case $\mathfrak{b}(t) = bt$, $\mathfrak{c}(t) = ct$ and $\mu(ds, dx) = \nu(dx)ds$, where b, c are fixed constants and ν is a fixed positive measure on $\mathbb{R}$ such that $a = \int_{\mathbb{R}\backslash\{0\}}[\![x]\!]^2\nu(dx) < \infty$. So we see than

$$\lambda(ds) = \kappa ds, \quad \text{where} \quad \kappa = (|b| + c + a),$$

$$\rho(ds) = \kappa^{-1}\nu(dx), \quad c(s) = \kappa^{-1}c, \quad b(s) = \kappa^{-1}b,$$

$$\overline{k}(x) = \kappa^{-1}\left(\int_{\mathbb{R}\backslash\{0\}}[\![xu]\!]^2\nu(du) + cx^2\right),$$

$$\overline{l}(x) = \kappa^{-1}\left(\int_{\mathbb{R}\backslash\{0\}}([\![xu]\!] - x[\![u]\!])\,\nu(du) + bx\right)$$

and

$$\psi(s,x) = \overline{\psi}(x) = \overline{k}(x) + \sup_{|y|\leq|x|}\overline{l}(y).$$

Since

$$\left|\frac{[\![xu]\!] - x[\![u]\!]}{x^2[\![u]\!]^2}\right| \leq 1 \quad \text{and, for each} \quad u \quad \lim_{|x|\to\infty}\frac{[\![xu]\!] - x[\![u]\!]}{x^2[\![u]\!]^2} = 0,$$

then from the Lebesgue Bounded Convergence Theorem it follows than

$$\lim_{|x|\to\infty}\frac{1}{x^2}\int_{\mathbb{R}\backslash\{0\}}([\![xu]\!] - x[\![u]\!])\,\lambda(du) = \lim_{|x|\to\infty}\int_{\mathbb{R}\backslash\{0\}}\frac{[\![xu]\!] - x[\![u]\!]}{x^2[\![u]\!]^2}[\![u]\!]^2\,\lambda(du) = 0.$$

Therefore, if process **X** has a nontrivial Gaussian component or, equivalently, if $c \neq 0$ then $\psi(s, x) = \tilde{\psi}(x) \sim c|x|^2$ for large x, which means that $\lim_{|x|\to\infty}\psi(x)/x^2 = c$. Thus in this case the class of Λ–integrable functions $L^{det}(\mathbf{X})$ coincides with $L^2(dt)$.

Example 4.7.2 α-stable Lévy motion.

Consider a process **X** with α–stable independent stationary increments, for $0 < \alpha \leq 2$. For $\alpha = 2$ (with $c \neq 0, \nu = 0$) the process is a Brownian motion and in this case $L^{det}(\mathbf{X}) = L^2(dt)$.

For $0 < \alpha < 2$, we have $c = 0$ and $\nu(dx) = g(x)\,dx$, where

$$g(x) = \begin{cases} r_1|x|^{-1-\alpha}, & \text{for } x > 0; \\ r_2|x|^{-1-\alpha}, & \text{for } x < 0, \end{cases}$$

for some $r_1, r_2 \geq 0$. Process **X** is symmetric if and only if $r_1 = r_2$. By elementary calculations we get

$$\overline{k}(x) = \frac{r_1 + r_2}{\kappa}\frac{2}{\alpha(2-\alpha)}|x|^\alpha.$$

Furthermore, for $\alpha \neq 1$ and $|x| \geq 1$,

$$\overline{l}(x) = \frac{r_1 + r_2}{\kappa}\left(\frac{|x|^\alpha}{\alpha} + \frac{|x|^{\alpha-1}}{1-\alpha} - \frac{|x|}{\alpha}\right)\text{sgn}(x) + \frac{b}{\kappa}x,$$

and, for $\alpha = 1$ and $|x| > 1$,

$$\bar{l}(x) = \frac{r_1 + r_2}{\kappa} x \log |x| + \frac{b}{\kappa} x.$$

Thus we see than only three essentially different situations arise when one tries to characterize spaces of integrable functions with respect to α–stable processes.

(A) Let $1 < \alpha \leq 2$, or $0 < \alpha \leq 1$ and $r_1 = r_2$, or $0 < \alpha < 1$, $r_1 \neq r_2$ and $b = (r_1 - r_2)/\alpha$. In this case

$$\overline{\psi}(x) \sim \beta |x|$$

for large $|x|$ and some constant β, and, as a consequence, $L^{det}(\mathbf{X}) = L^{\alpha}(dt)$.

(B) Let $\alpha = 1, r_1 \neq r_2$, and $b \neq (r_1 - r_2)/\alpha$. In this case

$$\overline{\psi}(x) \sim \beta |x|$$

for large $|x|$ and some constant β, so that $L^{det}(\mathbf{X}) = L^1(dt)$.

(C) Let $\alpha = 1, r_1 \neq r_2$. In this case

$$\overline{\psi}(x) \sim \beta |x| \log |x|$$

for large $|x|$ and some β, which implies that

$$L^{det}(\mathbf{X}) = \left\{ f : \int_{\mathbb{T}} |f| \log^+ |f| dt < \infty \right\}.$$

If we consider the integration with respect to stochastic measures generated by processes with independent increments, which have the finite p-th moment for a certain $p > 0$, it is natural to construct the integral with respect to a stochastic measure with values in the Banach space $L^p(\Omega, \mathcal{F}, P)$, rather than in $L^0(\Omega, \mathcal{F}, P)$.

To be more exact, for a given process $\mathbf{X}$ with independent increments, we are interested in the case when the associated additive set function Λ extends to a measure with values in $L^p(\Omega, \mathcal{F}, P)$. As before, for $L^0(\Omega, \mathcal{F}, P)$–valued stochastic measures, the fundamental question will be : what is the class of $\mathbf{X}$–integrable functions? In order to distinguish them from the elements of the class $L^0(\Omega, \mathcal{F}, P)$ we call such functions $(p, \mathbf{X})$–integrable functions. Obvious necessary condition for extendability of Λ to a stochastic measure with values in $L^p(\Omega, \mathcal{F}, P)$ are that Λ extends to a stochastic measure in $L^0(\Omega, \mathcal{F}, P)$, and that $E|\mathbf{X}(t)|^p < \infty$ for all $t \in \mathbb{T}$.

Before stating the main theorem characterizing $(p, \mathbf{X})$–integrable functions we need some preliminary results.

Proposition 4.7.1 *If* $\{X(t) : t \in \mathbb{T}\}$, *where* $\mathbb{T} = [0, T]$, *is a process with independent increments and cadlag paths, then the following three conditions are equivalent:*

(i) $E|X(T)|^p < \infty$;

(ii) $E(X^*)^p < \infty$;

(iii) $\int_{\mathbb{R}\setminus\{0\}} \int_{\mathbb{T}} |x|^p I_{\{|x|>1\}} \, \mu(ds, dx) < \infty$.

We shall assume that the process $\{X(t) : t \in \mathbb{T}\}$ satisfies any of the above equivalent conditions and that its Lévy characteristic $\mathfrak{b}$ is of bounded variation.

Let $h_p : \mathbb{R} \to \mathbb{R}^+$ be defined by the formula

$$h_p(x) = \begin{cases} x^2, & \text{if } |x| \leq 1, \\ |x|^p, & \text{if } |x| > 1. \end{cases}$$

With the quantities l, λ, ρ, k, ψ introduced at the beginning of this section we need now

$$k_p(s, x) = \int_{\mathbb{R}\setminus\{0\}} h_p(ux) \, \rho(s, du), \tag{4.7.5}$$

$$\psi_p(s, x) = l(s, x) + k_p(s, x). \tag{4.7.6}$$

The function ψ_p satisfies the conditions assuring that the functional

$$\Psi_{p,\mathbf{X}}(f) = \int_{\mathbb{T}} \psi_p(s, f(s)) \, \lambda(ds) \tag{4.7.7}$$

is a modular and step functions are dense in the defined above Musielak-Orlicz space $L^{\psi_p}(E, \mathcal{E}, \lambda)$. Thus, we have the following characterization of Λ-integrable functions.

Theorem 4.7.2 *An additive set function* Λ *extends to a stochastic measure with values in* $L^p(\Omega, \mathcal{F}, P)$ *if and only if the Lévy characteristic* $\mathfrak{b}$ *is of finite variation and* $E|\mathbf{X}(T)|^p < \infty$. *Furthermore,* $f : \mathbb{T} \to \mathbb{R}$ *is* $(p, \mathbf{X})$-*integrable if and only if* $\Psi_{p,\mathbf{X}}(f) < \infty$.

4.8 Integrals of Deterministic Functions with ID Integrators

Let $\Lambda = \{\Lambda(A) : A \in \mathcal{E}\}$ be an *infinitely divisible random measure* (in shortened notation, ID random measure), i.e. $\Lambda(A)$ is an ID random variable from $L^0(\Omega, \mathcal{F}, P)$ for every $A \in \mathcal{E}$. So its characteristic function can be written in the following Lévy form:

$$\widehat{\mathcal{L}}(\Lambda(A))(\theta) = \exp\{\psi(\theta)\} \tag{4.8.1}$$

with

$$\psi(\theta) = \left\{ i\theta\mathfrak{b}(A) - \frac{1}{2}\theta^2\mathfrak{c}(A) + \int_{\mathbb{R}\setminus\{0\}} \left(e^{i\theta x} - 1 - i\theta[\![x]\!] \right) \mu(A, dx) \right\},$$

where $-\infty < \mathfrak{b}(A) < \infty$, $0 \leq \mathfrak{c}(A) < \infty$ and $\mu(A, \cdot)$ is a Lévy measure on $\mathbb{R}$.

The next proposition (Urbanik and Woyczyński (1967)) states that there is a one–to–one correspondence between the class of ID random measures and the class of parameters $\mathfrak{b}$, $\mathfrak{c}$ and μ. This result provides also a measure λ, the control measure of Λ. Next, it follows that $\mu(\cdot, \cdot)$ determines a unique measure on $\mathcal{E} \times \mathcal{B}(\mathbb{R})$ which admits a factorization in terms of a family of Lévy measures $\rho(s, \cdot)$, $s \in E$ on $\mathbb{R}$ and the measure λ. This fact helps to derive another form of the characteristic function of $\mathcal{L}(\Lambda(A))$ in terms of the measures $\rho(s, \cdot)$ and λ. This form of the characteristic function allows us to obtain the characteristic function of the stochastic integral $\int_E f \, d\Lambda$.

Proposition 4.8.1 *(i) Let Λ be an ID random measure with the characteristic function given by 4.8.1. Then*

$\mathfrak{b} : \mathcal{E} \to \mathbb{R}$ *is a signed–measure;*

$\mathfrak{c} : \mathcal{E} \to [0, \infty)$ *is a measure;*

$\mu(A, \cdot)$ *is a Lévy measure on $\mathbb{R}$ for every $A \in \mathcal{E}$;*

$\mathcal{E} \ni A \to \mu(A, B) \in [0, \infty)$ *is a measure for every $B \in \mathcal{B}(\mathbb{R})$, whenever $0 \notin \overline{B}$.*

(ii) Let $\mathfrak{b}$, $\mathfrak{c}$ and μ satisfy the condition given in (i). Then there exists a unique (in the sense of finite–dimensional distributions) ID random measure Λ such that (4.8.1) holds.

(iii) Let $\mathfrak{b}$, $\mathfrak{c}$ and μ be as in (i) and define

$$\lambda(A) = |\mathfrak{b}|(A) + \mathfrak{c}(A) + \int_{\mathbb{R}\setminus\{0\}} \min\{1, x^2\} \, \mu(A, dx), \quad A \in \mathcal{E}.$$

Then

$\lambda : \mathcal{E} \to [0, \infty)$ *is a measure such that $\lambda(A_n) \to 0$ implies $\Lambda(A_n) \to 0$ in probability for every $\{A_n\} \subset \mathcal{E}$;*

if $\Lambda(A_n') \to 0$ in probability for every sequence $\{A_n'\} \subset \mathcal{E}$ such that $A_n' \subset A_n \in \mathcal{E}$, then $\lambda(A_n) \to 0$.

Remark 4.8.1 *Since $\lambda(E) < \infty$, $n = 1, 2, \ldots$, we can extend λ to a σ–finite measure on $(E, \mathcal{E})$; we call λ **the control measure** of Λ.*

Proposition 4.8.2 *Let μ be as in Proposition 4.8.1. Then there exists a function $\rho : E \times \mathcal{B}(\mathbb{R}) \to [0, \infty]$ such that*

(i) $\rho(s, \cdot)$ is a Lévy measure on $\mathcal{B}(\mathbf{R})$ for every $s \in E$,

(ii) $\rho(\cdot, B)$ is a Borel measurable function for every $B \in \mathcal{B}(\mathbb{R})$,

(iii)

$$\int_{E \times \mathbb{R}} h(s,x)\,\mu(ds,dx) = \int_E \left(\int_{\mathbb{R}} h(s,x)\rho(s,dx) \right)\,\lambda(ds),$$

for every $\mathcal{E} \times \mathcal{B}(\mathbb{R})$–measurable function $h : E \times \mathbb{R} \to [0,\infty]$.

Using previous results one can obtain a very useful form of the characteristic function of $\Lambda(A)$. Namely, it can be rewritten in the form

$$\widehat{\mathcal{L}}(\Lambda(A))(\theta) = \exp\left\{ \int_A K(\theta,s)\,\lambda(ds) \right\}, \quad A \in \mathcal{E}, \qquad (4.8.2)$$

where

$$K(\theta,s) = i\theta a(s) - \frac{1}{2}\theta^2 \sigma^2(s) + \int_{\mathbb{R}} \left(e^{i\theta x} - 1 - i\theta [\![x]\!] \right)\,\rho(s,dx),$$

$$a(s) = \frac{d\mathfrak{b}}{d\lambda}(s), \quad \sigma^2(s) = \frac{d\mathfrak{c}}{d\lambda}(s)$$

and ρ is given by Proposition 4.8.2. Moreover, we have

$$|a(s)| + \sigma^2(s) + \int_{\mathbb{R}} \min\{1,x^2\}\,\rho(s,dx) = 1 \quad \lambda - a.e.$$

Recall that, having a control measure λ of the random measure Λ, we can define a Λ–integrable function $f : (E,\mathcal{E}) \to (\mathbb{R},\mathcal{B}(\mathbb{R}))$ as such, for which there exists a sequence $\{f_n\}$ of simple functions such that

1. $f_n \to f$ $\lambda - a.e.$;

2. for any $A \in \mathcal{E}$ the sequence $\{\int_A f_n d\Lambda\}$ converges in probability as $n \to \infty$.

In this case we define

$$\int_A f\,d\Lambda = P - \lim_{n \to \infty} \int_A f_n\,d\Lambda.$$

The following proposition provides an explicit formula for the characteristic function of $\int_E f\,d\Lambda$.

Proposition 4.8.3 *If f is Λ–integrable, then $\int_E |K(\theta f(s),s)|\,\lambda(ds) < \infty$ and*

$$\widehat{\mathcal{L}}\left(\int_E f\,d\Lambda \right)(\theta) = \exp\left\{ \int_E K(\theta f(s),s)\,\lambda(ds) \right\}. \qquad (4.8.3)$$

We are now ready to formulate the theorem which provides necessary and sufficient conditions for the existence of $\int_E f\,d\Lambda$ in terms of the deterministic characteristics of Λ.

Theorem 4.8.1 *Let $f : E \to \mathbb{R}$ be a $\mathcal{E}$-measurable function. Then f is Λ-integrable if and only if the following three conditions hold:*

1. $\int_E |l(s, f(s))|\ \lambda(ds) < \infty;$

2. $\int_E |f(s)|^2 \sigma^2(s)\ \lambda(ds) < \infty;$

3. $\int_E k_0(s, f(s))\ \lambda(ds) < \infty,$

where

$$l(s, x) = \int_{\mathbb{R}\setminus\{0\}} (\llbracket xu \rrbracket - x\llbracket u \rrbracket)\ \rho(s, du) + x a(s),$$

$$k_0(s, x) = \int_{\mathbb{R}\setminus\{0\}} \min\{1, |xu|^2\}\ \rho(s, du).$$

Further, if f is Λ-integrable, then the characteristic function of $\int_E f d\Lambda$ can be written as

$$\widehat{\mathcal{L}} \left(\int_E f d\Lambda \right) (\theta) = \exp \left\{ i\theta a_f - \frac{1}{2}\theta^2 \sigma_f^2 + \int_{\mathbb{R}\setminus\{0\}} \left(e^{i\theta x} - 1 - i\theta \llbracket x \rrbracket \right)\ \mu(f, dx) \right\},$$

where

$$a_f = \int_E l(s, f(s))\ \lambda(ds), \quad \sigma_f^2 = \int_E |f(s)|^2 \sigma^2(s)\ \lambda(ds),$$

and

$$\mu(f, B) = \mu(\{(s, x) \in E \times \mathbb{R} : f(s)x \in B \setminus \{0\}\}), \quad B \in \mathcal{B}(\mathbb{R}).$$

As we already know, even in a more general case, one can identify the set of Λ-integrable functions as a certain Musielak–Orlicz modular space. We shall restrict ourselves here to the representation of ID processes and for such processes we shall determine the modular space and state the continuity of the mapping $f \to \int_E f d\Lambda$ from this modular space into $L^p(\Omega, \mathcal{F}, P)$. Under certain conditions the inverse of this map is also continuous.

Let q be a non–negative number such that

$$E|\Lambda(A)|^q < \infty, \quad \text{for all } A \in \mathcal{E}. \tag{4.8.4}$$

Further, we shall assume that this condition is satisfied and $q \in [0, \infty)$ is fixed (note that every Λ satisfies (4.8.4) for $q = 0$). Since, for ID distribution μ with Lévy measure G, the integral $\int_{\mathbb{R}\setminus\{0\}} |x|^q\ \mu(dx)$ is finite if and only if $\int_{\{|x|>1\}} |x|^q\ G(dx)$ is finite, thus

$$\int_A \left[\int_{\{|x|>1\}} |x|^q\ \rho(s, dx) \right] \lambda(ds) = \int_{\{|x|>1\}} |x|^q\ \mu(A, dx) < \infty,$$

for every $A \in \mathcal{E}$, where $\mu(A, \cdot)$ is the Lévy measure of $\mathcal{L}(\Lambda(A))$. Hence

$$\int_{\{|x|>1\}} |x|^q \rho(s, dx) < \infty \quad \lambda - -a.e.$$

Without loss of generality we can assume that the last inequality holds for all $s \in E$. Define for $0 \le p \le q$, $u \in \mathbb{R}$ and $s \in E$,

$$\psi_p(s, x) = k_p(s, x) + \sup_{|y| \le |x|} l(s, y),$$

where

$$k_p(s, x) = \int_{-\infty}^{\infty} \left(|ux|^p I_{\{y:|y|>1\}}(ux) + |ux|^2 I_{\{y:|y|\le 1\}}(ux) \right) \rho(s, du) + \sigma^2(s)x^2.$$

One can verify that the function $\psi_p(\cdot, \cdot)$ has the following properties

1. for every $s \in E$, $\psi_p(s, \cdot)$ is a continuous non–decreasing function on $[, \infty)$ with $\psi_p(s, 0) = 0$,

2. $\lambda(\{s : \psi_p(s, u) = 0 \quad \text{for some} \quad u = u(s) \ne 0\}) = 0$,

3. $\exists_{C>0} \forall_{u \ge 0} \forall_{s \in E} \quad \psi_p(s, 2u) \le C\psi_p(s, u)$.

Now we define the Musielak–Orlicz space

$$L^{\psi_p}(E, \mathcal{E}, \lambda) = \{f \in L^o(E, \mathcal{E}; \lambda) : \int_E \psi_p(s, f(s))\lambda(ds) < \infty\}.$$

The following properties $L^{\psi_p}(E, \mathcal{E}, \lambda)$ (once more we refer the reader to Musielak (1983)) for these and further facts concerning general Musielak–Orlicz spaces generated by functions which satisfy the above conditions) will be used in the sequel. The space $L^{\psi_p}(E, \mathcal{E}, \lambda)$ is a complete linear metric space with the F–norm defined by

$$\|f\|_{\psi_p} = \inf \left\{ c > 0 : \int_E \psi_p(s, c^{-1}f(s))\lambda(ds) \le c \right\}.$$

Simple functions are dense in $L^{\psi_p}(E, \mathcal{E}, \lambda)$ and the natural embedding of $L^{\psi_p}(E, \mathcal{E}, \lambda)$ into $L^o(E, \mathcal{E}, \lambda)$ is continuous, where $L^o(E, \mathcal{E}, \lambda)$ is equipped with the topology of convergence in measure λ on every set of finite λ–measure. Finally, $\|f_n\|_{\psi_p} \to 0$ if and only if $\int_E \psi_p(s, f_n(s))\lambda(ds) \to 0$.

Theorem 4.8.2 *Let* $0 \le p \le q$. *Then*

$$\{f : f \text{ is } \Lambda\text{–integrable and} \quad E \left| \int_E f d\Lambda \right|^p < \infty\} = L^{\psi_p}(E, \mathcal{E}, \lambda),$$

and the linear mapping

$$L^{\psi_p}(E, \mathcal{E}, \lambda) \ni f \to \int_E f d\Lambda \in L^p(\Omega, \mathcal{F}, P)$$

is continuous (here $L^{\psi_0}(E, \mathcal{E}, \lambda) = \{f : f \text{ is } \Lambda\text{–integrable}\}$*).*
Moreover, if

$$\exists_{0 \leq p \leq q} \exists_{C=C(p,q)>0} \forall_{u \geq 0} \quad |l(s,u)| \leq C \ k_p(s,u) \quad \lambda - a.e., \tag{4.8.5}$$

then the mapping $f \rightarrow \int_E f d\Lambda$ *is an isomorphism from* $L^{\psi_p}(E, \mathcal{E}, \lambda)$ *into the space* $L^p(\Omega, \mathcal{F}, P)$.

In this case

$$\left\{ \int_E f d\Lambda : f \in L^{\psi_p}(E, \mathcal{E}, \lambda) \right\} = \overline{\text{lin}} \{ \Lambda(A) : A \in \mathcal{E} \}_{L^p(\Omega, \mathcal{F}, P)}.$$

Notice that this construction of the stochastic integral generalizes all constructions presented and mentioned in Section 4.3. Also the space $L^{\psi_p}(E, \mathcal{E}, \lambda)$ of Λ–integrable functions does coincide with appropriate spaces of deterministic functions integrable with respect to an $S\alpha S$, symmetric $S(r, \alpha)$, or centered $S(r, \alpha)$ random measure Λ.

One can prove that the condition (4.8.5) is satisfied under any of the following two hypotheses on the ID random measure Λ and the real number p:

1. Λ is symmetric and $0 \leq p \leq q$ arbitrary,

2. $E\Lambda(A) = 0$ for all A and $1 \leq p \leq q$.

If Λ is a centered $S(\alpha)$ random measure, where $1 < \alpha < 2$, then $E|\Lambda(A)|^q < \infty$ for any $q < \alpha$, and $E\Lambda(A) = 0$ for every $A \in \mathcal{E}$. Hence the characteristic function of $\Lambda(A)$ is of the form

$$\widehat{\mathcal{L}}(\Lambda(A))(\theta) = \exp \left\{ -\frac{1}{2}\theta^2 \mathfrak{c}(A) + \int_{\mathbb{R} \backslash \{0\}} \left(e^{i\theta x} - 1 - i\theta x \right) \ \mu(A, dx) \right\}$$

$$= \exp \left\{ i\theta \mathfrak{b}(A) - \frac{1}{2}\theta^2 \mathfrak{c}(A) - \int_{\mathbb{R} \backslash \{0\}} \left(e^{i\theta x} - 1 - i\theta [\![x]\!] \right) \ \mu(A, dx) \right\},$$

where $\mathfrak{b}(A) = \int_{\mathbb{R} \backslash \{0\}} ([\![x]\!] - x) \ \mu(A, dx)$, $\mathfrak{c}(A) \equiv 0$ and $\mu(A, \cdot)$ is an $S(\alpha)$ Lévy measure.

If Λ is an $S(\alpha)$ random measure and $0 < \alpha < 1$, then

$$\widehat{\mathcal{L}}(\Lambda(A))(\theta) = \exp \left\{ \int_{\mathbb{R} \backslash \{0\}} \left(e^{i\theta x} - 1 \right) \ \mu(A, dx) \right\}$$

$$= \exp \left\{ i\theta \mathfrak{b}(A) + \int_{\mathbb{R} \backslash \{0\}} \left(e^{i\theta x} - 1 - i\theta [\![x]\!] \right) \mu(A, dx) \right\},$$

where $\mathfrak{b}(A) = \int_{\mathbb{R} \backslash \{0\}} [\![x]\!] \ \mu(A, dx)$ and $\mu(A, \cdot)$ is an $S\alpha S$ Lévy measure for every $A \in \mathcal{E}$. In view of the theorem we have

$$a(s) = \int_{\mathbb{R} \backslash \{0\}} [\![x]\!] \ \mu(A, dx) \quad \text{and} \quad l(s, x) = \int_{\mathbb{R} \backslash \{0\}} [\![ux]\!] \ \rho(s, du) \quad \lambda - a.e.$$

Finally, if Λ is a centered $S1S$ random measure, then Λ is symmetric and the characteristic function of $\Lambda(A)$ is given by (4.8.1) with $\mathfrak{b} \equiv \mathfrak{c} \equiv 0$ and $\mu(A, \cdot)$ is a symmetric $S1S$ Lévy measure for every $A \in \mathcal{E}$.

The following proposition states the fact that the conditional Lévy measures $\rho(s, \cdot)$ of an $S(\alpha)$ random measure Λ are $S(\alpha)$ measures.

Proposition 4.8.4 *(i) Let Λ be an $S(\alpha)$ random measure. Then*

$$\rho(s, dx) = c_1(s)I_{\mathbb{R}^+}(x)x^{-1-\alpha}\, dx + c_{-1}(s)I_{\mathbb{R}^-}(x)|x|^{-1-\alpha}\, dx, \quad \lambda - -a.s.,$$

where $c_1, c_{-1} : E \to [0, \infty)$ are measurable.

(ii) Let Λ be a $S(r, \alpha)$ random measure. Then for λ–almost all $s \in E$

$$\rho(s, B) = \sum_{-\infty}^{\infty} r^n \rho(s, (r^{\frac{n}{\alpha}} B) \cap \Delta)) \quad \text{for all} \ \ B \in \mathcal{B}(\mathbb{R}),$$

where $\Delta = \{x \in \mathbb{R} : r^{\frac{1}{\alpha}} < |x| \le 1\}$.
More generally, for λ–almost all $s \in S$, the following formulas hold

$$\int_{\mathbb{R}} f(x)\rho(s, dx) = \sum_{n=-\infty}^{\infty} r^n \int_{\Delta} f(r^{\frac{-n}{\alpha}} x)\rho(s, dx),$$

$$\int_{|x|>r^{\frac{k}{\alpha}}} f(x)\rho(s, dx) = \sum_{i=1}^{\infty} r^{-k+i} \int_{\Delta} f(r^{\frac{k-1}{\alpha}} x)\rho(s, dx),$$

$$\int_{|x|\le r^{\frac{k}{\alpha}}} f(x)\rho(s, dx) = \sum_{i=0}^{\infty} r^{-k-i} \int_{\Delta} f(r^{\frac{k+i}{\alpha}} x)\rho(s, dx),$$

for every Borel non–negative function f and an arbitrary integer k.

Proposition 4.8.5 *Let Λ be a centered $S(\alpha)$, or more generally, a centered $S(r, \alpha)$ random measure. Then condition (4.8.5) holds for $0 \le p < \alpha$ and we have $L^{\psi_p}(E, \mathcal{E}, \lambda) = L^{\alpha}(E, \mathcal{E}, \lambda)$ up to a renormalization for every $0 \le p < \alpha$. Consequently, there are positive constants C_1 and C_2 depending only on p, r and α such that*

$$C_1 \left(\int_E |f|^{\alpha}\, d\lambda \right)^{\frac{1}{\alpha}} \le \left(E \left| \int_E f\, d\Lambda \right|^p \right)^{\frac{1}{p}} \le C_2 \left(\int_E |f|^{\alpha}\, d\lambda \right)^{\frac{1}{\alpha}},$$

for every $f \in L^{\alpha}(E, \mathcal{E}, \lambda)$.

4.9 Integrals with Stochastic Integrands and ID Integrators

Now we are going to define $\mathbf{I} = \int \mathbf{Y} d\mathbf{X}$, where both the integrand and integrator are suitable stochastic processes. Of course, as we have already mentioned, we must limit the class of integrands and precisely describe the class of integrators.

Let $\{\mathcal{F}_t\}_{t \in \mathbb{T}}$ be a filtration. We say that the process $\{X(t) : t \in \mathbb{T}\}$ has *independent increments with respect to the filtration* $\{\mathcal{F}_t\}$ if it is $\mathcal{F}_t$–measurable for every $t \in \mathbb{T}$ and if, for each $0 < t < s < t_{\infty}$, the random variable $X(s) - X(t)$ is independent of $\mathcal{F}_t$.

Let $\mathcal{P}_0$ be the algebra of subsets of $\mathbb{T} \times \Omega$ generated by sets $(s, t] \times A$, where $s, t \in \mathbb{T}$ and $A \in \mathcal{F}_s$, and let $\mathcal{P}$ denote the σ–algebra generated by $\mathcal{P}_0$.

Definition 4.9.1 *The process* **Y** *is said to be a* **predictable process***, if it is* $\mathcal{P}$*– measurable as a function of* (t, ω)*. The process* **Y** *is said to be a* **predictable step process** *if it is a finite sum of processes of the form* $\xi I_{(s,t]}(t)$*,* $t \in \mathbb{T}$*, where the random variable* ξ *is* $\mathcal{F}_s$*-measurable.*

For such processes the definition of the integral is obvious. If

$$Y(t) = \sum_{k=1}^{n} \xi_k I_{(s_k, r_k]}(t),$$

then

$$\int_{\mathbb{T}} \mathbf{Y} d\mathbf{X} = \sum_{k=1}^{n} \xi_k [X(r_k) - X(s_k)].$$

Notice that not every predictable step process is a $\mathcal{P}_0$–step function. Nevertheless, each predictable step process on bounded $\mathbb{T} \times \Omega$ is a uniform limit of $\mathcal{P}_0$–processes. Therefore, for a predictable step process **Y** we can define

$$\rho_{\mathbf{X}}(\mathbf{Y}) = \sup_{\mathbf{V} \in \mathbf{P}_1} \| \int_{\mathbb{T}} \mathbf{V} \mathbf{Y} \, d\mathbf{X} \|_0,$$

where $\mathbf{P}_1$ denotes the class of all predictable step processes **V** such that $|\mathbf{V}| \leq 1$.

The σ–algebra $\mathcal{T} = \{A \times \Omega : A \in \mathcal{A}\}$ is contained in $\mathcal{P}$, so that each deterministic measurable process is predictable. Obviously, if the stochastic additive set function $\mathbf{\Lambda}$ generated by **X** on $\mathcal{P}_0$ extends also to a σ–additive stochastic measure on $\mathcal{P}$, then it also extends to a stochastic measure on $\mathcal{T}$, and therefore the Lévy characteristic $\mathfrak{b}$ is a function of bounded variation. It also follows that, conversely, if the Lévy characteristic $\mathfrak{b}$ is of bounded variation, then $\mathbf{\Lambda}$ extends to a σ–additive stochastic measure on P. Before we formulate a formal theorem providing the description of predictable processes which are **X**–integrable, let us recall that $\mathbf{L}^{det}(d\mathbf{X})$ denotes the class of deterministic functions which are integrable with respect to the process **X**. Analogously, by $\mathbf{L}^{rnd}(d\mathbf{X})$ we will denote the class of processes which are integrable with respect to **X**

Theorem 4.9.1 *Let* $\{X(t) : t \in \mathbb{T}\}$ *be a process with independent increments. Then the associated additive stochastic set function* $\mathbf{\Lambda}$ *extends to a* σ*– additive stochastic measure on P if and only if the Lévy characteristic* $\mathfrak{b}$ *is of bounded variation. In such case the following statements are equivalent*

1. *A predictable process* **Y** *is* **X***-integrable, i.e.* $\mathbf{Y} \in \mathbf{L}^{rnd}(d\mathbf{X})$*;*

2. *For almost every* $\omega \in \Omega$ *the deterministic function* $Y(\cdot, \omega)$ *is integrable with respect to* **X***, i.e.* $Y(\cdot, \omega) \in \mathbf{L}^{det}(d\mathbf{X})$*;*

3. *For almost every* $\omega \in \Omega$*,* $Y(\cdot, \omega) \in \mathbf{L}^{\phi}(d\nu)$*.*

Let us now make some remarks about stochastic integrals with respect to the Brownian motion. We begin with a fundamental result which is obtained by direct computation involving step processes and the standard approximation procedure.

Theorem 4.9.2 *Let* **W** *be a process with* $\{\mathcal{F}_t\}$-*independent increments which is also a Brownian motion process, and let* **F** *be an* $\{\mathcal{F}_t\}$-*predictable process such that* $E \int_{\mathbb{T}} F^2(t)dt < \infty$. *Then*

$$E \int_{\mathbb{T}} F d\mathbf{W} = 0,$$

and

$$E \left(\int_{\mathbb{T}} F d\mathbf{W} \right)^2 = E \int_{\mathbb{T}} F^2(t)\, dt.$$

Let us return now to Example (4.5.3). So, consider $(\Omega, \mathcal{F}, P)$ with the filtration $\{\mathcal{F}_t\}_{t \in [0,T]}$. Set $\mathbf{T} = [0, T] \times \Omega$ and let $\mathcal{P}_0$ be the "predictable" algebra generated by sets $(s, t] \cap \mathbf{T} \times A$, where $s, t \in [0, T]$ and $A \in \mathcal{F}_s$. If $\{X(t) : t \in [0, T]\}$ is an $\{\mathcal{F}_t\}$-adapted process, then the formula

$$\Lambda((s, t] \times A) = (X(t) - X(s))I_A,$$

defines an additive stochastic set function on $\mathcal{P}_0$ with values in $L^o(\Omega, \mathcal{F}, P)$. Further, we shall assume that the filtration is right continuous and that all the sample paths of **X** are in D(**T**).

Definition 4.9.2 *We shall say that the process* **X** *has property (B) if for each* $\varepsilon > 0$ *there exists an* $s > 0$ *such that for any sequence* $0 = t_0 < t_1 < ... < t_n = T$ *we have*

$$\left(\sum_{k=1}^{n} | E([[X(t_k) - X(t_{k-1})]] | \mathcal{F}_{t_{k-1}}) | > s \right) \leq \varepsilon.$$

Let $\pi^n = \{t_k^n\}$, where $k = 1, 2, ..., k_n$ and $n = 1, 2, ...$, be a sequence of nested, normal partitions of the interval **T**. To define characteristics of a process **X**, which we shall call *Jacod–Grigelionis characteristics* and which are analogous to Lévy 's characteristics for processes with independent increments, put

$$\mathfrak{b}_n(t) = \sum_{k:t_k^n \leq t} E\left([[d_k^n]] \mid \mathcal{F}_{k-1}^n\right), \tag{4.9.1}$$

$$\mathfrak{w}_n(t) = \sum_{k:t_k^n \leq t} E\left([[d_k^n]]_2 \mid \mathcal{F}_{k-1}^n\right), \tag{4.9.2}$$

$$\mathfrak{p}_n(t) = \sum_{k:t_k^n \leq t} E\left(f(d_k^n) \mid \mathcal{F}_{k-1}^n\right), \tag{4.9.3}$$

where we use the abbreviation $\mathcal{F}_k^n = \mathcal{F}_{t_k^n}$, and where f is, throughout the remainder of this section, a fixed function from

$$\mathcal{R}_0^1 = \{f \in \mathcal{R}_0 : |f(x) - f(y)| < \kappa |[[x - y]]| \text{ for some } \kappa > 0\},$$

with $\mathcal{R}_0$ defined by (4.6.2). As before,

$$d_k^n = X(t_k) - X(t_{k-1}),$$

Recall that a sequence of processes $\{Z_n(t) : t \in \mathbb{T}\}$ is **uniformly convergent in probability** if there exists a process $\mathbf{Z}$ such that

$$\lim_{n \to \infty} \|(Z_n - Z)^*(T)\|_0 = 0,$$

where

$$(Z_n - Z)^*(T) = \sup_{t \in \mathbb{T}} |(Z_n(t) - Z_m(t)|,$$

which is equivalent to

$$\lim_{n,m \to \infty} \|(Z_n - Z_m)^*(T)\|_0 = 0.$$

Theorem 4.9.3 *If process* $\mathbf{X}$ *has property (B) and is quasi–left continuous then, for each* $t \in \mathbb{T}$, *limits in probability*

$$\mathfrak{b}(t) = \lim_{n \to \infty} \mathfrak{b}_n(t), \tag{4.9.4}$$

$$\mathfrak{w}(t) = \lim_{n \to \infty} \mathfrak{w}_n(t), \tag{4.9.5}$$

$$\mathfrak{p}(t) = \lim_{n \to \infty} \mathfrak{p}_n(t), \tag{4.9.6}$$

exist and the convergence in probability is uniform on $\mathbb{T}$.
   Moreover, if $f \in \mathcal{R}_0^1$ *is such that* $|f(x)| \le c$ *for some* $c > 0$ *and* $f(x) = 0$ *for* $|x| < r$ *and some* $r > 0$ *then, for each* $s > 0$ *and* $n \in \mathbb{N}$,

$$P(\mathfrak{p}_n^* > s) \le \frac{2s + c}{s} P(X^* > \frac{r}{2}).$$

Remark 4.9.1   *1. In the proof of the uniform convergence in probability of the sequence* $\{\mathfrak{p}_n(t)\}$ *the fact that process* $\mathbf{X}$ *satisfies assumption (B) is not used.*

2. The limit processes $\mathfrak{b}$, $\mathfrak{w}$, *and* $\mathfrak{p}$ *have a.s. continuous sample paths.*

3. Property (B) implies that the limiting process $\mathfrak{b} = \{\mathfrak{b}(t) : t \in \mathbb{T}\}$ *has a.a. sample paths of bounded variation.*

Now we can define the Jacod–Grigelionis characteristics of a process $\mathbf{X}$.

Jacod–Grigelionis characteristics. The process $\mathfrak{b}$ defined by (4.9.4) is called the **first Jacod–Grigelionis characteristic** of the process $\mathbf{X}$. Hence it is a predictable process with sample paths that are a.s. continuous and of bounded variation.
   The **second Jacod–Grigelionis characteristic** is a random measure μ defined as follows. For a fixed family of functions $f \in \mathcal{R}_0^1$ which separate points,

choose subsequences of sequences $\{\mathfrak{b}_n\}$, $\{\mathfrak{w}_n\}$ and $\{\mathfrak{p}_n\}$ so that they are convergent a.s. uniformly on $\mathbf{T}$. One can prove, using Theorem 4.9.3 that if function f is as above, then there exists a unique measure μ on $\mathbf{T} \times (\mathbb{R} \setminus \{0\})$ such that

$$\lim_{n \to \infty} \sum_{k : t_k^n \le t} E\left(f(d_k^n)|\mathcal{F}_{k-1}^n\right)(\omega) = \int_{\mathbb{R}\setminus\{0\}} \int_0^t f(x)\mu(ds, dx, \omega).$$

Moreover, the process on the right hand side is a.s. predictable, its a.a. sample paths are continuous and, additionally,

$$\int_{\mathbb{R}\setminus\{0\}} \int_0^t [\![x]\!]^2 \mu(ds, dx, \omega) < \infty \quad \text{a.e.}$$

The predictability of the process defined by the above double integral justifies labeling the measure μ *predictable* since, for each Borel set $A \subset \mathbb{R} \setminus \{0\}$ the process $\{\mu((0,t] \times A : t \in \mathbf{T}\}$ is predictable.

The *third Jacod–Grigelionis characteristic* is a process $\mathfrak{c} = \{\mathfrak{c}(t)\}$ defined by equation

$$\mathfrak{c}(t) = \mathfrak{w}(t) - \int_{\mathbb{R}\setminus\{0\}} \int_0^t f(x) \, \mu(ds, dx),$$

which is a predictable process with nondecreasing sample paths.

The following result summarizes the properties of the Jacod–Grigelionis characteristics. (The class $\mathcal{R}$ appearing below was introduced in Section 4.6.)

Theorem 4.9.4 *For each* $f \in \mathcal{R}$ *we have*

$$\lim_{n \to \infty} \sum_{k : t_k^n \le t} E\left(f(d_k^n)|\mathcal{F}_{k-1}^n\right)$$

$$= f(0)\mathfrak{b}(t) + \frac{1}{2}\ddot{f}(0)\mathfrak{c}(t) + \int_{\mathbb{R}\setminus\{0\}} \int_0^t f(x)\mu(ds, dx).$$

As we observed before, the Lévy characteristics determine the distribution of a process with independent increments. This is no longer true for the Jacod–Grigelionis characteristics and general processes. The only exception is the situation when all the characteristics $\mathfrak{b}$, μ and $\mathfrak{c}$ are deterministic. Then the process $\mathbf{X}$ has $\{\mathcal{F}_t\}$–independent increments. This fact is known as the Grigelionis Theorem. In general, the best analog of the Lévy–Khinchine Formula that can be obtained states that the process described by

$$E e^{i\theta X(t)} = \exp\left\{-i\theta\mathfrak{b}(t) - \theta^2\mathfrak{c}(t) - \int_{\mathbb{R}\setminus\{0\}} \left(e^{i\theta x} - 1 - i\theta x\right) \mu(dt, dx)\right\}$$

is a local L^2–martingale.

By the Bichteler–Dellacherie Theorem, a quasi–left continuous process $\mathbf{X}$ has property (B) if and only if it is a semimartingale in the classical sense, i.e. it is a sum of two processes with sample paths in $D(\mathbf{T})$: a local L^2–martingale and a process that has a.a. sample paths of bounded variation.

4.10   Diffusions Driven by Brownian Motion

In this section we will discuss general stochastic differential equations driven by a Brownian motion process. Next, we give an explicit formula for solutions of linear systems and present an Ornstein–Uhlenbeck process as an example of such solution.

General theorem on existence of solutions.   Let $f = f(x,t)$ and $g = g(x,t)$ be measurable functions on $\mathbb{R}^n \times [0,T]$ with values in $\mathbb{R}^n$ and $\mathcal{L}(\mathbb{R}^m, \mathbb{R}^n)$, respectively. Let $\{W(t)\}$ denote the standard m–dimensional Wiener process.

We are looking for a stochastic process $\{X(t) : t \in [0,T]\}$ which solves a *stochastic differential equation* of the form

$$dX(t) = f(X(t),t)\,dt + \langle g(X(t),t), dW(t)\rangle.$$

Of course, if $g(x,t) \equiv 0$, then this equation becomes a well–known ordinary differential equation

$$\frac{dX}{dt}(t) = f(X(t),t).$$

Uniqueness of the solution is insured by an initial condition.

To be more precise, we have to introduce an increasing family $\{\mathcal{F}_t\}$ of σ–algebras, with respect to which the process $\{W(t)\}$ is an $\{\mathcal{F}_t\}$ martingale and to formulate the problem with the use of stochastic integrals introduced in Section 4.2.

Definition 4.10.1 *The process $\{X(t)\}$ with values in $\mathbb{R}^n$ satisfying the conditions*

- *$\{X(t)\}$ is adapted to $\{\mathcal{F}_t\}$;*

- *$\{X(t)\}$ is continuous;*

- *and solving the equation*

$$X(t) = \xi + \int_0^t f(X(s),s)\,ds + \int_0^t \langle g(X(s),s), dW(s)\rangle, \quad t \in [0,T] \quad (4.10.1)$$

  for a given $\mathcal{F}_0$–measurable random variable ξ

*will be called a **solution of the stochastic differenctial equation** (4.10.1).*

We will (as is commonly understood) also write

$$dX(t) = f(X(t),t)\,dt + \langle g(X(t),t), dW(t)\rangle, \quad t > 0; \quad X(0) = \xi. \quad (4.10.2)$$

Let us formulate and prove a theorem on existence and uniqueness of a solution to this problem.

Theorem 4.10.1 *We suppose that*

$$\forall_{x,y \, \in \mathbb{R}^n} \, \forall_{t \, \in [0,T]} \, |f(x,t) - f(y,t)| + |g(x,t) - g(y,t)| \leq L|x - y|, \quad (4.10.3)$$

$$\forall_{x \, \in \mathbb{R}^n} \, \forall_{t \, \in [0,T]} \, |f(x,t)|^2 + |g(x,t)|^2 \leq K^2(1 + |x|^2), \quad (4.10.4)$$

$$E|\xi|^2 < \infty. \quad (4.10.5)$$

Then there exists a solution $\{X(t) : t \in [0,\infty)\}$ *of the stochastic differential equation 4.10.1, and such that*

$$E \sup_{0 \leq t \leq T} |X(t)|^2 \leq C \left(1 + E|\xi|^2\right), \quad (4.10.6)$$

where the constant C *depends only on* K *and* T. *This solution is uniquely determined.*

PROOF. *Uniqueness.* If $\{X_1(t)\}$ and $\{X_2(t)\}$ are two solutions of (4.10.1) then for all $t \in [0,T]$ we have

$$X_1(t) - X_2(t) = \int_0^t [f(X_1(s),s) - f(X_2(s),s)] \, ds \quad (4.10.7)$$

$$+ \int_0^t \langle g(X_1(s),s) - g(X_2(s),s), dW(s)\rangle$$

and thus

$$|X_1(t) - X_2(t)|^2 \leq 2\left\{\int_0^t [f(X_1(s),s) - f(X_2(s),s)] \, ds\right\}^2$$

$$+ \, 2\left\{\int_0^t \langle g(X_1(s),s) - g(X_2(s),s), dW(s)\rangle\right\}^2.$$

By applying inequality (4.2.16) we obtain

$$E|X_1(t) - X_2(t)|^2 \leq 2t \int_0^t E|f(X_1(s),s) - f(X_2(s),s)|^2 \, ds \quad (4.10.8)$$

$$+ 2 \int_0^t E|g(X_1(s),s) - g(X_2(s),s)|^2 \, ds.$$

If we define

$$\phi(t) = E|X_1(t) - X_2(t)|^2$$

and make use of (4.10.3), then we have

$$\phi(t) \leq 2(T + 1) \, L^2 \int_0^t \phi(s) \, ds$$

and we derive from the Gronwall Inequality that $\phi(t) \equiv 0$ and

$$\forall_{t \, \in [0,T]} |X_1(t) - X_2(t)| = 0, \quad \text{a.s.,}$$

which gives $X_1(t) = X_2(t)$ a.s. for all $t \in [0, T]$, thanks to the continuity of the processes in question.

Existence. We define the sequence

$$X^0(t) \equiv \xi, \tag{4.10.9}$$

$$X^n(t) = \xi + \int_0^t f(X^{n-1}(s), s) \, ds + \int_0^t \langle g(X^{n-1}(s), s), dW(s) \rangle, \tag{4.10.10}$$

which is correct since $\{g(X^{n-1}(t), t)\}$ defines a process from $L^2_W(0, T; \mathbb{R}^n)$.

The argument from the first part of the proof gives

$$E|X^{n+1}(t) - X^n(t)|^2 \leq M \int_0^t E|X^n(s) - X^{n-1}(s)|^2 \, ds, \tag{4.10.11}$$

where $M = 2(T + 1)L^2$.

By iterating (4.10.11) we obtain

$$E|X^{n+1}(t) - X^n(t)|^2 \leq M^n \int_0^t \frac{(t-s)^{n-1}}{(n-1)!} E|X^1(s) - \xi|^2 \, ds. \tag{4.10.12}$$

But

$$E|X^1(s) - \xi|^2 \leq MK^2 T(1 + E|\xi|^2),$$

hence

$$E|X^{n+1}(t) - X^n(t)|^2 \leq C \frac{T^n}{n!}. \tag{4.10.13}$$

On the other hand,

$$\sup_{0 \leq t \leq T} |X^{n+1}(t) - X^n(t)| \leq \int_0^T |f(X^n(s), s) - f(X^{n-1}(s), s)| \, ds$$

and thanks to (4.2.23) we get

$$E \sup_{0 \leq t \leq T} |X^{n+1}(t) - X^n(t)|^2$$

$$\leq 2TL^2 E \int_0^T |X^n(s) - X^{n-1}(s)|^2 \, ds + M E \int_0^T |X^n(s) - X^{n-1}(s)|^2 \, ds$$

$$\leq M \frac{T^{n-1}}{(n-1)!}.$$

Thus, we have

$$\sum_{n=1}^\infty P\left\{ \sup_{0 \leq t \leq T} |X^{n+1}(t) - X^n(t)| > \frac{1}{n^2} \right\} \leq \sum_{n=1}^\infty n^4 C_1 \frac{T^{n-1}}{(n-1)!}. \tag{4.10.14}$$

The series on the right hand side of this inequality converges, which means that the series

$$\xi + \sum_{n=0}^\infty \left(X^{n+1}(t) - X^n(t) \right)$$

converges a.s. uniformly with respect to t. There exists a continuous process $\{X(t)\}$ such that

$$X^n(t) \to X(t) \quad \text{uniformly with respect to } t. \tag{4.10.15}$$

It remains to show that this process solves our problem. From (4.10.10) we have $X(0) = \xi$. Observe that

$$E|X^n(t)|^2 \le 3\left\{ E|\xi|^2 + KT \int_0^t \left(1 + E|X^{n-1}(s)|^2\right) \, ds \right.$$
$$\left. + K^2 \int_0^t \left(1 + E|X^{n-1}(s)|2\right) \, ds \right\},$$

$$E|X^n(t)|^2 \le \left(1 + E|\xi|^2\right) + M \int_0^t E|X^{n-1}(s)|^2 \, ds,$$

and by repeating this argument recursively, we get finally

$$E|X^n(t)|^2 \le \left(1 + E|\xi|^2\right) \left(M + M^2 t + \ldots + M^{n-1}\frac{t^n}{n!}\right)$$
$$\le M \left(1 + E|\xi|^2\right).$$

This allows us to pass to the limit with $n \to \infty$ in (4.10.10) and gives (4.10.1), and also (4.10.6).   □

Ornstein–Uhlenbeck process.

Let us start with the observation, that the Ornstein–Uhlenbeck process $\{X(t); \; t \ge 0\}$ defined by

$$X(t) = e^{-\lambda t} X(0) + \mu \int_0^t e^{-\lambda(t-s)} \, dB(s), \qquad \lambda > 0, \quad \mu > 0,$$

with a fixed $X(0) = X_0$, and with a given Brownian motion $\{B(t); \; t \ge 0\}$, is a diffusion process, i.e., that it can be considered as a solution to the equation (4.10.1) with drift and dispersion coefficients defined as $f(s, X(s)) = -\lambda X(s)$ and $g(s, X(s)) \equiv \mu$. Figures 4.10.1 – 4.10.3 show four approximate trajectories of the Ornstein–Uhlenbeck processes $\{X(t); \; t \in [0,1]\}$ with $X(0) \sim \mathcal{N}(0,1)$, for three different values of λ: $\lambda = 1$, $\lambda = 2$ and $\lambda = 4$ with the same value of $\mu = 2$. In all cases the trajectories are included in the same rectangle $(t, X(t)) \in [0,1] \times [-2,2]$. As in all figures in this section the trajectories are represented by thin lines. The two pairs of quantile lines defined by $p_1 = 0.1$ and $p_2 = 0.25$ show that only for the case $\lambda = 2$ and $\mu = 2$ (Fig. 4.10.2) they are "parallel". This means that the quantile lines are time invariant, demonstrating the stationarity of the corresponding Ornstein–Uhlenbeck process (only in this case we have $Var \; X_0 = 1 = \frac{\mu^2}{2\lambda}$). Moreover, the field of directions of the ordinary differential equation $dx = -\lambda x(t) \, dt$ inserted in these figures helps to estimate the proper value of the parameter λ assuring stationarity of the process (in Fig. 4.10.1 it is too small, in Fig. 4.10.3 it is too large).

This observation helps to estimate this parameter in more complicated situations, e.g., when we work with α–stable Ornstein–Uhlenbeck processes, assuming μ and X_0 to be fixed.

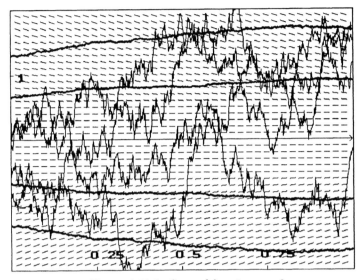

Figure 4.10.1. Case of $\lambda = 1$, $\mu = 2$.

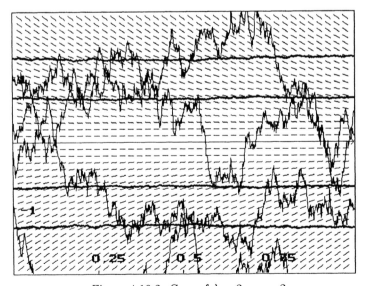

Figure 4.10.2. Case of $\lambda = 2$, $\mu = 2$.

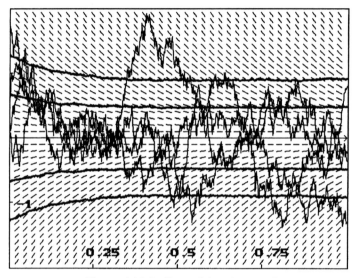

Figure 4.10.3. Case of $\lambda = 4$, $\mu = 2$.

Systems of linear stochastic equations.

A very important sub-class of $\mathbb{R}^d$–valued diffusion processes $\{X(t) : t \in [0, T]\}$ driven by an $\mathbb{R}^p$–valued Brownian motion process $\{B(t) : t \geq 0\}$ is defined by a system of d scalar linear stochastic equations of the form

$$dX(t) = (f(t) + g(t)X(t))\, dt + h(t)\, dB(t), \qquad (4.10.16)$$

with $X(0) = X_0$, where f, g, h are given vector–, or matrix–valued functions with values in $\mathbb{R}^d$, $\mathbb{R}^d \times \mathbb{R}^d$ and $\mathbb{R}^d \times \mathbb{R}^p$, respectively.

Let us mention a few of features of linear stochastic differential equations that make them so important.

- The class of stationary Gaussian processes can be derived as a family of solutions of appropriately defined linear stochastic equations.

- Assuming some regularity of coefficient functions (to simplify the exposition let us assume them to be continuous on $[0, T]$), solutions of such equations can be expressed in the following explicit form involving stochastic integrals

$$X(t) = G(t) \left(X(0) + \int_0^t G^{-1}(s)f(s)\, ds + \int_0^t G^{-1}(s)h(s)\, dB(s) \right),$$

for $t \in [0, T]$, where $G = G(t)$ is an $\mathbb{R}^d \times \mathbb{R}^d$–valued function and is defined as a fundamental matrix of the linear deterministic system

$$\frac{d}{dt}G(t) = g(t)G(t), \quad \text{for } t \in [0, T],$$

i.e., its solution with the following initial condition

$$G(0) = I.$$

- It is possible to get additional information on a solution. For example, using the Itô Formula one can derive deterministic differential equations describing the evolution in time of first and second moments of diffusions in question.

- It is quite easy to extend some methods and facts concerning the equation (4.10.16) to linear stochastic equations with respect to other classes of stochastic measures.

- A large variety of physical problems can be modeled by means of such systems of stochastic equations.

4.11   Diffusions Driven by α–Stable Lévy Motion

We have already defined an α–stable stochastic measure on $\mathbb{R}$ and an α–stable stochastic integral

$$I(f) = \int_A f(u) \, dL_\alpha(u).$$

Thus, we are in a position to describe diffusions driven by α–stable Lévy motion in terms of such integrals.

We are looking for a real–valued stochastic process $\{X(t) : t \in [0,T]\}$ such that

$$dX(t) = a(t, X(t)) \, dt + c(t) \, dL_\alpha(t), \quad X(0) = X_0, \tag{4.11.1}$$

with $t \in [0,T]$ and X_0 – a given stable random variable.

Strictly speaking, it means that the following "integral" equation should be satisfied

$$X(t) = X_0 + \int_0^t a(s, X(s-)) \, ds + \int_0^t c(s) \, dL_\alpha(s). \tag{4.11.2}$$

The more general problem, involving stochastic integrands and integrators, can be written in the following setting

$$X(t) = X_0 + \int_0^t a(s, X(s-)) \, ds + \int_0^t c(s, X(s-)) \, dL_\alpha(s). \tag{4.11.3}$$

Notice that more general than equations (4.11.2) and (4.11.3) are the so called stochastic differential equations with jumps, involving stochastic integrals with respect to Poisson random measures of suitable point processes with given deterministic intensity measures (see Ikeda and Watanabe (1981)). In turn, all these stochastic equations are special cases of general stochastic differential equations driven by semimartingales, i.e. equations of the form

all these stochastic equations are special cases of general stochastic differential equations driven by semimartingales, i.e. equations of the form

$$X(t) = X_0 + \int_0^t f(X(s-))\, dY(s), \qquad (4.11.4)$$

where $\{Y(t)\}$ stands for a given semimartingale process.

There is a vast literature concerning this topic (see for example Protter (1990) and the bibliography therein).

To see that the differential equation (4.11.3) driven by α–stable Lévy motion is a special case of the equation (4.11.4) with a semimartingale as an integrator it is enough to notice that α–stable Lévy motion can serve as an example of a semimartingale.

Remark 4.11.1 *The Lévy motion can serve as an example of a semimartingale. Regarding processes having independent increments without a Brownian component, i.e., processes with the Laplace exponent of the form* $E[e^{-zX(t)}] = e^{-\kappa(z)t}$, *we have*

$$\kappa(z) = bz + \int_{\mathbb{R}} \left(1 - zx I_{\{x:\ |x| \le 1\}}(x) - e^{-zx}\right)\, \nu(dx),$$

where ν is the Lévy measure satisfying the inequality $\int_{\mathbb{R}} (x^2 \wedge 1)\, \nu(dx) < +\infty$, b being the (non–canonical) drift. In the case of α–stable process $\{X(t)\}$ with $1 < \alpha < 2$ we have $\nu(dx) = c\, \alpha\, x^{-\alpha-1} dx / \Gamma(1 - \alpha)$ and thus $\kappa(z) = \bar{c} \int_0^\infty (1 - zx - e^{-zx})\, x^{-\alpha}\, dx$, where $\bar{c}$ is a positive constant and the drift is defined by $b = \bar{c} \int_1^\infty x^{-\alpha}\, dx$. Moreover, $\int (|x| \wedge 1)\, \nu(dx)$ is infinite, so the process $\{X(t) : t \ge 0\}$ is a totally discontinuous martingale with infinite variance (for detailed information we refer the reader to Jacod (1979), Protter (1990) or Kwapień and Woyczyński (1992)).

This fact allows us to obtain theorems on existence of solutions of stochastic differential equations (4.11.2) or (4.11.3), driven by stable measures. It is enough to employ corresponding theorems concerning semimartingales.

Often we will be interested in a linear version of this equation, i.e. we will discuss the α–stable diffusions described by the equation

$$X(t) = X_0 + \int_0^t \left(a(s) + b(s)X(s)\right)\, ds + \int_0^t c(s)\, dL_\alpha(s). \qquad (4.11.5)$$

Equation (4.11.5) is of independent interest because, as is easily seen, the general solution belongs to the class of α–stable processes. It may be expressed in the following form

$$X(t) = \Phi(t,0)X_0 + \int_0^t \Phi(t,s)\, a(s)\, ds + \int_0^t \Phi(t,s)\, c(s)\, dL_\alpha(s),$$

where $\Phi(t,s) = \exp\left\{\int_s^t b(u)\, du\right\}$.

α–stable Ornstein–Uhlenbeck process.

Let us consider a simplified version of Equation (4.11.5), namely a linear stochastic equation of the form

$$dV_\alpha(t) = -\lambda \, V_\alpha(t) \, dt + dL_\alpha(t), \quad V_\alpha(0) = V_0. \tag{4.11.6}$$

Analogously to the Gaussian case, the solution of Equation (4.11.6) can be given in the following explicit form

$$V_\alpha(t) = e^{-\lambda t} \, V_\alpha(0) + \int_0^t e^{-\lambda(t-s)} \, dL_\alpha(s).$$

Remark 4.11.2 *In the case of $\alpha = 2$, a solution to (4.11.6) is represented by an Ornstein–Uhlenbeck process with Gaussian distribution. From the point of view of stochastic modeling such a system response is Gaussian and a fluctuation–dissipation relation can be satisfied (see Gardner (1985)). If $\{L_\alpha(t) : t \geq 0\}$ is a non–Gaussian stable process ($\alpha \in (1, 2)$) then the response of the system is of the α–stable form. In this case the system response has infinite variance, which corresponds to the situation when the particle with the velocity $V_\alpha(t)$ and subject to linear damping λ has infinite kinetic energy. Thus, the fluctuations supply an infinite amount of energy, which can not be balanced by the linear dissipation. This means that the fluctuation must be regarded as external and no fluctuation–dissipation relation can be imposed on the system. (See West and Seshardi (1982) for more details.)*

Moreover, in the case of Equation (4.11.6) it is possible to describe the characteristic function of the response of the system $\phi = \phi(t, \theta)$ as a solution of the following simple partial differential equation of evolution

$$\frac{\partial \phi}{\partial t} + \lambda \, \theta \, \frac{\partial \phi}{\partial \theta} \, \phi = -\sigma^\alpha \, |\theta|^\alpha \, \phi.$$

From these solutions one can derive an explicit formula for the probability density in the Gaussian ($\alpha = 2$) and Cauchy ($\alpha = 1$) cases.

Chapter 5

Spectral Representations of Stationary Processes

5.1 Introduction

In this chapter we discuss some classes of stochastic processes which can be represented by stochastic integrals of deterministic functions with respect to α–stable or infinitely divisible stochastic measures. The results presented here are extensively applied in Chapter 10 to the study of ergodic properties of α–stable and infinitely divisible processes.

Definition 5.1.1 *We say that a process* $\mathbf{X}$ *has the* **representation** $\mathbf{Y}$ *if* $\mathbf{X} \overset{\mathrm{d}}{=} \mathbf{Y}$, *i.e. if the processes* $\mathbf{X}$ *and* $\mathbf{Y}$ *have the same finite–dimensional distributions.*

When discussing spectral integral representations of stochastic processes we will rely mainly on the results of Hardin (1982) and Rajput and Rosiński (1989).

In general the problem is to find for a given stochastic process $\mathbf{X} = \{X(t) : t \in \mathbb{T}\}$ a measurable space $(E, \mathcal{E})$, a random measure $\boldsymbol{\Lambda}$ and measurable functions f_t such that

$$\{X(t) : t \in \mathbb{T}\} \overset{\mathrm{d}}{=} \{\textstyle\int f_t d\boldsymbol{\Lambda} : t \in \mathbb{T}\}. \tag{5.1.1}$$

Definition 5.1.2 *The map* $t \to f_t$ *defined by (5.1.1) is called a* **spectral representation** *of the process* $\mathbf{X}$.

There are two main reasons that make such representations useful in answering various questions about the process $\mathbf{X}$.

(a) Many problems of interest about $\mathbf{X}$ can be reformulated in terms of non–random functions f_t and the measure $\boldsymbol{\Lambda}$ (or in terms of certain parameters characterizing $\boldsymbol{\Lambda}$, e.g. its Lévy characteristics).

(b) These reformulated questions can be effectively answered by making use of the rich structure of the metric linear space of functions generated by $\{f_t\}$

and the fact that Λ possesses properties very similar to $\mathbf{X}$, but admits much simpler probabilistic structure.

Many authors constructed spectral representations for special subclasses of infinitely divisible processes (see Urbanik (1968), Maruyama (1970), Schilder (1970), Kuelbs (1973), Hardin (1982), Rosiński (1986), Rajput and Rama-Murthy (1987)). Finally, Rajput and Rosiński (1989) established that for any given infinitely divisible process $\mathbf{X}$ there exist a family of non–random functions f_t and an infinitely divisible random measure Λ such that (5.1.1) holds and the following conditions are satisfied:

(i) the measure Λ retains properties similar to $\mathbf{X}$, for example, if $\mathbf{X}$ belongs to a known class of processes, such as α–stable or self–decomposable processes, then Λ belongs to the corresponding class of random measures;

(ii) the functions f_t belong to a linear topological space, which has a similar structure as the linear space of the process $\mathbf{X}$.

In addition to the above representations, which are valid only in law, they also obtained spectral representations which are valid almost surely.

Of particular interest is a vast class of stationary processes (e.g. *moving averages, harmonizable processes,* etc.), so important from the point of view of ergodic theory. Thus, at the end of this section let us recall this basic definition.

Definition 5.1.3 *An arbitrary stochastic process* $\{X(t) : t \in \mathbf{T}\}$ *is called* ***stationary*** *if and only if the joint distribution functions*

$$F_{t_1,\ldots,t_n}(x_1, \ldots, x_n) = P\{X(t_1) \leq x_1, \ldots, X(t_n) \leq x_n\} \qquad (5.1.2)$$

have the property

$$F_{t_1,\ldots,t_n} = F_{t_1+h,\ldots,t_n+h}$$

for all $n = 1, \ldots$ *and all* $t_1, \ldots, t_n$ *such that* $t_1 < \ldots < t_n$ *and all* $t_i,\ t_i + h \in \mathbf{T}$ *with any* $h > 0$. *In other words, the joint distribution function of* $X(t_1), \ldots, X(t_n)$ *depends only on the differences* $t_2 - t_1, t_3 - t_2, \ldots, t_n - t_{n-1}$.

5.2   Gaussian Stationary Processes

In this section we briefly discuss Gaussian stationary Markov processes.

Definition 5.2.1 *Let* $\lambda > 0$ *and let* $\{B(t) : t \in \mathbb{R}\}$ *be a Brownian motion process defining corresponding stochastic Gaussian measure on* $\mathbb{R}$. *The process*

$$X(t) = \int_{-\infty}^{t} e^{-\lambda(t-s)}\, dB(s), \quad t \in \mathbb{R}, \qquad (5.2.1)$$

is called a ***Gaussian Ornstein–Uhlenbeck process***.

According to Definition 5.1.3, stationarity of a stochastic process implies stationarity in the wide sense for an L^2-process, i.e., the process with finite second moments, since

$$\begin{aligned} E[X(s)X(t)] &= \int\int xy\, dF_{s,t}(x,y) = \int\int xy\, dF_{s+h,t+h}(x,y) \\ &= E[X(s+h)X(t+h)], \end{aligned}$$

and, similarly,

$$E[X(t)] = E[X(t+h)].$$

The converse is not true in general; however, it holds for stationary Gaussian processes. This follows directly from the formula for the probability density of an n-tuple of jointly Gaussian and linearly independent variables, or from the form of the joint characteristic function, which for jointly Gaussian variables $\mathbf{X} = [X(t_1), X(t_2), ..., X(t_n)]$ is given by

$$\Phi_{\mathbf{X}}(\boldsymbol{\theta}) = E\{e^{i(\mathbf{X},\boldsymbol{\theta})}\} = exp\{i(\boldsymbol{\theta}, \mathbf{m_X}) - \frac{1}{2}(\boldsymbol{\theta}, \mathbf{K_X}(\boldsymbol{\theta}))\}.$$

This formula depends only on the mean vector $\mathbf{m_X}$ and the covariance matrix $\mathbf{K_X}$. Moreover, if they are translation–invariant then the probability density and the characteristic function are translation–invariant.

Example 5.2.1 *A thermal–noise voltage.*

As discussed in Gardner(1986), a thermal–noise voltage is appropriately modeled as a stationary Gaussian process assuming that the environment of noisy resistance is time–invariant. Furthermore, it is shown there that the covariance function is given by

$$K_X(\tau) = N_0 \int_{-\infty}^{\infty} h(t+\tau)h(t)\, dt,$$

where $h(t)$ is the impulse response of the voltmeter used to measure the noise voltage waveform and $N_0 = 2KTR$, where K is the Boltzmann constant, T is the temperature (in degrees Kelvin) of the resistor, B is the bandwidth (in Hertz) of the voltmeter, and R is the resistance (in ohms). This result is based on the physical assumption that the response time of the voltmeter is much larger than the mean relaxation time of free electrons within the resistance. If

$$h(t) = \begin{cases} e^{-\alpha t} & \text{if } t > 0, \\ 0 & \text{otherwise,} \end{cases}$$

then $\{X(t) : t \geq 0\}$ is the Ornstein–Uhlenbeck process.

Now let us prove the Doob Theorem characterizing stationary Markov Gaussian processes (cf., also Itô (1944b)).

Theorem 5.2.1 *If $\{X(t) : t \in \mathbb{R}\}$ is a real–valued stationary Gaussian process with mean 0 and variance 1, then it is a Markov process if and only if its covariance function has the form*

$$r(t) = K_X(t) = \exp(-\lambda|t|), \quad \lambda > 0. \tag{5.2.2}$$

PROOF. *Necessity.* For the process $\{X(t)\}$ with joint densities W we have

$$W(x_1, t_1; x_2, t_2; x_3, t_3)W(x_2, t_2) = W(x_1, t_1; x_2, t_2)W(x_2, t_2; x_3, t_3), \tag{5.2.3}$$

and these four densities can be easily computed. First,

$$W(x_2, t_2) = \frac{1}{\sqrt{2\pi}} e^{x_2^2/2}.$$

Next, since $X(t)$ has mean 0 and variance 1, the two–dimensional density takes the form

$$W(x_1, t_1; x_2, t_2) = \frac{\sqrt{\det \mathbf{A}}}{2\pi} e^{-(a_{11}x_1^2 + 2a_{12}x_1 x_2 + a_{22}x_2^2)/2},$$

where

$$\mathbf{A} = [r(t_i - t_j)]^{-1} = \begin{bmatrix} 1 & r_{12} \\ r_{12} & 1 \end{bmatrix}^{-1} = \begin{bmatrix} \frac{1}{1-r_{12}^2} & \frac{-r_{12}}{1-r_{12}^2} \\ \frac{-r_{12}}{1-r_{12}^2} & \frac{1}{1-r_{12}^2} \end{bmatrix},$$

which means that

$$W(x_1, t_1; x_2, t_2) = \frac{1}{2\pi\sqrt{1-r_{12}^2}} \exp\left(-\frac{1}{2}\frac{x_1^2 - 2r_{12}x_1 x_2 + x_2^2}{1-r_{12}^2}\right)$$

and

$$W(x_2, t_2; x_3, t_3) = \frac{1}{2\pi\sqrt{1-r_{23}^2}} \exp\left(-\frac{1}{2}\frac{x_2^2 - 2r_{23}x_2 x_3 + x_3^2}{1-r_{23}^2}\right).$$

It is also clear that

$$W(x_1, t_1; x_2, t_2; x_3, t_3) = \frac{\sqrt{\det \mathbf{B}}}{\sqrt{(2\pi)^3}} \exp\left(-\frac{1}{2}\langle \mathbf{x}, \mathbf{Bx}\rangle\right)$$

with the matrix $\mathbf{B}$ of the form

$$\mathbf{B} = \begin{bmatrix} 1 & r_{12} & r_{13} \\ r_{12} & 1 & r_{23} \\ r_{13} & r_{23} & 1 \end{bmatrix}^{-1}$$

and

$$\det \mathbf{B} = \frac{1}{1 - r_{23}^2 - r_{12}^2 - 2_{13}^2 + 2r_{12}r_{13}r_{23}}.$$

Substituting all these densities into (5.2.3) with

$$x_1 = x_2 = x_3 = 0$$

we get

$$(1 - r_{12}^2(1 - r_{23}^2)) = 1 - r_{23}^2 - r_{12}^2 - r_{13}^2 + 2r_{12}r_{13}r_{23},$$

or

$$(r_{13} - r_{12}r_{23})^2 = 0,$$

or else

$$r_{12}r_{23} = 0.$$

This, by setting $s = t_2 - t_1$ and $t = t_3 - t_2$, becomes

$$r(s)r(t) = r(s + t).$$

This functional equation has only one continuous solution of the form (5.2.2).

Sufficiency. In order to complete the proof it is enough to check that the family of Gaussian densities with mean 0 and the covariance matrix

$$(r_{ij}) = \left(e^{-\lambda|t_i - t_j|}\right)$$

satisfies the conditions

$$W(x_1, t_1; ...; x_{n+1}, t_{n+1}) = P(x_{n+1}, t_{n+1}|x_n, t_n) \cdots P(x_2, t_2|x_1, t_1)W(x_1, t_1)$$

and

$$\int_{-\infty}^{\infty} P(x_{i+1}, t_{i+1}|x_i, t_i)P(x_i, t_i|x_{i-1}, t_{i-1}) \, dx_i = P(x_{i+1}, t_{i+1}|x_{i-1}, t_{i-1}),$$

which is not a complicated task (see e.g. Iranpour and Chacon (1987), p. 176). This ends the proof. □

Remark 5.2.1 *By replacing the process $\{B(t)\}$ in (5.2.1) by an α–stable Lévy motion $\{L_\alpha(t) : t \in \mathbb{R}\}$ we obtain the definition of an α–stable Ornstein–Uhlenbeck process. The Doob Theorem cannot be extended to the case when $\alpha \in (0, 2)$. The example below shows that in this case there are two such processes: the Ornstein–Uhlenbeck and the reverse Ornstein–Uhlenbeck processes which are stationary and Markovian.*

Example 5.2.2 *Two $S\alpha S$ Ornstein–Uhlenbeck processes.*

Let $\lambda > 0$ and M be a $S\alpha S$ random measure, $0 < \alpha < 2$, with Lebesgue control measure. The process

$$X(t) = \int_{-\infty}^{t} e^{-\lambda(t-x)} M(dx), \quad -\infty < t < \infty$$

is called an $S\alpha S$ *Ornstein–Uhlenbeck process.* $\{X(t)\}$ is a moving average process with $f(x) = \exp(-\lambda x)I_{[0,\infty)}(x)$, see pages 104–5, and hence it is stationary. To

To see this observe that for any $t_1, ... t_n, h \in \mathbb{R}$ and real $a_1, ..., a_n$ we have

$$
\begin{aligned}
\| \sum_{j=1}^{n} a_j X(t_j + h) \|_{\alpha}^{\alpha} &= \int_{-\infty}^{\infty} \left| \sum_{j=1}^{n} a_j f(t_j + h - x) \right|^{\alpha} dx \\
&= \int_{-\infty}^{\infty} \left| \sum_{j=1}^{n} a_j f(t_j - y) \right|^{\alpha} dy \\
&= \| \sum_{j=1}^{n} a_j X(t_j) \|_{\alpha}^{\alpha}.
\end{aligned}
$$

So the finite dimensional distributions of the process are shift invariant. Note also that for any fixed $s < t$,

$$
X(t) - e^{-\lambda(t-s)} X(s) = \int_s^t e^{-\lambda(t-x)} \, M(dx)
$$

is an $S\alpha S$ random variable independent of $\sigma(X(u) : u \leq s)$. This implies that the $S\alpha S$ Ornstein–Uhlenbeck process is also a Markov process. The Doob Theorem shows that in the Gaussian case ($\alpha = 2$) the Ornstein–Uhlenbeck process is the only stationary Markov Gaussian process. However, in the case of $0 < \alpha < 2$, there is another stationary Markov $S\alpha S$ process. It can be given by the following formula

$$
X(t) = \int_t^{\infty} e^{-\lambda(x-t)} \, M(dx), \quad -\infty < t < \infty.
$$

It is called the *reverse $S\alpha S$ Ornstein–Uhlenbeck process*. Using the same argument as above, we conclude that it is also a Markov process. The Doob Theorem shows that for $\alpha = 2$ the Ornstein–Uhlenbeck process and the reverse Ornstein–Uhlenbeck process are identical, which can also be verified by computing the covariances. It turns out that, in contrast to the Gaussian case, these two processes are different when $0 < \alpha < 2$. To see this, let $\mathbf{X}_1$ and $\mathbf{X}_2$ be the Ornstein–Uhlenbeck and the reverse Ornstein–Uhlenbeck processes, respectively. Fix $s < t$ and evaluate the spectral measure Γ_1 of the $S\alpha S$ random vector $(X_1(s), X_1(t))$ and the spectral measure Γ_2 of the $S\alpha S$ random vector $(X_2(s), X_2(t))$ and observe that they are different (cf., Samorodnitsky and Taqqu (1993), Chapter 3). The uniqueness of the spectral measure implies that $(X_1(s), X_1(t)) \neq (X_2(s), X_2(t))$. Therefore, the Ornstein–Uhlenbeck and the reverse Ornstein–Uhlenbeck processes are different for $0 < \alpha < 2$.

5.3   Representation of α–Stable Stochastic Processes

We are concerned here with the spectral representations of $S\alpha S$ processes $\mathbf{X} = \{X(t) : t \in \mathbf{T}\}$ in the form of α–stable stochastic integrals.

Definition 5.3.1 *A stochastic process* **X** *is called* **symmetric** α-**stable** *or* **Lévy SαS** *or, shortly,* **SαS process** *for* $\alpha \in (0, 2]$, *if for every* $n \in \mathbb{N}$ *and any* $a_1, ..., a_n \in \mathbb{R}$, $t_1, ..., t_n \in \mathbb{R}$, *the random variable* $Y = \sum_{i=1}^{n} a_i X(t_i)$ *has a symmetric stable distribution with index* α.

Let **X** be an $S\alpha S$ process, $\alpha \in (0, 2]$. For an $S\alpha S$ random variable Y, set $\|Y\|_\alpha = c_Y^{1/\alpha}$. Then $\| \cdot \|_\alpha^{1 \wedge \alpha}$ defines a norm in the case $1 \leq \alpha \leq 2$ and a quasi–norm in the case $0 < \alpha < 1$ on the space $lin\{X(t) : t \in \mathbb{R}\}$, metrizing the convergence in probability. Then, for $Y \in lin\{X(t) : t \in \mathbb{R}\}$ we have

$$\mathbb{E} e^{i\theta Y} = \exp(-|\theta|^\alpha \|Y\|_\alpha^\alpha).$$

Taking the closure of the linear span $lin\{X(t) : t \in \mathbb{R}\}$ with respect to the norm (quasi–norm) $\| \cdot \|_\alpha$ in the space $L_0(\mathbf{X}) \subseteq L^o(\Omega, \mathcal{F}, P)$ we obtain the space $L_\alpha(\mathbf{X})$.

Definition 5.3.2 *We say that the process* **X** *is* **separable in probability** *if there exists a countable set* $\mathbb{T}_0 \subset \mathbb{T}$ *such that the set of random variables* $\{X(t) : t \in \mathbb{T}_0\}$ *is a dense subset of* $\{X(t) : t \in \mathbb{T}\}$ *with respect to the topology of convergence in probability.*

It is not difficult to show that **X** is separable if and only if $L_\alpha(\mathbf{X})$ is separable. Here we will consider only stochastic processes that are separable in probability.

Theorem 5.3.1 *Let* $\mathbf{X} = \{X(t) : t \in \mathbb{T}\}$ *be an* $S\alpha S$ *process and let* $\{M(s) : s \in [0, 1]\}$ *be an* α-*stable Lévy process. Then there exists a set of functions* $\{f_t : t \in \mathbb{T}\} \subseteq L^\alpha[0, 1]$ *such that the process*

$$\left\{ \int_0^1 f_t(s) \ dM(s) \right\}_{t \in \mathbb{T}} \tag{5.3.1}$$

is stochastically equivalent to **X**.

PROOF. It is shown in Schilder (1970) that each finite dimensional subspace of $L_\alpha(\mathbf{X})$ imbeds linearly and isometrically into $L^\alpha[0, 1]$. This implies that there exists a measure space $(E, \mathcal{E}, m)$ and a linear isometric imbedding of $L_\alpha(\mathbf{X})$ into $L^\alpha(E, \mathcal{E}, m))$, (see Bretagnolle, Dacunha–Castelle and Krivine (1966)). The space $L_\alpha(\mathbf{X})$ is separable since **X** is separable and therefore we may choose $(E, \mathcal{E}, m)$ to be $[0, 1]$ with the Lebesgue measure. It follows from the fact that under separability condition of Definition 5.3.2, the space $L^\alpha(E, \mathcal{E}, m)$ is isometric to either $L^\alpha[0, 1]$, $(L^\alpha[0, 1] \oplus \ell_n^\alpha)_\alpha$ or $(L^\alpha[0, 1] \oplus \ell^\alpha)_\alpha$, (cf., Lacey (1974), p. 128). Consequently, we may represent the characteristic function of the process **X** as

$$\mathbb{E} \exp\left(i \sum_{j=1}^{d} a_j X(t_j) \right) = \exp\left(- \| \sum_{j=1}^{d} a_j f_{t_j} \|^\alpha \right), \tag{5.3.2}$$

where $\{f_t : t \in \mathbf{T}\} \subseteq L^{\alpha}[0,1]$. Conversely, the Kolmogorov Theorem implies that for any choice of functions $\{f_t\}$ in $L^{\alpha}[0,1]$, formula (5.3.1) defines an $S\alpha S$ process which is a representation of $\{X(t) : t \in \mathbf{T}\}$, so the map $t \to f_t$ is its spectral representation.   □

The spectral representations were the subject of many works. If we do not require the separability condition then we lose the stochastic integral representation and the luxury of working exclusively in $L^{\alpha}[0,1]$, but there still exist a measure space $(S, \mathcal{B}_S, \mu)$ and a spectral representation $t \to f_t \in L^{\alpha}(S, \mathcal{B}_S, \mu)$ satisfying (5.3.2). The concept of a *minimal spectral representation* introduced by Hardin (1982) (see also Rosinski (1993)) leads to the uniqueness of the spectral representation.

Definition 5.3.3 *Let $\mathbf{F_X}$ denote the closed linear span of $\{f_t : t \in \mathbf{T}\}$ in $L^{\alpha}[0,1]$ and let $\rho(F)$ denote the "ratio" σ-field of $\mathbf{F_X}$ defined by the formula $\rho(\mathbf{F_X}) = \sigma\{f/g : f,g \in \mathbf{F_X}\}$. We let $|A|$ denote the Lebesgue measure of the Borel set $A \subseteq [0,1]$, considered with the induced σ-field. A spectral representation $t \to f_t$ is called a **minimal spectral representation** if the following conditions are satisfied:*

(M_1) there is no set B of positive measure such that $f_t = 0$ a.e. on B for all t;

(M_2) to every Borel set B which is almost disjoint from the atoms of $\rho(\mathbf{F_X})$, there corresponds a set $B' \in \rho(\mathbf{F_X})$ such that $|B \triangle B'| = 0$;

(M_3) whenever B is an atom of $\rho(\mathbf{F_X})$, then f_t is a.e. constant on B for all t.

Theorem 5.3.2   *(i) Every $S\alpha S$ process has a minimal spectral representation.*

(ii) If $t \to f_t$ and $t \to g_t$ are minimal spectral representations for a given non-Gaussian $S\alpha S$ process then there exists an isometric automorphism A of $L^{\alpha}[0,1]$ such that $Af_t = g_t$ for all $t \in \mathbf{T}$.

PROOF. (1) Let $\{g_t : t \in \mathbf{T}\}$ be the spectral representation of the process in question. Let $\mathbf{G_X} = \overline{\mathrm{lin}}\{g_t : t \in \mathbf{T}\}$ and let g be a normalized function which has full support in $\mathbf{G_X}$ (see Hardin (1981)). Define a new measure space $(E_0, \mathcal{E}_0, \mu)$ by setting $E_0 = \mathrm{supp}(g)$, $\mathcal{E}_0 = \{A \cap E_0 : I_A \in \rho(\mathbf{G_X})\}$ and $d\mu(s) = |g(s)|^p \, ds$. $(E_0, \mathcal{E}_0, \mu)$ has at most countably many atoms, say $A_1, A_2, \dots$. Set $A = \cup_{n=1}^{\infty} A_n$. We define a measure algebra isomorphism T of $(E_0, \mathcal{E}_0, \mu)$ into $([0,1], \mathcal{B}([0,1]), \lambda)$. Start by setting $x_0 = 0, x_n = \sum_{k=1}^{n} \mu(A_k)$, and define $T(A_n) = [x_{n-1}, x_n)$. We can extend T to $\sigma(A_1, A_2, \dots)$ in an obvious way. Consider a measure space $(E', \mathcal{E}', \mu')$ where $E' = E \backslash A, \mathcal{E}' = \{B \backslash A : B \in \mathcal{E}\}$ and μ' denotes μ restricted to $E \backslash A$. It is a non-atomic separable measure space with total mass $1 - \sum_{n \geq 1} \mu(A_n) \equiv 1 - a$, and hence (by Halmos (1950)) there exists a measure algebra isomorphism T of $(E', \mathcal{E}', \mu')$ into $([a,1], \mathcal{B}([a,1]), \lambda)$, which we use to define T for sets in $\mathcal{E}'$. For general sets $B \in \mathcal{E}$ we set $TB = T(B \cap A) \cup T(B \backslash A)$. It is easy to see that T thus defined is a measure algebra isomorphism of $(E_0, \mathcal{E}_0, \mu)$ onto $([0,1], \sigma\{[x_{n-1}, x_n), B : n \geq 1, B \in \mathcal{B}([a,1])\}, \lambda)$.

Now define $f_t = T(g_t/g)$, where T denotes the canonical map on measurable functions induced by our isomorphism (Doob (1953)). It appears that $\{f_t : t \in \mathbb{T}\}$ is a minimal representation. It is a spectral representation since T is measure–preserving. For $g \in \mathbf{G_X}$ we have $I_E \in \overline{\text{lin}}\{g_t/g\}_{L^\alpha(E_0, \mathcal{E}_0, \mu)}$, and thus $I_{[0,1]} = T I_{E_0} \in \mathbf{F_X}$ which shows that (M_1) is satisfied. Condition (M_2) is satisfied since $\rho(\mathbf{F_X}) = \sigma(\mathbf{F_X}) = T(\mathcal{E}) = \sigma\{[x_{n-1}, x_n), B : n \geq 1, B \in \mathcal{B}([a,1])\}$. Finally, (M_3) holds since if B is an atom of $\rho(\mathbf{F_X})$, then $f = f/I_{[0,1]}$ must be a.e. constant when restricted to B.

(2) With the previous notations let us consider the map $f_t \to g_t$. This map induces an isometry U of $\mathbf{F_X}$ onto $\mathbf{G_X}$ which admits a unique isometric extension $\overline{U}$ to the subspace $\{g \in L^\alpha([0,1]) : g = rf, f \in \mathbf{F_X}, r$ is $\rho(\mathbf{F_X})$–measurable$\}$. Further $\overline{U}$ must have the form $\overline{U}(rf) = (Tr)(Uf)$, where T is a regular set isomorphism of $\rho(\mathbf{F_X})$ onto $\rho(\mathbf{G_X})$ (see Hardin (1981)). Now, let h have full support in $\mathbf{F_X}$. To extend $\overline{U}$ to the desired isometric automorphism we need only to extend T to the regular set isomorphism $\overline{T}$ of $\mathcal{B}([0,1])$ onto itself such that R defined by $Rf = \overline{T}(f/h)Uh$ is isometric. One can do this as follows. Let $A_1, A_2, \ldots$ be all atoms of $\rho(\mathbf{G_X})$. Since T is a regular set isomorphism, setting $B_i = T A_i$ gives a listing of the atoms of $\rho(\mathbf{G_X})$. Then (Halmos (1950)) there exists a measure algebra isomorphism between $(A_i, \mathcal{B}(A_i), \frac{1}{\lambda(A_i)}\lambda)$ and $(B_i, \mathcal{B}(B_i), \frac{1}{\lambda(B_i)}\lambda)$. Let $A_0 = [0,1]\backslash\bigcup_{i\geq 1} A_i$, $B_0 = [0,1]\backslash\bigcup_{i\geq 1} B_i$. Note that $T(A_0) = B_0$ and that, by (M_3), T maps A_0 onto B_0. Now define $\overline{T}$ by $\overline{T}(B) = T(B \cap A_0) \cup [\bigcup_{i\geq 1} T_i(B \cap A_i)]$. It is easy to see that $\overline{T}$ is a regular set isomorphism of $\mathcal{B}([0,1])$ onto itself which extends T. Now one has only to show that R is isometric. It follows from (M_2) that it is enough to show that for arbitrary Borel set B in A_0 the equality $\|RI_B\|_\alpha = \|I_B\|_\alpha$ holds. Let $i \geq 0$. By (M_3), h must be a.e. equal to some constant c_i on A_i, Uh must be a.e. equal to some constant d_i on B_i and the relationship between c_i and d_i is expressed by

$$c_i^\alpha \lambda(A_i) = \|I_{A_i} h\|_\alpha^\alpha = \left\|\overline{U}(I_{A_i} h)\right\|_\alpha^\alpha = \|(T I_{A_i})(Uh)\|_\alpha^\alpha + d_i^\alpha \lambda(B_i).$$

For $B \subset A_i$, $i \geq 1$, it is easy to check that

$$\begin{aligned}
\|RI_B\|_\alpha^\alpha &= \left\|\overline{T}(I_B/h) \cdot Uh\right\|_\alpha^\alpha = \left\|\frac{1}{c_i} \cdot T I_B \cdot Uh\right\|_\alpha^\alpha \\
&= \frac{\lambda(A_i)}{\lambda(B_i)}\left\|\frac{1}{d_i} \cdot I_{\overline{T}B} \cdot d_i\right\|_\alpha^\alpha = \frac{\lambda(A_i)}{\lambda(B_i)}\lambda(\overline{T}B) = \lambda(B) = \|I_B\|_\alpha^\alpha.
\end{aligned}$$

This ends the proof. □

Example 5.3.1 *The $S\alpha S$ Lévy motion.*

It is easy to check that each $S\alpha S$ Lévy motion has the minimal representation $t \to I_{(0,t]}$. For this, let

$$X(t) = \int_0^\infty I_{(0,t]}(x) \, dM(x) = \int_0^t dM(x), \quad t \geq 0,$$

where M is $S\alpha S$ on $[0, \infty)$, with control measure $dm(x) = dx$. Then

$$X(0) = 0 \quad \text{a.s.}$$

and

$$X(t) - X(s) = \int_s^t dM(x) = M([s,t]) \sim S_\alpha(|t-s|^{1/\alpha}, 0, 0).$$

If $0 \le t_1 < t_2 < ... < t_n$, then

$$(X(t_2) - X(t_1), X(t_3) - X(t_2), ..., X(t_n) - X(t_{n-1})) =$$

$$\left(\int_{t_1}^{t_2} dM(x), \int_{t_2}^{t_3} dM(x), ..., \int_{t_{n-1}}^{t_n} dM(x) \right).$$

The components of this random vector are independent because the integrands have disjoint supports. Hence, $\{X(t) : t \ge 0\}$ is a process which starts at 0, has stationary independent increments, and is $S\alpha S$–distributed. Therefore, it is the $S\alpha S$ Lévy motion.

Example 5.3.2 *Mean zero Gaussian processes.*

Each mean zero Gaussian process $\{X(t) : t \in \mathbf{T}\}$ on the probability space $(\Omega, \mathcal{F}, P)$ can be represented by

$$\tilde{X}(t) = \int_\Omega X(t, \omega) \, dG(\omega),$$

where G is the canonical independently scattered Gaussian random measure on the underlying probability space. Since the characteristic functions of $\{X(t) : t \in \mathbf{T}\}$ and $\{\tilde{X}(t) : t \in \mathbf{T}\}$ coincide, these processes are stochastically equivalent. To see this observe that

$$E \exp\left\{ i \sum_j a_j \tilde{X}(t_j) \right\} = \exp\left\{ -\frac{1}{2} \text{Var}\left[\sum_j a_j \tilde{X}(t_j) \right] \right\}$$

$$= \exp\left\{ -\frac{1}{2} \sum_{j,k} a_j a_k \text{Cov}\left(X(t_j), X(t_k) \right) \right\}.$$

If the process $\{X(t) : t \in \mathbf{T}\}$ satisfies the separability condition then the underlying probability space can be chosen separable, cf., Breiman (1968) for such a construction. This space must be also nonatomic since Gaussian variables take on no constant value with positive probability. By the isomorphism theorem, one may find $\{f_t : t \in \mathbf{T}\} \subseteq L^2[0,1]$ stochastically equivalent to $\{X(t) : t \in \mathbf{T}\}$. The calculation above shows that the spectral representation of the original Gaussian process $\{X(t) : t \in \mathbf{T}\}$ has the form $t \to f_t \in L^2[0,1]$. Thus, the representation of $\{X(t) : t \in \mathbf{T}\}$ by the mapping $t \to f_t \equiv X(t)$ is not very useful.

Example 5.3.3 *Sub–stable processes.*

An $S\alpha S$ process $\mathbf{X} = \{X(t) : t \in \mathbb{T}\}$ will be called sub–stable if there exist $\alpha' > \alpha$ and a $S\alpha'S$ process $\mathbf{Y} = \{Y(t) : t \in \mathbb{T}\}$, called a governing process, such that

$$-\log \mathbf{E} \exp \left(i \sum_j a_j X(t_j) \right) = \left[-\log \mathbf{E} \exp \left(i \sum_j a_j Y(t_j) \right) \right]^{\alpha/\alpha'},$$

for all a_j and t_j. If $\alpha' = 2$ then $\mathbf{X}$ is called sub–Gaussian.

Sub–stable processes exist in great profusion. Let $\{Y(t) : t \in \mathbb{T}\}$ be any $S\alpha'S$ process, where $0 < \alpha < \alpha'$, and let A be a positive p–stable random variable independent of the process $\mathbf{Y}$ and having the Laplace transform $\mathbf{E} \exp(-uA) = \exp(-|u|^p)$, where $p = \alpha/\alpha'$. Set

$$X(t) = A^{1/\alpha'} Y(t), \quad t \in \mathbb{T}.$$

Then the process $\mathbf{X}$ is a sub–stable process with the governing process $\mathbf{Y}$. Indeed, for any $d \geq 1$, $t_1, ..., t_d \in \mathbb{T}$, $b_1, ..., b_d$ real, θ real,

$$
\begin{aligned}
\mathbf{E} \exp \left\{ i\theta \sum_{j=1}^d X(t_j) \right\} &= \mathbf{E} \, \mathbf{E} \left[\exp \left\{ i\theta A^{1/\alpha'} \sum_{j=1}^d b_j Y(t_j) \right\} | A \right] \\
&= \mathbf{E} \exp \left\{ -A \| \theta \sum_{j=1}^d b_j Y(t_j) \|_{\alpha'}^{\alpha'} \right\} \\
&= \exp \left\{ -|\theta|^\alpha \| \sum_{j=1}^d b_j Y(t_j) \|_{\alpha'}^{\alpha'} \right\}.
\end{aligned}
$$

To find its spectral representation, let $(\Omega, \mathcal{F}, P)$ denote the probability space on which $\mathbf{Y}$ is defined, let M be an $S\alpha S$ random measure on $(\Omega, \mathcal{F})$ with control measure P, and let $c_{\alpha,\alpha'}$ be a constant depending only on α and α' such that if A is an α' random variable with characteristic function

$$\mathbf{E} \exp(i\theta A) = \exp(-k|\theta|^{\alpha'}),$$

then ($\mathbf{E}|A|^\alpha)^{1/\alpha} = c^{-1} k^{1/\alpha'}$. Hence

$$
\begin{aligned}
-\log \mathbf{E} \exp \left(i \sum_j \theta_j X(t_j) \right) &= \left[-\log \mathbf{E} \exp \left(i \sum_j \theta_j Y(t_j) \right) \right]^{\alpha/\alpha'} \\
&= k^{\alpha/\alpha'} = \left[c \left(\mathbf{E}| \sum_j^d \theta_j Y(t_j)|^\alpha \right) 1/\alpha \right]^\alpha \\
&= \mathbf{E}| \sum_j (cY(t_j))|^\alpha.
\end{aligned}
$$

The calculation above shows that

$$\mathbf{E} \exp \left(i \sum_j \theta_j X(t_j) \right) = \exp \left(-\| \sum_j \theta_j (cY(t_j)) \|^\alpha \right),$$

which establishes the following spectral representation

$$\{X(t) : t \in \mathbf{T}\} = \left\{ c \int_\Omega Y(t,\omega)dM(\omega) : \quad t \in \mathbf{T} \right\}.$$

Here, similarly as in Example 5.3.2, one can use the separability condition and, by the Isomorphism Theorem, find $\{f_t, t \in \mathbf{T}\} \subseteq L^\alpha[0,1]$ stochastically equivalent to $\mathbf{Y}$.

However, this spectral representation is not minimal! Although $\rho(F)$ is nonatomic, it is properly contained in $\sigma(F)$, and we are not able to map $\rho(F)$ isomorphically onto the Borelian σ-field on $[0,1]$ without altering the distribution of $\{f_t, t \in \mathbf{T}\}$.

If $\alpha = 2$ then the sub-Gaussian process $X(t) = A^{1/2}G(t)$, $t \in \mathbf{T}$, where $G(t)$ is a Gaussian process and A is a $S\frac{\alpha}{2}S$ random variable independent from the process $G(t)$, has the following representation:

$$\{X(t) : t \in \mathbf{T}\} = \left\{ \frac{1}{d\sqrt{2}} \int_\Omega G(t,\omega)dM(\omega) : \quad t \in \mathbf{T} \right\},$$

where $d = (\ E|N(0,1)|^\alpha)^{1/\alpha}$.

Now we are interested in the problem of uniqueness of the spectral representation under an additional measurability condition. The following theorem constitutes one of the main results in Rosinski (1993).

Theorem 5.3.3 *Let* $\mathbf{T}$ *be a separable metric space. Let* $\{f_t^{(i)}\}_{t \in \mathbf{T}}$ *contained in* $L^\alpha(S_i, \mathcal{B}_{S_i}, \mu_i)$ *(i = 1,2) be two representations of the same SαS real (complex) process* $\{X(t)\}_{t \in \mathbf{T}}$. *Assume that the mapping* $(t,s) \rightarrow f_t^{(i)}(s)$ *is Borel measurable,* $i = 1,2$.

Then, for every σ-*finite Borel measure* λ *on* $\mathbf{T}$, *there exist Borel functions* $\phi : S_2 \rightarrow S_1$ *and* $h : S_2 \rightarrow \mathbf{R}\ (\mathbf{C})$ *such that*

$$f_t^{(2)}(s) = h(s)f_t^{(1)}(\phi(s)) \qquad \lambda \otimes \mu_2 - a.e. \qquad (5.3.3)$$

Hardin (1981) introduced functions with full support to establish equimeasurability of certain mappings. Since we will use a similar method in the proof, we recall the definition and some basic facts on functions with full support. Let F be a collection of real (or complex) functions on some measure space (S, μ). We say that a function g has *full support* in F if $g \in F$ and $\mu(\{s : f(s) \neq 0 \text{ and } g(s) = 0\}) = 0$ for each $f \in F$.

From Lemmas 3.2 and 3.4 of Hardin (1981) one can derive the following

Lemma 5.3.1 *Let* F *be a closed subspace of* $L^p(\mu)$, $0 < p < \infty$.

(i) *If* F *is separable or if* μ *is* σ-*finite then there exists a function of full support in* F.

(ii) *Let* $U : F \rightarrow L^p(\nu)$ *be a linear isometry. If* $0 < p \leq 2$ *and* g *has full support in* F *then* Ug *has full support in* $U(F)$.

PROOF of Theorem 5.3.3. Let F_i be the closure in $L^\alpha(S_i, \mathcal{B}_{S_i}, \mu_i)$ of $\lin\{f_t^{(i)}:$ $t \in \mathbb{T}\}$, $i = 1, 2$. Since the families $\{f_t^{(i)}\}_{t \in \mathbb{T}}$ $(i = 1, 2)$ represent the same $S\alpha S$ process, the mapping $Uf_t^{(1)} = f_t^{(2)}$, $t \in \mathbb{T}$, extends to a linear isometry of F_1 onto F_2. By Lemma 5.3.1 there exists a function $g_1 \in F_1$ with full support in F_1 and $g_2 = Ug_1$ has full support in F_2. Let

$$S_i^0 = \{s \in S_i : g_i(s) \neq 0\}, \qquad i = 1, 2.$$

Without loss of generality, we may assume that λ is a finite measure. Define Borel maps $\xi : S_1^0 \to L^o(\mathbb{T}, \mathcal{B}_\mathbb{T}, \lambda)$ and $\eta : S_2 \to L^o(\mathbb{T}, \mathcal{B}_\mathbb{T}, \lambda)$ by

$$[\xi(s)](t) = \frac{f_t^{(1)}(s)}{g_1(s)}, \qquad s \in S_1^0, \ t \in \mathbb{T},$$

and

$$[\eta(s)](t) = \begin{cases} \dfrac{f_t^{(2)}(s)}{g_2(s)}, & \text{if } s \in S_2; \\ 0, & \text{otherwise,} \end{cases}$$

where $t \in \mathbb{T}$. We will prove that

$$\nu_1 \circ \xi^{-1} = \nu_2 \circ \eta^{-1}, \qquad (5.3.4)$$

where $\nu_i(ds) = |g_i(s)|^\alpha \mu_i(ds)$. Since $\nu_2(S_2 - S_2^0) = 0$, it is enough to show that for every $n \geq 1$, $t_1, \ldots, t_n \in \mathbb{T}$, $B \in \mathcal{B}_{\mathbb{R}^n}$,

$$\nu_1(\{s \in S_1^0 : (f_{t_1}^{(1)}(s)/g_1(s), \ldots, f_{t_n}^{(1)}(s)/g_1(s)) \in B\})$$

$$= \nu_2(\{s \in S_2^0 : (f_{t_1}^{(2)}(s)/g_2(s), \ldots, f_{t_n}^{(2)}(s)/g_2(s)) \in B\}). \qquad (5.3.5)$$

We have, for every $n \geq 1$, $a_1, \ldots a_n \in \mathbb{R}$ ($\mathbb{C}$),

$$\int_{S_1^0} \left| 1 + \sum_{j=1}^n a_j \frac{f_{t_j}^{(1)}(s)}{g_1(s)} \right|^\alpha \nu_1(ds) = \int_{S_1} \left| g_1(s) + \sum_{j=1}^n a_j f_{t_j}^{(1)}(s) \right|^\alpha \mu_1(ds)$$

$$= \left\| U\left(g_1 + \sum_{j=1}^n a_j f_{t_j}^{(1)}\right) \right\|_\alpha^\alpha$$

$$= \int_{S_2} \left| g_2(s) + \sum_{j=1}^n a_j f_{t_j}^{(2)}(s) \right|^\alpha \mu_2(ds)$$

$$= \int_{S_2^0} \left| 1 + \sum_{j=1}^n a_j \frac{f_{t_j}^{(2)}(s)}{g_2(s)} \right|^\alpha \nu_2(ds).$$

Thus, (5.3.5) follows from the Rudin Theorem (see, e.g., Hardin (1981)) which proves (5.3.4). Applying Theorem 2.2 from Rosiński (1993) for $\mathcal{A} = \sigma(\xi)$ we infer that there exists a Borel function $\phi : S_2 \to S_1^0$ such that for ν_2-almost all $s \in S_2$,

$$\eta(s) = \xi(\phi(s)) \qquad \text{in } L^o(\mathbb{T}, \mathcal{B}_\mathbb{T}, \lambda).$$

By the Fubini Theorem we get

$$f_t^{(2)}(s) = \frac{g_2(s)}{g_1(\phi(s))} f_t^{(1)}(\phi(s)) \qquad \lambda \otimes \nu_2 - a.e. \tag{5.3.6}$$

Put $h(s) = g_2(s)/g_1(\phi(s))$ if $s \in S_2^0$, and $h(s) = 0$ otherwise. Since μ_2 is equivalent to ν_2 on S_2^0, (5.3.6) implies (5.3.3) on $\mathbb{T} \times S_2^0$. Since $f_t^{(2)}(s) = 0$ $\lambda \otimes \mu_2$–a.e. on $\mathbb{T} \times (S_2 - S_2^0)$, (5.3.3) holds true on $\mathbb{T} \times (S_2 - S_2^0)$. This completes the proof of Theorem 5.3.3.   □

Condition (5.3.3) implies that there exists a set $\mathbb{T}_0 \subset \mathbb{T}$ with $\mu(\mathbb{T} - \mathbb{T}_0) = 0$ such that for every $t \in \mathbb{T}_0$

$$f_t^{(2)}(s) = h(s) f_t^{(1)}(\phi(s)) \qquad \mu_2 - a.e. \tag{5.3.7}$$

A natural question is whether one can remove the auxiliary measure λ from the statement of Theorem 5.3.3 to have (5.3.7) valid for every $t \in \mathbb{T}$. The answer is no. Indeed, let $S_1 = [0,1]$, $\mu_1 = Leb$, $f_t^{(1)}(s) = I_{[0,1]\backslash\{t\}}(s)$, $t \in \mathbb{T} = [0,1]$; $S_2 = \{0\}$, $\mu_2 = \delta_0$, and $f_t^{(2)}(0) = 1$. Then (5.3.7) fails for $t = \phi(0)$ and any choice of ϕ. However, under some separability conditions on the spectral representation, formula (5.3.7) is proven for every $t \in \mathbb{T}$ in Rosinski (1993) as Theorem 4.1.

Theorem 5.3.3 can be used to distinguish various classes of stable processes or to identify processes within the same class. Its usefulness for this purpose will be demonstrated in the following examples taken from Rosinski (1993).

Example 5.3.4 *Disjointness of moving averages and harmonizable processes.*

Suppose that there is an $S\alpha S$ process which possesses both harmonizable and moving average representations, i.e., $\int_{-\infty}^{\infty} e^{its} Z(ds)$ and $\int_{-\infty}^{\infty} f(t-s)\, dM(s)$, where the $S\alpha S$ measure $Z(\cdot)$ is necessarily complex–valued. (A complex random variable is $S\alpha S$ if its real and imaginary parts are jointly $S\alpha S$). Using Theorem 5.3.3 for $f_t^{(1)}(s) = e^{its}$, $f_t^{(2)}(s) = f(t-s)$, with $t, s \in \mathbb{R}$ and $f \in L^\alpha(\mathbb{R})$, we get

$$f(t-s) = h(s) e^{it\phi(s)} \qquad \lambda \otimes \lambda - a.e.,$$

where λ denotes the Lebesgue measure on $\mathbb{R}$. Integrating the α-th power of the magnitude of both sides of this equation with respect to t over $\mathbb{R}$, we get the left–hand side finite and the right–hand side infinite, unless $h(s) = 0$. Therefore, the process in question must be a zero process.

Example 5.3.5 *Kanter Theorem and its generalizations.*

Suppose that an $S\alpha S$ process possesses two moving average representations, $f_t^{(1)}(s) = f(t+s)$ and $f_t^{(2)}(s) = g(t+s)$, $f, g \in L^\alpha(\mathbb{R})$. In view of Theorem 5.3.3,

$$g(t+s) = h(s) f(t + \phi(s)) \qquad \lambda \otimes \lambda - a.e.$$

Fix $s = s_0$ for which the above equation holds for λ-almost all t, and put $u = t + s_0$. Then

$$g(u) = h(s_0)f(u + u_0) \quad \text{for} \quad \lambda - \text{almost all} \ u,$$

where $u_0 = \phi(s_0) - s_0$. Therefore

$$\int_{\mathbb{R}} |g(u)|^\alpha \, du = |h(s_0)|^\alpha \int_{\mathbb{R}} |f(u - u_0)|^\alpha \, du = |h(s_0)|^\alpha \int_{\mathbb{R}} |g(u)|^\alpha \, du,$$

which yields $|h(s_0)| = 1$. This proves the Kanter Theorem (e.g., Kanter (1975)) in both real and complex cases. Notice that the above argument works for arbitrary metrizable separable locally compact Abelian groups (instead of $\mathbb{R}$) with Haar measure λ. The non–Abelian case is also simple. Using the same argument and multiplicative notation, we get

$$g(u) = h(s_0)f(uu_0) \quad \text{for} \quad \lambda - \text{almost all} \ u,$$

where λ denotes the left invariant Haar measure on the group, say, G. We have

$$\begin{aligned}
\int_G |g(u)|^\alpha \lambda(du) &= |h(s_0)|^\alpha \int_G |f(uu_0)|^\alpha \ \lambda(du) \\
&= |h(s_0)|^\alpha [\Delta(u_0^{-1})]^{-1} \int_G |f(u)|^\alpha \ \lambda(du) \\
&= |h(s_0)|^\alpha [\Delta(u_0^{-1})]^{-1} \int_G |g(u)|^\alpha \ \lambda(du),
\end{aligned}$$

where Δ is the modular function on G determined by $\Delta(x)\lambda(Bx) = \lambda(B)$ for every $B \in \mathcal{B}_G$. Hence, $h(s_0) = \varepsilon[\Delta(u_0^{-1})]^{1/\alpha}$ with $|\varepsilon| = 1$. Consequently,

$$g(u) = \varepsilon[\Delta(u_0^{-1})]^{1/\alpha}f(uu_0) \quad \text{for} \quad \lambda - \text{almost all} \ u.$$

This generalization of the Kanter Theorem was obtained earlier (without the assumption on metrizability or separability of G) by Hardin and Pitt (1983) by more complicated methods.

Example 5.3.6 *Generalized moving averages.*

Let ν be a σ–finite measure on $\mathcal{X}$. Given $f \in L^\alpha(\mathcal{X} \times \mathbb{R}^d, \nu \otimes \lambda)$, where λ is Lebesgue measure on $\mathbb{R}^d$, put

$$X(t) = \int_{\mathcal{X} \times \mathbf{R}^d} f(x, t - s) \, M(dx, ds),$$

where M is an $S\alpha S$ random measure with the control measure $\nu \otimes \lambda$. Such process $\{X(t) : t \in \mathbb{R}^d\}$ is called a generalized moving average (see Surgailis *et al.* (1992)). Suppose that we are given two representations of generalized moving average processes, $f_t^{(1)}(x, s) = f(x, t - s)$ and $f_t^{(2)}(y, s) = g(y, t - s)$, where $f \in L^\alpha(\mathcal{X} \times \mathbb{R}^d, \nu \otimes \lambda)$ and $g \in L^\alpha(\mathcal{Y} \times \mathbb{R}^d, \rho \otimes \lambda)$, and $\mathcal{X}$ and $\mathcal{Y}$ are Borel subsets of certain separable complete metric spaces. Define two measurable maps $\xi : \mathcal{X} \to L^\alpha(\mathbb{R})$ and $\eta : \mathcal{Y} \to L^\alpha(\mathbb{R})$ by

$$\xi(x) = \frac{f(x, \cdot)}{\|f(x, \cdot)\|_\alpha}$$

and

$$\eta(x) = \frac{g(y, \cdot)}{\|g(y, \cdot)\|_\alpha},$$

where $\|k\|_\alpha = \{\int_{\mathbb{R}^d} |k(u)|^\alpha \, du\}^{1/\alpha}$, $k \in L^\alpha(\mathbb{R}^d)$. Let $\nu_f(dx) = \|f(x, \cdot)\|_\alpha^\alpha \nu(dx)$ and $\rho_g(dy) = \|g(y, \cdot)\|_\alpha^\alpha \rho(dy)$. Surgailis *et al.* (1992) have shown that $\{f_t^{(1)}\}$ and $\{f_t^{(2)}\}$ represent the same process if and only if

$$\nu_f \circ \xi^{-1} = \rho_g \circ \eta^{-1} \qquad (5.3.8)$$

on shift invariant symmetric Borel subsets of $L^\alpha(\mathbb{R}^d)$. Using Theorem 5.3.3, we can get more explicit relationship between f and g.

Namely, if $\{f_t^{(1)}\}$ and $\{f_t^{(2)}\}$ represent the same process, then by Theorem 5.3.3 there exist $\phi : \mathcal{Y} \times \mathbb{R}^d \to \mathcal{X} \times \mathbb{R}^d$, $\phi(y, s) = (\phi_1(y, s), \phi_2(y, s))$, and $h : \mathcal{Y} \times \mathbb{R}^d \to \mathbb{R}$ ($\mathbb{C}$, in the complex case) such that

$$g(y, t - s) = h(y, s) f(\phi_1(y, s), t - \phi_2(y, s))$$

for $\lambda \otimes \rho \otimes \lambda$–almost all t, y, s. Choose s_0 such that for $\rho \otimes \lambda$–almost all (y, t)

$$g(y, t - s_0) = h(y, s_0) f(\phi_1(y, s_0), t - \phi_2(y, s_0)).$$

Define $a(y) = \phi_2(y, s_0) - s_0$, $b(y) = \phi_1(y, s_0)$, and $c(y) = h(y, s_0)$. Then

$$g(y, u) = c(y) f(b(y), u - a(y)) \qquad \rho \otimes \lambda - a.e.$$

Integrating with respect to u, we obtain $c(y) = \varepsilon(y)\|g(y, \cdot)\|_\alpha\|f(b(y), \cdot)\|_\alpha^{-1}$ with $|\varepsilon(y)| = 1$. Therefore,

$$\frac{g(y, u)}{\|g(y, \cdot)\|_\alpha} = \varepsilon(y)\frac{f(b(y), u - a(y))}{\|f(b(y), \cdot)\|_\alpha} \qquad \rho \otimes \lambda - a.e. \qquad (5.3.9)$$

Using (5.3.8), we get

$$\nu_f \circ \xi^{-1} = \rho_g \circ (\xi \circ b)^{-1} \qquad (5.3.10)$$

on shift invariant symmetric Borel subsets of $L^\alpha(\mathbb{R}^d)$. Conversely, if there exist measurable functions $a : \mathcal{Y} \to \mathbb{R}^d$, $b : \mathcal{Y} \to \mathcal{X}$, and $\varepsilon : \mathcal{Y} \to \mathbb{R}$ ($\mathbb{C}$) with $|\varepsilon(y)| = 1$ such that (5.3.9) and (5.3.10) hold then $\{f_t^{(1)}\}$ and $\{f_t^{(2)}\}$ represent the same process.

Example 5.3.7 *Series and integral representations of stable processes.*

Let $\mathbb{T} = \{1, 2\}$. It is well–known (see, e.g., Linde (1987)) that there are $S\alpha S$ processes (i.e., 2–dimensional $S\alpha S$ random variables) which do not admit the following series representation:

$$\{X_i\}_{i=1,2} \stackrel{\mathrm{d}}{=} \left\{ \sum_{j=1}^{\infty} a_{ij}\theta_j \right\}_{i=1,2} \tag{5.3.11}$$

where θ_j are i.i.d. standard $S\alpha S$ random variables and $a_{ij} \in \mathbb{R}$. It can be interesting to examine this fact in the light of Theorem 5.3.3. Suppose that we are given a stochastic process

$$X_i = \int_0^1 f_i(s)\, dM(s), \quad i = 1, 2, \tag{5.3.12}$$

where M is a real independently scattered $S\alpha S$ random measure with the control Lebesgue measure μ. If the process (5.3.12) admits a representation (5.3.11) then by Theorem 5.3.3 (for $\lambda = \delta_1 + \delta_2$) there exist $\phi : [0,1] \to \mathbb{N}$ and $h : [0,1] \to \mathbb{R}\ (\mathbb{C})$ such that

$$f_i(s) = h(s)a_{i\phi(s)}, \quad i = 1, 2$$

for each $s \in [0,1] - A_0$, where $\mu(A_0) = 0$. Let $A_j = \{s \in [0,1] : \phi(s) = j\}$, $j \geq 1$. Therefore, for every $j \geq 1$ and $s \in A_j \cap A_0^c$,

$$f_2(s) = h(s)a_{2j} = f_1(s)\frac{a_{2j}}{a_{1j}},$$

provided $a_{1j} \neq 0$. Thus f_1 and f_2 restricted to $A_j \cap A_0^c$ are linearly dependent for each $j \geq 1$. We obtain the following conclusion.

Proposition 5.3.1 *A process $\mathbf{X}$ defined by (5.3.12) admits a series representation (5.3.11) if and only if there exists a partition $\{\pi_j\}_{j\geq 0}$ of $[0,1]$ such that f_1 and f_2 are linearly dependent on each π_j, $j \geq 1$, and $\mu(\pi_0) = 0$.*

Any process defined by (5.3.12), and such that functions f_1, f_2 are linearly independent and analytic in some open subinterval of $(0,1)$ does not admit series representation (5.3.11).

5.4   Structure of Stationary Stable Processes

Now we present a theorem characterizing minimal representations due to Hardin (1982).

Theorem 5.4.1 *A non–Gaussian $S\alpha S$ process satisfying the separability condition is stationary if and only if it has a minimal representation of the form $t \to f_t = P_t\phi$, where ϕ is a fixed function in $\mathbf{L}^\alpha[0,1]$ and $\{P_t : t \in \mathbb{T}\}$ is a group of isometries on $\mathbf{L}^\alpha[0,1]$.*

PROOF. Assume that the process is stationary and that $t \to f_t$ is its minimal representation. Stationarity implies that for any fixed $s \in \mathbf{T}$ the map $t \to f_{s+t}$ is another minimal representation for $\mathbf{X}$, and also that the closed linear extension of the map $U_s : f_t \to f_{s+t}$ is an isometry of $\mathbf{F}$ onto $\mathbf{F}$. By Hardin (1981), Theorem 4.2, each U_s has a unique isometric extension $\bar{U}_s$ to the space

$$\bar{F} = \{g \in \mathbf{L}^\alpha[0,1] : g = rf, \ f \in \mathbf{F}, \ \text{and } r \text{ is } \rho(\mathbf{F})\text{-measurable}\},$$

and

$$\bar{U}_s(g) = (T_s r)(U_s f),$$

where T_s is a regular set isomorphism of $\rho(\mathbf{F})$ onto itself. It is easily seen that $\{U_t\}$ has the group property on $\mathbf{F}$, hence $\{\bar{U}_t\}$ has the group extension on $\bar{\mathbf{F}}$. From this one can conclude that $\{T_t\}$ is a group of regular set isomorphism on $\rho(\mathbf{F})$, i.e., $T_s T_t A = T_{s+t} A$ for each $A \in \rho(\mathbf{F})$ since $\bar{U}_t h$ has full support in $\mathbf{F}$.

To extend the group $\{\bar{U}_t\}$ on $\bar{F}$ to a group of isometries P_t on all of $\mathbf{L}^\alpha[0,1]$, we will extend $\{T_t\}$ to a group of regular set isomorphism $\{\bar{T}_t\}$ on the Borel σ-field on $[0,1]$ and define

$$P_t f = \bar{T}_t(f/h)U_t h.$$

If we define $\{P_t\}$ as above, then it clearly extends $\{\bar{U}_t\}$. To see that $\{P_t\}$ is isometric it suffices to show that $\| P_t I_A \| = \| I_A \|$ for an arbitrary Borel set A contained in some atom $A_i, i \geq 0$ of $\rho(\mathbf{F})$. The identity can be checked by using the second and third condition in the definition of the minimal representation. It is also clear that $\{P_t\}$ is a group. The required representation is then $t \to P_t \phi$, where $\phi = f_0$. The "if" part of the theorem is obvious.   □

Example 5.4.1 *Doubly stationary processes.*

Let $(S, \mathcal{B}, \mu)$ be an arbitrary (finite or infinite) measure space and let $\{F_t : t \in \mathbf{T}\}$ be a collection of measurable functions on S. Now $\mathbf{T}$ is some group, usually understood to be $\mathbf{Z}$ or $\mathbf{R}$. Call $\{f_t\}$ *stationary* if the μ-distribution of the vector $(f_{t_1+s}, ..., f_{t_n+s})$ is independent of $s \in G$ for each fixed choice of n and $t_j \in G$. An $S\alpha S$ process will be called *doubly stationary* if it has the same distribution as some process

$$\left\{ X(t) = \int_S f_t(u)dZ(u) : t \in G \right\},$$

where $\{f_t\} \subseteq \mathbf{L}^\alpha(S, \mathcal{B}, \mu)$ is stationary and $\mathbf{Z}$ is the canonical independently scattered random measure on $(S, \mathcal{B}, \mu)$. So they are, loosely speaking, those $S\alpha S$ processes whose spectral representations are themselves stationary, see Cambanis, Hardin and Weron (1987).

It is obvious by checking characteristic functions that doubly stationary $S\alpha S$ processes are also stationary. Example (iv) below shows the converse does not hold. For stationary $\{f_t\}$ we may find, just as in the case of stationary processes, a group of measure-preserving transformations U_t of $S = \sigma\{f_t\}$ such that $f_t = U_t f_0$. (We also denote by U_t the induced map on measurable functions.) Conversely, any group of measure-preserving set transformations defines stationary functions $\{U_t f_0\}$ for arbitrary measurable f_0. Thus an $S\alpha S$ process is doubly stationary if and only if it has a spectral representation as in Theorem 5.4.1, where the group $\{P_t\}$ is induced by such group $\{U_t\}$. This equivalent definition will be more useful for us, if not as picturesque. Now, we will illustrate this class of stationary processes with four special examples.

(i) *Every mean–zero stationary Gaussian process is doubly stationary.* To see this, let $\{X(t)\}$ be a mean–zero Gaussian process on $(\Omega, \mathcal{F}, P)$ and let $\mathbf{Z}$ be the canonical independently scattered Gaussian measure on $(S, \mathcal{B}, \mu) = (\Omega, \mathcal{F}, P)$. Then $\{Y(t) = \int_\Omega X(t, \omega) dZ(\omega)\}$ is seen (by checking characteristic functions as in Example 5.3.2) to have the same distribution as $\{X(t)\}$. Hence it is doubly stationary.

(ii) *Every stationary sub–Gaussian process is doubly stationary.* Let $\{X(t)\}$ be α–sub–Gaussian on $(\Omega, \mathcal{F}, P)$, represented as $X(t) = A^{1/2} G(t)$, as in Example 5.3.3. As was seen there, $\{X(t)\}$ is distributed as $\{Y(t) = \int_\Omega c G(t, \omega) dZ(\omega)\}$, where $\mathbf{Z}$ is the canonical independently scattered $S\alpha S$ random measure on $(\Omega, \mathcal{F}, P)$ and c is a constant depending on α. The process $\{G(t)\}$ is stationary since $\{X(t)\}$ is. Thus $\{X(t)\}$ is doubly stationary.

(iii) *All $S\alpha S$ generalized moving averages are doubly stationary.*

In order to show that the $S\alpha S$ generalized moving averages defined in Example 5.3.6 are also doubly stationary, it is enough to check that their spectral kernel $f(\cdot, t - \cdot)$ is stationary with respect to the control measure $\nu \otimes \lambda$, i.e., the measure $(\nu \otimes \lambda) \circ (f(\cdot, t_1 + \tau - \cdot), ..., f(\cdot, t_n + \tau - \cdot))^{-1}$ on $\mathbb{R}^n$ is independent of τ. Indeed, for all $B \in \mathcal{B}(\mathbb{R}^n)$ using the fact that λ is the Lebesgue measure we have

$$(\nu \otimes \lambda)\{(x, s) : (f(x, t_1 + \tau - s), ..., f(x, t_n + \tau - s)) \in B\}$$

$$= \int_X \lambda\{s : (f(x, t_1 + \tau - s), ..., f(x, t_n + \tau - s)) \in B\} \nu(dx)$$

$$= \int_X \lambda\{u : (f(x, t_1 - u), ..., f(x, t_n - u)) \in B\} \nu(dx)$$

$$= (\nu \otimes \lambda)\{(x, s) : (f(x, t_1 - s), ..., f(x, t_n - s)) \in B\}.$$

Let us also observe that in the Gaussian case ($\alpha = 2$) the generalized moving averages coincide with the usual moving averages since

$$\begin{aligned} EX(t)X(t') &= \int_X \left\{ \int_{\mathbb{R}} f(x, t - s) f(x, t' - s) ds \right\} \nu(dx) \\ &= \int_X \left\{ \frac{1}{\sqrt{2\pi}} \int_{\mathbb{R}} \exp i(t - t', u) |F(x, u)|^2 du \right\} \nu(dx) \\ &= \frac{1}{\sqrt{2\pi}} \int_{\mathbb{R}} \exp i(t - t', u) f_*(u) du \\ &= \int_{\mathbb{R}} g(t - s) g(t' - s) ds, \end{aligned}$$

where $F(x, u)$ is the L^2–Fourier transform of $f(x, \cdot)$, a function $f_*(u)$ defined
by $f_*(u) = \int_X |F(x, u)|^2 \nu(dx)$ belongs to $L^1(d\lambda)$ since we have $\int_{\mathbb{R}} f_*(u)du = \sqrt{2\pi} \int_{X \times \mathbb{R}} |f(x, t)|^2 dt\nu(dx) < \infty$, and where $g \in L^2(d\lambda)$ is the L^2–Fourier trans-
form of $(f_*)^{1/2} \in L^2(d\lambda)$; so that

$$\{X(t) : t \in \mathbb{R}\} \overset{d}{=} \left\{ \int_{\mathbb{R}} g(t - s)\xi(ds) : t \in \mathbb{R} \right\},$$

where ξ stands for a Gaussian white noise.

In the non–Gaussian stable case ($0 < \alpha < 2$) the generalized moving averages
form a larger class than the usual moving averages since, as it was shown in
Surgailis *et al.* (1992) sums of independent $S\alpha S$ usual moving averages are
distinct from the usual moving averages.

(iv) *There exists a stationary $S\alpha S$ process which is not doubly stationary.*

For simplicity we take $\alpha = 1$, although this example may be altered easily
to work for each $\alpha \in (0, 2)$. Define $P_t : L^1[0, 1] \to L^1[0, 1]$ for real t by

$$(P_t f)(x) = 2^t x^{2^t - 1} f(x^{2^t}).$$

It is easily checked that $\{P_t\}$ is a strongly continuous group of linear isometries,
so that

$$\left\{ X_t = \int_0^1 P_t I_{[0,1]} d\mathbf{Z} : t \in \mathbb{R} \right\}$$

is a stationary α–stable process continuous in probability. Here, $\mathbf{Z}$ is the Cauchy
motion on $[0, 1]$ (the canonical $S1S$ independent increments process on $[0, 1]$).
We claim that $\{X(t)\}$ is not doubly stationary.

Thus, if $\{X(t)\}$ were doubly stationary, we could find a measure space
(S, Σ, μ), a group of measure–preserving set maps $U_t : \Sigma \to \Sigma$ and a function $g(t) \in L^\alpha(\mu)$ such that $t \to U_t g$ is a spectral representation for $\{X(t)\}$.
Since $t \to P_t l I_{[0,1]}$ is also a spectral representation for $\{X(t)\}$, so we have

$$\left\| \sum a_j P_{t_j} f_0 \right\|_{L^\alpha[0,1]} = \left\| \sum a_j U_{t_j} g \right\|_{L^\alpha(\mu)}$$

for all choices of a_j and t_j. Hence the map $P_t I_{[0,1]} \to U_t g$ extends to a linear
isometry of $lin\{P_t I_{[0,1]}\}_{L^\alpha[0,1]}$ onto $lin\{U_t g\}_{L^\alpha(\mu)}$. This isometry in fact extends
to all of $L^\alpha[0, 1]$ by Hardin (1981), Corollary 4.3, since $P_0 I_{[0,1]} = I_{[0,1]}$ and
$P_1 I_{[0,1]}(x) = 2x$ are both in $sp\{P_t I_{[0,1]}\}$. Call this extension V. Again by Hardin
(1981), Corollary 4.3, V has the form

$$(Vf)(x) = h(x)(\psi f)(x),$$

where ψ is induced by a regular set isomorphism of the Borel σ–field of $[0, 1]$
with Lebesgue measure to (Σ, μ). Since $V P_t I_{[0,1]} = U_t g$, we have

$$U_t g = V P_t I_{[0,1]} = V(2^t id^{2^t - 1}) = h\psi(2^t - 1) = 2^t h[\psi(id)]^{2^t - 1},$$

where $id(x) = x$. Since $0 < id < 1$ a.e., thus $0 < \psi(id) < 1$ μ–a.e. If $0 < x < 1$
then $2^t x^{2^t - 1} \to 0$ as $t \to \infty$. But $U_t g$ must be equidistributed for all t (since U_t

is measure–preserving), and $2^t h[\psi(id)]^{2^t-1}$ is not by the above, since by choosing t large enough we may, for any $\epsilon > 0$, force $\mu\{|2^t h[\psi(id)]^{2^t-1}| < \epsilon\}$ as close to $\mu(S)$ as desired if $\mu(S) < \infty$. Thus the process $\{X(t)\}$ has no doubly stationary representation on a space of finite measure.

However, it has a moving average (and hence doubly stationary) representation on $\mathbb{R}$. Namely, let $(E_1, \mathcal{E}_1, \lambda_1)$ and $(E_2, \mathcal{E}_2, \lambda_2)$ denote Lebesgue measure on $[0,1]$ and $\mathbb{R}$, respectively, with the corresponding Borel σ–fields. Define $\Phi_1^t : (E_1, \mathcal{E}_1, \lambda_1) \to (E_1, \mathcal{E}_1, \lambda_1)$ by

$$\Phi_1^t x = x^{2^t}, \qquad t \in \mathbb{R}, x \in [0,1].$$

Then $(\phi_1^t)_{t \in \mathbb{R}} \stackrel{\mathrm{df}}{=} ((\Phi_1^t)^{-1})_{t \in \mathbb{R}}$ is a group of regular set isomorphisms which induces the group (U^t) of isometries as described in Lamperti (1958). Similarly, define $\Phi_2^t : (E_2, \mathcal{E}_2, \lambda_2) \to (E_2, \mathcal{E}_2, \lambda_2)$ by

$$\Phi_2^t x = x + t, \qquad t \in \mathbb{R}, x \in \mathbb{R};$$

then $(\phi_1^t)_{t \in \mathbb{R}} \stackrel{\mathrm{df}}{=} ((\Phi_2^t)^{-1})_{t \in \mathbb{R}}$ is a group of regular set isomorphisms which preserve Lebesgue measure. Define $\Psi : \mathbb{R} \to [0,1]$ by

$$\Psi x = 2^{-2^x}, \qquad t \in \mathbb{R}, x \in \mathbb{R},$$

and define the regular set isomorphism $\psi = \Psi^{-1}$. One can check directly that

$$\Phi_1^t \Psi = \Psi \Phi_2^t, \qquad t \in \mathbb{R},$$

and that therefore ϕ_1, ϕ_2 and ψ satisfy

$$\psi \phi_1 = \phi_2 \psi. \tag{5.4.1}$$

Now define the positive isometries induced by the set isomorphisms by

$$U_i f = (\phi_i')^{1/\alpha} \cdot (\phi_i f), \qquad i = 1, 2,$$

$$W f = (\psi')^{1/\alpha} \cdot (\psi f).$$

(Note that $(U_1, 1)$ represents the $S\alpha S$ process defined in this example.) Then for all I_A in $L^p(\mu_1)$,

$$W U_1 I_A = [\phi_1(\psi') \cdot \phi_1']^{1/\alpha} \cdot I_{\psi\phi_1 A}$$

$$U_2 W I_A = [\psi(\phi_2') \cdot \psi']^{1/\alpha} \cdot I_{\phi_2\psi A}$$

It now follows from Equation (5.4.1) and an application of the chain rule that

$$W U_1 = U_2 W.$$

Thus $(U_2, W1)$ represents the same $S\alpha S$ process as $(U_1, 1)$. But U_2 is the group of shift operators, i.e.

$$U_2^t f(x) = f(x + t), \qquad t \in \mathbb{R}, f \in L^1(\mathbb{R}),$$

so the process given in the example is a moving average.

The above discussion shows that the question of whether there exists a stationary $S\alpha S$ process which do not admit doubly stationary representation has a measure theoretical character. Gross and Weron (1993) constructed an example of such process (actually a class of examples for every $\alpha \in (0,2)$) by using an example of Ornstein (1960) of a nonsingular transformation on $[0,1]$ which does not admit a σ-finite invariant measure equivalent to Lebesque measure. This example is constructed in the following four steps.

1. Establish a relation between the σ-field $\rho(L_1)$ and the Radon-Nidodym derivatives of a regular set isomorphism.

2. Describe a nonsingular point transformation τ on $[0,1]$ with Lebesgue measure m, due to Ornstein (1960), which does not admit a σ-finite, invariant, equivalent measure; then, using Step 1, show that this particular example can be taken so that $\rho(L_1) = \mathcal{B}_{[0,1]}$.

3. Use a result of Hardin (1981) to show that, if the process is represented by another isometry induced by a measure-preserving transformation T on some measure space $(E, \mathcal{E}, \lambda)$, then there is a regular set isomorphism $\psi : \mathcal{B}_{[0,1]} \to \mathcal{E}$ such that $\psi\tau^{-1} = T\psi$.

4. Conclude that the $S\alpha S$ sequence represented by the isometry U^n induced by τ and the function $f = 1$,

$$Ug = h \cdot (g \circ \tau)$$

where h is any function satisfying $|h|^\alpha = d(m \circ \tau^{-1})/dm$, cannot be doubly stationary.

Example 5.4.2 *Automorphisms defining stationary processes.*

With a given $(S, \mathcal{B}_S, \mu)$, let

$$U_t : S \to S, \qquad t \in \mathbb{R} \tag{5.4.2}$$

be a collection of Borel maps such that the following conditions hold

$$U_0 = I \qquad \mu - a.e.$$

$$U_u \circ U_v = U_{u+v} \qquad \mu - a.e., \quad \text{for every} \ \ u, v \in \mathbb{R} \tag{5.4.3}$$

$$\mu \circ U_t^{-1} \sim \mu, \quad \text{for every} \ \ t \in \mathbb{R}.$$

Let

$$\varepsilon_t : S \to \{-1, 1\} \qquad (\{|z| = 1\}, \ \text{resp.}), \quad t \in \mathbb{R},$$

be a collection of Borel maps such that

$$\varepsilon_0 = 1 \qquad \mu - a.e.,$$

$$\varepsilon_{u+v} = \varepsilon_u \varepsilon_v \circ U_u \quad \mu - a.e., \quad \text{for every} \quad u, v \in \mathbb{R}. \tag{5.4.4}$$

Let $f_0 \in L^{\alpha}(S, \mathcal{B}_S, \mu)$. It is easy to verify that

$$f_t(s) = \varepsilon_t(s) \left\{ \frac{d(\mu \circ U_{-t}^{-1})}{d\mu}(s) \right\}^{1/\alpha} f_0(U_t(s)) \tag{5.4.5}$$

is a representation of a stationary $S\alpha S$ process $\{X(t)\}_{t \in \mathbb{R}}$. Representation (5.4.5), in conjunction with the uniqueness results, indicates how rich the class of stationary $S\alpha S$ processes actually is. Each group of automorhisms $\{U_t\}$ which preserve sets of measure 0 and a family of multipliers $\{\varepsilon_t\}$ produce a (virtually different) class of stationary process. The classes of moving average and harmonizable processes correspond to the cases $U_t(s) = t + s$, $\varepsilon_t \equiv 1$, and $U_t = I$, $\varepsilon_t(s) = e^{its}$, respectively; thus these two classes cover very little of the spectrum of possible stationary stable processes. The class of doubly stationary processes corresponds to the case when the automorphism $\{U_t\}$ is a measure–preserving transformation and then formula (5.4.5) takes the form $f_t(s) = \varepsilon_t(s)f_0(U_t(s))$. Hence, it is clear why moving average and harmonizable processes are doubly stationary.

Now we prove a theorem which shows that each stationary $S\alpha S$ process has such representation as described in Example 5.4.2.

Theorem 5.4.2 *An $S\alpha S$ stochastic process $\{X(t)\}_{t \in \mathbb{R}}$ is stationary if and only if it admits spectral representation (5.4.5).*

PROOF. In view of Theorem 5.4.1 (Hardin (1982)) it is enough to prove that every group of linear isometries $(P_t)_{t \in \mathbb{R}}$ on $L^{\alpha}(S, \mathcal{B}_S, \mu)$ has the form

$$P_t f(s) = \varepsilon_t(s) \left\{ \frac{d(\mu \circ U_{-t}^{-1})}{d\mu}(s) \right\}^{1/\alpha} f(U_t(s)). \tag{5.4.6}$$

Since P_t is an isometry, the collections of functions $\{P_t f\}_{f \in L^{\alpha}}$ and $\{f\}_{f \in L^{\alpha}}$ can be viewed as representations of the same $S\alpha S$ process indexed by $L^{\alpha} = L^{\alpha}(S, \mathcal{B}_S, \mu)$. Let $u, v \in \mathbb{R}$ be fixed. From Theorem 4.1 of Rosinski (1993) it follows that there exist functions $U_u, U_v, U_{u+v} : S \to S$, and $h_u, h_v, h_{u+v} : S \to \mathbb{R}$ ($\mathbb{C}$) such that for every $f \in L^{\alpha}$

$$P_u f = h_u f(U_u), \quad P_v f = h_v f(U_v), \quad \text{and} \quad P_{u+v} f = h_{u+v} f(U_{u+v}), \tag{5.4.7}$$

modulo μ. For every $f \in L^{\alpha}$, we have (modulo μ)

$$
\begin{aligned}
h_{u+v} f(U_{u+v}) &= P_{u+v} f \\
&= P_v(P_u f) = P_v[h_u f(U_u)] \\
&= h_v h_u(U_v) f(U_u(U_v)).
\end{aligned}
$$

Let $f = I_B$, where $B \in \mathcal{B}_S$, $\mu(B) < \infty$. From the above equalities we get

$$h_{u+v} I_B(U_{u+v}) = h_v h_u(U_v) I_B(U_u(U_v)) \quad \mu - a.e. \tag{5.4.8}$$

Notice that $h_{u+v} \neq 0$, $\mu - a.e.$ Indeed, if $\mu(\{s : h_{u+v}(s) = 0\}) > 0$, then for every $A \subset \{s : h_{u+v}(s) = 0\}$ with $0 < \mu(A) < \infty$, $I_A \notin P_{u+v}(L^\alpha)$ by (5.4.7). This contradicts that P_{u+v} is onto. Therefore, (5.4.8) implies

$$\mu\left(\phi_{u+v}^{-1}(B) - (U_u \circ U_v)^{-1}(B)\right) = 0.$$

Since

$$\{s : U_{u+v}(s) \neq (U_u \circ U_v)(s)\} = \bigcup_n \bigcup_i \left[\phi_{u+v}^{-1}(B_i \cap K_n) - (U_u \circ U_v)^{-1}(B_i \cap K_n)\right],$$

where $\{B_i\}$ is a sequence of balls in S with centers in a countable dense set and with positive rational radii, and $\{K_n\}$ is a sequence of sets of finite μ–measure such that $S = \bigcup_n K_n$, we infer that

$$U_{u+v} = U_u \circ U_v \qquad \mu - a.e. \tag{5.4.9}$$

Since P_0 is the identity map, an argument similar to the above gives $U_0(s) = s$ and $h_0(s) = 1$, $\mu - a.e.$ Using (5.4.8) with $B = K_n$, and (5.4.9), and letting $n \to \infty$, we get

$$h_{u+v} = h_u h_v(U_u) \qquad \mu - a.e. \tag{5.4.10}$$

Now, for every $B \in \mathcal{B}_S$ with $\mu(B) < \infty$ we have

$$\begin{aligned}
\mu(U_{-t}^{-1}(B)) &= \int_S I_B(U_{-t}) \, d\mu \\
&= \|I_B(U_{-t})\|_\alpha^\alpha = \|P_t I_B(U_{-t})\|_\alpha^\alpha \\
&= \|h_t I_B(U_{-t} \circ U_t)\|_\alpha^\alpha = \int_B |h_t|^\alpha \, d\mu.
\end{aligned}$$

This completes the proof of (5.4.3) and establishes the relation

$$\frac{d(\mu \circ U_{-t}^{-1})}{d\mu} = |h_t|^\alpha.$$

Put

$$\varepsilon_t(s) = \frac{h_t(s)}{|h_t(s)|}.$$

Then (5.4.10) yields (5.4.4) and ends the proof of (5.4.6). $\qquad \square$

5.5 Self–similar α–Stable Processes

Self–similar (ss) processes are processes that are invariant under suitable translations of time and scale. It is well known that the Brownian motion is ss and the Gaussian Ornstein–Uhlenbeck process is its corresponding stationary process. The first paper giving a rigorous treatment of general ss processes is Lamperti (1962), where a fundamental relationship between stationary and ss processes was proved. Self–similar processes are also related to limit theorems, random walks in random scenery, statistical physics, and fractals. Interested reader is referred to a bibliographical guide by Taqqu (1986) and to a survey by Maejima (1989).

Definition 5.5.1 *Let* $\mathbb{T}$ *be* $(-\infty, \infty)$, $[0, \infty)$ *or* $[0, 1]$. *A real-, or complex-valued stochastic process* $\mathbf{X} = \{X(t)\}_{t \in \mathbb{T}}$ *is said to be* **H-self-similar (H-ss)** *if all finite–dimensional distributions of* $\{X(ct)\}$ *and* $\{c^H X(t)\}$ *are the same for every* $c > 0$, *and to have stationary increments* (**si**) *if the finite–dimensional distributions of* $(X(t + b) - X(b))$ *do not depend on* $b \in \mathbb{T}$.

According to Lamperti (1962), there is the following relationship between stationary and ss–processes.

Proposition 5.5.1 *If* $\{Y(t)\}$ *is a continuous in probability stationary process and if for some* $H > 0$

$$X(t) = t^H Y(\log t), \quad for \quad t > 0, \quad X(0) = 0,$$

then $X(t)$ *is* H–*ss. Conversely, every nontrivial ss–process with* $X(0) = 0$ *is obtained in this way from some stationary process* $\{Y(t)\}$.

One instance of this relationship has been used by Doob (1953) to deduce properties of the stationary Ornstein–Uhlenbéck velocity process from those of the Brownian motion. The attractive feature of the above proposition is that this relationship preserves the Markov property but not stationarity. Thus from the viewpoint of stationary $S\alpha S$ processes it is interesting to know examples of $S\alpha S$ ss–processes other than the Brownian motion.

Examples. (1) $H > \max(1, 1/\alpha)$. Such H–ss si $S\alpha S$ processes do not exist. This is a consequence of the following two properties of ss si non–degenerate processes:

(i) if $0 < \gamma < 1$ and $E|X(1)|^\gamma < \infty$, then $H < 1/\gamma$ (see Maejima (1986));

(ii) if $E|X(t)| < \infty$, then $H < 1$. If we apply these to α–stable processes $\mathbf{X}$, for which $E|X(1)|^\gamma < \infty$ for all $\gamma < \alpha$, we see that $H < \max(1, 1/\alpha)$.

(2) $H = 1/\alpha$.
(2.1) $\alpha = 2$. The Brownian motion: $\{B(t)\}_{t \geq 0}$.
(2.2) $0 < \alpha < 2$. The α–stable Lévy motion: $\{Z_\alpha(t)\}_{t \geq 0}$. Here, by α–stable Lévy motion we mean the $1/\alpha$–ss si α–stable process with independent increments.
(2.3) $1 < \alpha < 2$. The log–fractional stable process (see Kasahara, Maejima and Vervaat (1988)):

$$X_1(t) = \int_{-\infty}^{\infty} \log \left| \frac{t-s}{s} \right| \, dZ_\alpha(s).$$

(2.4) $1 < \alpha < 2$. The sub–Gaussian process (see Kasahara, Maejima and Vervaat (1983)):
$X_2(t) = Z^{1/2} B_{1/\alpha}(t)$, where Z is a positive strictly $\alpha/2$–stable random variable and $\mathbf{B}_H$ is an H–ss si 2–stable (Gaussian) process independent of Z. In fact, $\mathbf{B}_H$ is the fractional Brownian motion defined below.

(2.5) $1 < \alpha < 2$. The (complex–valued) harmonizable fractional stable process (see Cambanis and Maejima (1989)):

$$X_3(t) = \int_{-\infty}^{\infty} \frac{e^{it\lambda} - 1}{i\lambda} |\lambda|^{1-2/\alpha} d\tilde{M}_\alpha(\lambda),$$

where $\tilde{M}_\alpha$ is a complex $S\alpha S$ motion.

(2.6) $\alpha = 1$. A linear function with random slope: $X_4(t) = tX(1)$.

(3) $0 < H < 1$, $0 < \alpha < 2$, $H \neq 1/\alpha$.

(3.1) $\alpha = 2$. The fractional Brownian motion (see Mandelbrot and Van Ness (1968)):

$$
\begin{aligned}
B_H(t) \;=\; C \Bigg[&\int_{-\infty}^{0} \left\{ (t-s)^{H-1/2} - (-s)^{H-1/2} \right\} dB(s) \\
+ &\int_{0}^{t} (t-s)^{H-1/2} \, dB(s) \Bigg],
\end{aligned}
$$

where $\mathbf{B}$ is the standard Brownian motion and C is a normalizing constant assuring $E|B_H(t)|^2 = 1$.

(3.2) $0 < \alpha < 2$. The linear fractional stable process (see e.g., Maejima (1983)):

$$
\begin{aligned}
X_5(t) \;=\; \int_{-\infty}^{\infty} \Bigg[&a \left\{ (t-s)_+^{H-1/\alpha} - (-s)_+^{H-1/\alpha} \right\} \\
+ &b \left\{ (t-s)_-^{H-1/\alpha} - (-s)_-^{H-1/\alpha} \right\} \Bigg] \, dZ_\alpha(s),
\end{aligned}
$$

where $a, b \in \mathbb{R}$ with $ab \neq 0$, $x_+ = \max(x, 0)$, $x_- = \max(-x, 0)$ and $\mathbf{Z}_\alpha$ is an α–stable motion.

(3.3) $0 < \alpha < 2$. The (complex–valued) harmonizable fractional stable process (see Cambanis and Maejima (1989)):

$$X_6(t) = \int_{-\infty}^{\infty} \frac{e^{it\lambda} - 1}{i\lambda} \left(a\lambda_+^{1-H-1/\alpha} + b\lambda_-^{1-H-1/\alpha} \right) \, d\tilde{M}_\alpha(\lambda),$$

where $a, b \in \mathbb{R}$ and $\tilde{M}_\alpha$ is the same as in (2.5).

(4) $0 < H < 1/\alpha$, $0 < \alpha < 2$.

(4.1) Substable processes (see Hardin (1982)):

$$X_7(t) = Z^{1/\beta} Y(t),$$

where $(Y(t))$ is an $H-ss$ si symmetric β–stable process $(0 < \beta < 2)$ and Z is a positive strictly α/β–stable random variable (so $0 < \alpha < \beta$), independent of $(Y(t))$. As special cases we have

$$X_8(t) = Z^{1/2} B_H(t) \quad \text{and} \quad X_9(t) = Z^{1/\beta} Z_\beta(t),$$

where $\mathbf{B}_H$ is the fractional Brownian motion and $\mathbf{Z}_\beta$ is the β–stable motion.

(4.2) Recently, Takenaka (1992) has constructed a class of H-ss si α-stable processes, $\mathbf{X}_{10}$, say, for any $0 < H < 1/\alpha$, $0 < \alpha < 2$, by means of integral geometry. His process can also be constructed by using the stable integral in the following way.

Let $\mathbb{R}_+^2 = \{(x,y) : x > 0, y \in \mathbb{R}\}$, $\mathcal{F} = \{$ bounded Borel sets in $\mathbb{R}_+^2\}$ and let $\{M(A) : A \in \mathcal{F}\}$ denote a family of $S\alpha S$ random measures such that

(i) $E[\exp\{i\theta M(A)\}] = \exp\{-m(A)|\theta|^\alpha\}$, where $m(dxdy) = x^{H\alpha-2}dxdy$,

(ii) $M(A_j), j = 1, \cdots, n$, are mutually independent if $A_j \cap A_k = \phi, j \neq k$, and

(iii) $M(\cup_j A_j) = \sum_j M(A_j)$ a.s. for any disjoint family $\{A_j, j = 1, 2, \cdots\}$.

Let further $c(t; x, y)$ be the indicator function of $\{(x,y) : |y| < x\} \triangle \{(x,y) : |y - t| < x\}$, where $\triangle$ means the symmetric difference of two sets.

Then

$$X_{10}(t) = \int_{\mathbb{R}_+^2} c(t; x, y) \, dM(x, y) \qquad (5.5.1)$$

is an H-ss si $S\alpha S$ process. This class seems new, or at least different from the other examples above. This fact has been proved by Sato (1989), who determined the supports of their Lévy measures.

Sample path properties. If an H-ss si $S\alpha S$ process $\{X(t)\}_{t \in \mathbb{T}}$ satisfies $1/\alpha < H \leq 1$, then it has a sample continuous version, which can be shown by Kolmogorov's moment criterion.

The sample path properties examined in the literatures can be listed as follows:

Property I : There exists a sample continuous version.

Property II : Property I does not hold, but there is a version whose sample paths are right–continuous and have left limits.

Property III : Any version of the process is nowhere bounded, i.e., unbounded on every finite interval.

The examples in the previous section are classified as follows:

Property I : $\mathbf{B}, \mathbf{X}_2, \mathbf{X}_3, \mathbf{X}_4, \mathbf{B}_H, \mathbf{X}_5$ for $1/\alpha < H < 1$, $\mathbf{X}_6, \mathbf{X}_8$.

Property II : $\mathbf{Z}_\alpha, \mathbf{X}_9, \mathbf{X}_{10}$.

Property III : $\mathbf{X}_1, \mathbf{X}_5$ for $0 < H < 1/\alpha$.

Proofs are needed to justify the classifications of $\mathbf{X}_1, \mathbf{X}_3, \mathbf{X}_5$ for $0 < H < 1/\alpha, \mathbf{X}_6$ and $\mathbf{X}_{10}$. They can be based on Theorem 5.5.1 below, due to Nolan (1989).

Let an $S\alpha S$ process $\mathbf{X} = \{X(t)\}_{t \in \mathbb{T}}$ be given by

$$X(t) = \int_U f(t, u) \, dW_m(u),$$

where $\mathbb{T}$ is a finite interval, $(U, \mathcal{U}, m)$ is some σ–finite measure space, $f : \mathbb{T} \times U \to \mathbb{R}$ is a function with the property that for each $t \in \mathbb{T}$, $f(t, \cdot) \in L^\alpha(U, \mathcal{U}, m)$, and W_m is an $S\alpha S$ random measure with control measure m such that

$$E[\exp\{i\theta W_m(A)\}] = \exp\{-m(A)|\theta|^\alpha\}, \quad A \in \mathcal{U}.$$

We assume **X** is continuous in probability and take a separable version of **X**. A kernel $f_0(t, u)$ is a modification of $f(t, u)$ if for all $t \in \mathbf{T}$, $f_0(t, \cdot) = f(t, \cdot)$ $m - a.e.$ on U. Then $\mathbf{X}_0 = (X_0(t) = \int_U f_0(t, u) \, dW_m(u))$ is a version of **X**.

When $1 < \alpha < 2$, define β by $1/\alpha + 1/\beta = 1$. For $\varepsilon > 0$ and d a metric or pseudo–metric on $\mathbf{T}$, let

$$H_\beta(d; \varepsilon) = \begin{cases} (\log N(d; \varepsilon))^{1/\beta}, & 2 < \beta < \infty; \\ \log^+ \log N(d; \varepsilon), & \beta = \infty, \end{cases}$$

where $N(d; \varepsilon) = N(\mathbf{T}, d; \varepsilon)$ is the minimum number of d–balls of radius ε with centers in $\mathbf{T}$ that cover $\mathbf{T}$. Let

$$d_X(t, s) = (-\log[\boldsymbol{E} e^{i(X(t)-X(s))}])^{1/\alpha}.$$

We consider three conditions on the kernel $f(t, u)$ and one condition on $H_\beta(d_X; \varepsilon)$:

(C1) f has a modification f_0 such that for every $u \in U$, $f_0(t, u)$ is continuous;

(C2) $f^*(u) = \sup_{t \in \mathbf{T}_0} |f(t, u)|$ is in $L^\alpha(U, \mathcal{U}, m)$, where $\mathbf{T}_0 \subset \mathbf{T}$ is a countable separant for **X** that is dense in $\mathbf{T}$;

(C3)

$$\int \left(\sup_{s, t \in \mathbf{T}} \frac{|f(t, u) - f(s, u)|}{d_X(t, s)} \right)^\alpha dm(u) < \infty;$$

(C4)

$$\int_0^\infty H_\beta(d_X; \varepsilon) \, d\varepsilon < \infty.$$

According to Nolan (1989), we obtain the following theorem.

Theorem 5.5.1 *Let* $0 < \alpha < 1$.

(i) **X** *has Property I if and only if (C1) and (C2) hold.*

(ii) **X** *has a version with discontinuous, bounded sample paths if and only if (C1) fails to hold and (C2) holds.*

(iii) **X** *has Property III if and only if (C2) fails to hold.*

Let $1 < \alpha < 2$.
(iv) *If (C1), (C2), (C3) and (C4) are fulfilled, then* **X** *has Property I.*

Following Kono and Maejima (1991), we give now the proofs for the above classification of the listed examples.

PROOFS. (1) The fact that $\mathbf{X}_1$ and $\mathbf{X}_5$ for $0 < H < 1/\alpha$ have Property III is verified by Theorem 5.5.1 (iii) above, or by Theorem 4 of Rosinski (1989).

(**2**) Recall

$$X_6(t) = \int_{-\infty}^{\infty} f(t, \lambda) d\tilde{M}_\alpha(\lambda), \quad 0 < t < 1,$$

where

$$f(t, \lambda) = \frac{e^{it\lambda} - 1}{i\lambda}(a\lambda_+^{1-H-1/\alpha} + b\lambda_-^{1-H-1/\alpha}).$$

(When $H = 1/\alpha$, $\mathbf{X}_6 = \mathbf{X}_3$.) Let $f(t, \lambda) = g(t, \lambda) + ih(t, \lambda)$ and $\tilde{M}_\alpha = M_\alpha^{(1)} + iM_\alpha^{(2)}$. Then

$$X_6(t) = \int (g dM_\alpha^{(1)} - h dM_\alpha^{(2)}) + i \int (h dM_\alpha^{(1)} + g dM_\alpha^{(2)}).$$

Obviously h and g satisfy (C1). Observe that g^* and h^* are in $L^\alpha(\mathbb{R}, \mathcal{B}, dx)$, satisfying (C2). Hence, when $0 < \alpha < 1$, it follows from Theorem 5.5.1 (i) that $\mathbf{X}_6$ has Property I.

If $\mathbf{X}$ is H–ss si $S\alpha S$, then

$$d_X(t, s) = C|t - s|^H$$

for some positive constant C, and thus, when $\mathbb{T} = [0, 1]$,

$$N(d_X; \varepsilon) = \begin{cases} C([\varepsilon^{-1/H}] + 1) & \text{if} \quad \varepsilon < 1, \\ C & \text{if} \quad \varepsilon \geq 1. \end{cases}$$

Hence, when $1 < \alpha < 2$, we have (C4). Condition (C3) is also satisfied. Thus, by Theorem 5.5.1 (iv), we conclude that $\mathbf{X}_6$ has Property I.

(**3**) When $0 < \alpha < 1$, we can apply Theorem 5.5.1 (ii) to show that $\mathbf{X}_{10}$ has Property II.

Recall that

$$X_{10}(t) = \int_{\mathbb{R}_+^2} c(t; x, y) dM(x, y).$$

Applying now Theorem 5.5.1 (ii) with $U = \mathbb{R}_+^2$, $u = (x, y)$, $dm(u) = x^{H\alpha-2} dx dy$, $W_m = M$ we see that $c(\cdot; x, y)$ does not satisfy (C1). However, since

$$\sup_{t \in [0,1]} c(t; x, y) = c(1; x, y),$$

(C2) is fulfilled. Hence, when $0 < \alpha < 1$, $\mathbf{X}_{10}$ has a version with discontinuous, bounded sample paths. By an observation due to Vervaat (1985), H–ss si processes with such versions have Property II. In the case $1 < \alpha < 2$, we arrive at the same conclusion if we represent the stochastic integral $X_{10}(t)$ by the pathwise integral after the integration by parts.   □

Remark 5.5.1 *For $1/\alpha < H < 1$, the processes $\mathbf{X}_5$ and $\mathbf{X}_8$ cannot be discriminated by Property I. More delicate path properties exhibit them as different. Takashima (1989) has recently proved that for any $\varepsilon > 0$,*

$$\limsup_{t \downarrow 0} \frac{|X_5(t)|}{t^H (\log 1/t)^{1/\alpha-\varepsilon}} = \infty \quad a.s.$$

However, $\mathbf{X_8}$ has the form $\mathbf{X_8} = Z^{1/2}\mathbf{B}_H$, where Z and $\mathbf{B}_H$ are independent. Hence, it follows from the law of the iterated logarithm for $\mathbf{B}_H$ (cf., Marcus (1968)) that

$$\limsup_{t\downarrow 0} \frac{|X_8(t)|}{t^H(2\log\log 1/t)^{1/2}} < \infty \qquad a.s.$$

Let

$$
\begin{aligned}
A_1 &= \{(H,\alpha) : \alpha = 2,\ 0 < H < 1/2\}, \\
A_2 &= \{(H,\alpha) : 0 < \alpha < 2,\ 0 < H < 1/2\}, \\
B_1 &= \{(H,\alpha) : \alpha = 2,\ 1/2 < H < 1\}, \\
B_2 &= \{(H,\alpha) : 1/H < \alpha < 2,\ 1/2 < H < 1\}, \\
B_3 &= \{(H,\alpha) : \alpha = 1/H,\ 1/2 < H < 1\}, \\
B_4 &= \{(H,\alpha) : 0 < \alpha < 1/H,\ 1/2 < H < 1\}, \\
C_1 &= \{(H,\alpha) : \alpha = 1/H,\ H > 1\} \\
C_2 &= \{(H,\alpha) : 0 < \alpha < 1/H,\ H > 1\}.
\end{aligned}
$$

Following Kono and Maejima (1991) one can systematize the different contingencies of α and H, existence of appropriate processes, and validity of path properties by the following picture of the (H,α) plane.

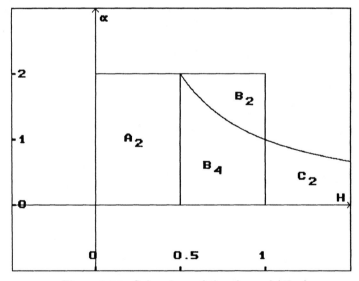

Figure 5.5.1. Subregions of the plane of (H,α).

Chapter 6

Computer Approximations of Continuous Time Processes

6.1 Introduction

The aim of this chapter is to provide some constructive computer methods designed to approximate and visualize the main classes of stochastic processes discussed in this book. It turns out that with the use of suitable statistical estimation techniques, computer simulation procedures and numerical discretization methods it is possible to construct approximations of stochastic integrals with stable measures as integrators. As a consequence we obtain an effective, general method giving approximate solutions for a wide class of stochastic differential equations involving such integrals. Application of computer graphics provides interesting quantitative and visual information on those features of stable variates which distinguish them from their commonly used Gaussian counterparts. It is possible to demonstrate evolution in time of densities with heavy tails of appropriate processes, to visualize the effect of jumps of trajectories, etc.

We try to demonstrate that stable variates can be very useful in stochastic modeling of problems of different kinds, arising in science and engineering, which often provide better description of real life phenomena than their Gaussian counterparts.

First, we focus our attention on diffusion processes described as solutions of stochastic differential equations with respect to both Gaussian and stable measures. As a particular case we obtain approximate representations of stochastic integrals.

Next, we describe the method of approximation and simulation of stochastic processes represented by stochastic integrals of deterministic functions with respect to stochastic measures. We present moving averages processes as examples of stationary $S\alpha S$ processes, which are so important from the point of view of ergodic theory of stochastic processes.

Computer methods of construction of stochastic processes involve at least two kinds of discretization techniques: discretization of the time parameter and an approximate representation of random variates with the aid of artificially produced finite–time series data sets or statistical samples. Statistical methods of data analysis such as constructions of empirical cumulative distribution functions or kernel probability density estimates, regression analysis, etc., provide powerful and very useful tools for practical investigation of interesting properties of different classes of stochastic processes. By applying IBM PC graphics we attempt to demonstrate the good results they can provide. (Keeping in mind technical abilities of such computers we restrict ourselves to $\mathbb{R}$–, or $\mathbb{R}^2$–valued processes.)

6.2 Approximation of Diffusions Driven by Brownian Motion

Now our aim is to present some methods of numerical approximation and computer simulation of solutions of Itô–type stochastic differential equations. Applying some statistics we describe a method of computer visualization of stochastic processes that satisfy these equations.

Let us start the discussion of methods leading to computer construction of diffusion processes with the simplest, but very important in applications, particular case.

Approximation of Itô–type stochastic integral of deterministic function. We are interested in a method of construction of an univariate stochastic process $\{X(t) : t \in [0, T]\}$ defined in the following way

$$X(t) = \int_0^t f(s) \, dB(s), \qquad t \in [0, T], \tag{6.2.1}$$

where $f = f(t)$ is a given, say, continuous function on $[0, T]$ and $\{B(t) : t \in [0, +\infty)\}$ is the Brownian motion.

Let us fix a positive integer $I \in \mathbb{N}$ and introduce a regular mesh of points $\{t_i\}$ discretizing the time interval $[0, T]$

$$t_i = i\tau, \qquad \text{for} \quad i = 0, 1, ..., I, \quad \text{where} \quad \tau = T/I. \tag{6.2.2}$$

Definition 6.2.1 *We approximate the process $\{X(t) : t \in [0, T]\}$ described by (6.2.1) by a discrete–time process $\{X_{t_i}^\tau\}_{i=0}^I$ defined in the following way:*

1. $X_0^\tau = 0$ a.s.;

2. for $i = 1, 2, ..., I$ define

$$X_{t_i}^\tau = X_{t_{i-1}}^\tau + f(t_{i-1})\Delta B_i^\tau, \tag{6.2.3}$$

where ΔB_i^τ denotes the stochastic (Gaussian) measure of an interval $[t_{i-1}, t_i)$, i.e., in this special case, simply the Gaussian random variable defined by

$$\Delta B_i^\tau = B([t_{i-1}, t_i)) = B(t_i) - B(t_{i-1}) \sim \sqrt{\tau}\mathcal{N}(0, 1) = \mathcal{N}(0, \tau); \tag{6.2.4}$$

3. the sequence $\{\Delta B_i^\tau; \; i = 1, ..., I\}$ *is an i.i.d. sequence.*

In particular we have

$$I = \int_0^T f(s) \; dB(s) = X(T) \sim \sum_{i=1}^I f(t_{i-1}) \Delta B_i^\tau.$$ (6.2.5)

Someone looking for continuous time approximation of the process $\{X(t)\}$ can derive from the above definition the following formula

$$X^\tau(t) = X_{t_{i-1}}^\tau \quad \text{for } t \in [t_{i-1}, t_i),$$ (6.2.6)

with $i = 1, 2, ..., I$.

Simulation of univariate stochastic processes.

Now we describe rather general technique of computer simulation of univariate stochastic processes $\{X(t) : t \in [0, T]\}$ with independent identically distributed increments by a discrete time process of the form $\{X_{t_i}^\tau\}_{i=0}^I$, defined by the formula

$$X_{t_i}^\tau = X_{t_{i-1}}^\tau + Y_i^\tau,$$ (6.2.7)

with a given X_0^τ and where Y_i^τ's form a sequence of i.i.d. random variables.

In computer calculations each random variable $X_{t_i}^\tau$ defined by (6.2.7) is represented by its N independent realizations, i.e. a random sample $\{X_i^\tau(n)\}_{n=1}^N$. So, let us fix $N \in \mathcal{N}$ large enough. The algorithm consists in the following:

1. simulate a random sample $\{X_0^\tau(n)\}_{n=1}^N$ for X_0^τ;

2. for $i = 1, 2, ..., I$ simulate a random sample $\{\Delta L_{\alpha,i}^\tau(n)\}_{n=1}^N$ for α–stable random variable $\Delta L_{\alpha,i}^\tau$;

3. for $i = 1, 2, ..., I$ compute the random sample $\{Y_i^\tau(n)\}_{n=1}^N$ of Y_i^τ and $X_i^\tau(n) = X_{i-1}^\tau(n) + Y_i^\tau(n), \quad n = 1, 2, ..., N$;

4. construct kernel density estimators $f_i = f_i^{I,N} = f_i^{I,N}(x)$ of the densities of $X(t_i)$, using for example the optimal version of the Rosenblatt–Parzen method, and their distribution functions $F_i = F_i^{I,N} = F_i^{I,N}(x)$.

Observe that we have produced N finite time series of the form $\{X_i^\tau(n)\}_{i=0}^I$ for $n = 1, 2, ..., N$. We regard them as "good" approximations of the trajectories of the process $\{X(t); \; t \in [0, T]\}$.

Visualization of univariate stochastic processes.

In order to obtain a graphical approximation of the integral $\{X(t); \; t \in [0, T]\}$ defined by (6.2.1) we propose the following:

1. fix a rectangle $[0, T] \times [c, d]$ that should include the trajectories of $\{X(t)\}$;

2. for each $n = 1, 2, ..., n_{max}$ (with fixed $n_{max} \ll N$) draw the line segments determined by the points $(t_{i-1}, X_{i-1}^r(n))$ and $(t_i, X_i^r(n))$ for $i = 1, 2, ..., I$, constructing n_{max} approximate trajectories of the process X (thin lines on all figures);

3. fix a few values of a "probability parameter" p_j from $(0, 1/2)$ for $j = 1, 2, ..., J$ and for each of them compute 2 quantiles: $q_{min}^{i,j} = F_i^{-1}(p_j)$ and $q_{max}^{i,j} = F_i^{-1}(1 - p_j)$ for all $i = 1, 2, ..., I$; then draw the line segments determined by the points $(t_{i-1}, q_{min}^{i-1,j})$, $(t_i, q_{min}^{i,j})$ and $(t_{i-1}, q_{max}^{i-1,j})$, $(t_i, q_{max}^{i,j})$ for $i = 1, 2, ..., I$ and $j = 1, 2, ..., J$, constructing J (varying in time) prediction intervals (thick lines on all figures) that determine subdomains of $\mathbb{R}^2$ to which the trajectories of the approximated process should belong with probabilities $1 - 2p_j$ at any fixed moment of time $t = t_i$. Each line obtained in this way we call *quantile line*.

Remark 6.2.1 *On the computer screen an interval $[0, T]$ is represented by a few hundreds of pixels, i.e. computer screen "points" (working with VGA graphics cart one has exactly 640 pixels). Computer experiments with Brownian motion processes $\{X(t)\}$ proved that simulating them on a finite interval $[0, T]$ with the mesh consisting of 1000, 10000, 100000 subintervals (some number of time steps is performed within 1 pixel size) the pictures of approximate trajectories look very much alike. An argument acceptable for statisticians, physicists and engineers, even for mathematicians, is that graphical approximations of such strange functions as the trajectories of Brownian motion are acceptable for a human eye.*

Experiments with meshes of 100 or 300 subintervals usually give poor results.

It is an obvious observation that some features of graphs of trajectories constructed on a computer heavily depend on a proper scaling of both axes. For example, self-similarity of trajectories of Brownian motion can be seriously distorted in the case of an interval $[0, T]$ with T much too small or much too large in comparison with the chosen intervals to represent the values of $X(t)$.

Examples of stochastic integrals. In Chapter 2 we have already presented a few examples of computer visualizations of stochastic processes obtained by this technique and we believe they provide some interesting quantitative and qualitative information on stochastic processes.

Here we present 3 examples. Each of them is described by 2 figures. The first includes a number of approximate trajectories (20, 20, 50, respectively) and 3 pairs of quantile lines (our dynamical histogram) defined by $p_1 = 0.05$, $p_2 = 0.15$, $p_3 = 0.25$ and 50 trajectories, the second – the histogram and kernel density estimator (see Chapter 3) of the last calculated value X(T) of the process. In all cases we have chosen I=2000, N=2000.

Example 6.2.1 *(Integral (i).)*

In order to obtain the Brownian motion it is enough to notice that

$$B(t) = \int_0^t dB(s), \quad t \in [0, T]. \tag{6.2.8}$$

Invoking the next two examples we want to demonstrate that there is a vast class of very interesting processes defined very simply. Their existence is not acknowledged in applications. Computer methods provide a deeper insight into their structure.

Example 6.2.2 *(Integral (ii).)*

$$X(t) = \int_0^t \cos(s)dB(s), \quad t \in [0, T]. \tag{6.2.9}$$

Example 6.2.3 *(Integral (iii).)*

$$X(t) = \int_0^t e^{-s} \, dB(s), \quad t \in [0, T]. \tag{6.2.10}$$

One can observe how quickly this process starts to behave like a stationary process or even trivially stationary (all trajectories practically become constant) for t large enough.

Approximation of stochastic integrals of random functions.

Here we are going to point out some well–known but very important and in some sense surprising facts concerning constructions of the"Riemann type" integral sums approximating stochastic integrals of a random function, i.e. processes $\{X(t) : t \in [0, T]\}$ defined by

$$X(t) = \int_0^t f(s, B(s)) \, dB(s), \quad t \in [0, T], \tag{6.2.11}$$

where $f = f(t, x)$ is a given, say, continuous function on $[0, T] \times \mathbb{R}$ and $\{B(t) : t \in [0, +\infty)\}$ is the standard Brownian motion, or by

$$X(t) = \int_0^t f(s, B^{(2)}(s)) \, dB^{(1)}(s), \quad t \in [0, T], \tag{6.2.12}$$

where $\{B^{(1)}(t) : t \in [0, +\infty)\}$ and $\{B^{(2)}(t) : t \in [0, +\infty)\}$ are two independent copies of the standard Brownian motion.

When constructing appropriate sums approximating such stochastic integrals one has to carefully distinguish between

(I) the *Itô stochastic integral* or *forward integral* that is obtained as a limit of sums

$$X_{t_i}^\tau = X_{t_{i-1}}^\tau + f(t_{i-1}, B(t_{i-1})) \, \Delta B_i^\tau \tag{6.2.13}$$

with respect to a given infinite sequence of appropriately chosen meshes,

(b) the *backward integral* that is obtained as a limit of sums

$$X_{t_i}^\tau = X_{t_{i-1}}^\tau + f(t_i, B(t_i))\, \Delta B_i^\tau \tag{6.2.14}$$

- the *Stratonovich stochastic integral* defined by

$$X_{t_i}^\tau = X_{t_{i-1}}^\tau + \frac{f(t_{i-1}, B(t_{i-1})) + f(t_i, B(t_i))}{2}\, \Delta B_i^\tau. \tag{6.2.15}$$

One can even try convex linear sums of the form

$$X_{t_i}^\tau = X_{t_{i-1}}^\tau + (\, (1-\lambda)f(t_{i-1}, B(t_{i-1})) + \lambda f(t_i, B(t_i))\,)\, \Delta B_i^\tau,$$

with $\lambda \in [0,1]$.

Example 6.2.4 *(Integral (iv).)*

We are going to demonstrate how stochastic integrals of random functions depend on definitions of their partial sums.

It is well known that in the limit we obtain

(I) $\int_0^t B(s)\, dB(s) = \frac{1}{2}(B^2(t) - t)$,

(b) $\int_0^t B(s)\, dB(s) = \frac{1}{2}(B^2(t) + t)$,

(S) $\int_0^t B(s)\, dB(s) = \frac{1}{2}B^2(t)$,

respectively.

We have chosen this example to test the correctness and accuracy of the developed method of computer simulation and visualization of stochastic integrals and diffusion processes. It is not difficult to check that the random variable $\frac{1}{2}B^2(t)$ has gamma distribution, i.e. its density is defined by

$$f(x) = \begin{cases} \frac{1}{\sqrt{\pi t}} x^{-1/2} e^{-x/t} & \text{if } x > 0, \\ 0 & \text{otherwise,} \end{cases}$$

for $t > 0$. The density is unbounded, so this is a hard test of "classical" versions of kernel density estimators. Figures 6.2.7–6.2.10 present graphically the Itô and Stratonovich integrals from Example 6.2.4.

Remark 6.2.2 *The convergence of processes $\{X^\tau(t)\}$ to $\{X(t)\}$ for $t \in [0, T]$ when $\tau \to 0$ follows from the definition of the Itô stochastic integral. It follows also from the theorems on convergence of approximations for diffusion processes that are presented below.*

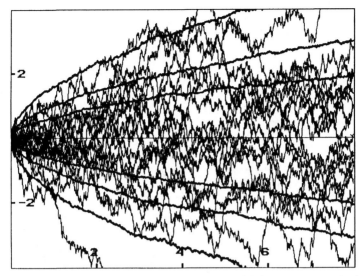

Figure 6.2.1. Visualization of the integral (i) (Brownian motion).

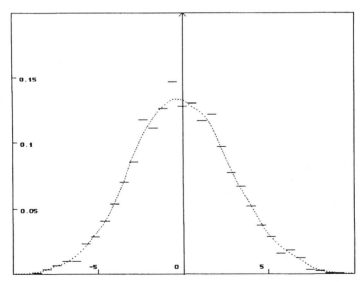

Figure 6.2.2. Histogram and kernel density estimator of B(8) (integral (i)).

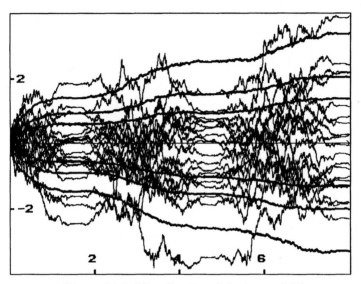

Figure 6.2.3. Visualization of the integral (ii).

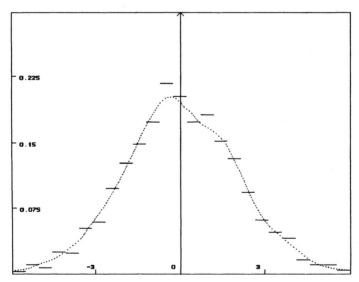

Figure 6.2.4. Histogram and kernel density estimator of X(8) for integral (ii).

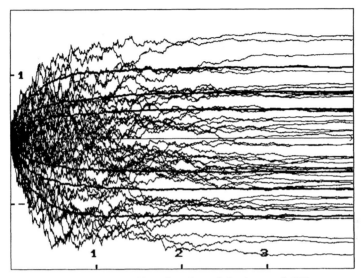

Figure 6.2.5. Visualization of the integral (iii).

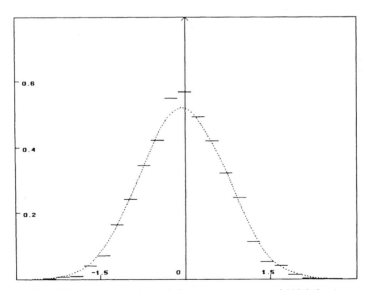

Figure 6.2.6. Histogram and kernel density estimator of X(4) for integral (iii).

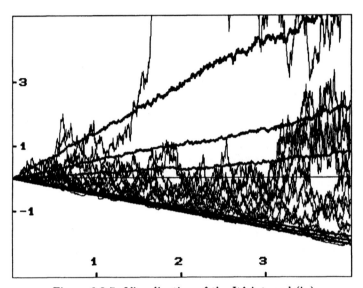

Figure 6.2.7. Visualization of the Itô integral (iv).

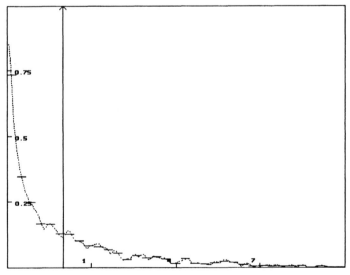

Figure 6.2.8. Estimators of X(4) of (iv) for Itô integral.

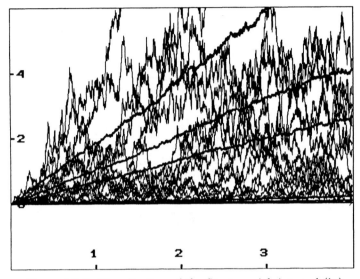

Figure 6.2.9. Visualization of the Stratonovich integral (iv).

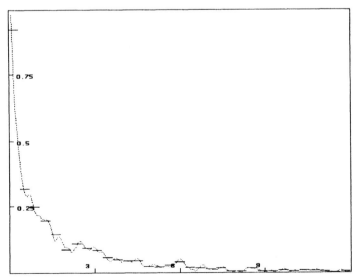

Figure 6.2.10. Estimators of X(4) for Stratonovich integral (iv).

Remark 6.2.3 *In Section 3.5 we made a remark on four possible sources of errors of computer calculations and simulations. Now we have two new types of errors induced by approximation:*

1. *discretization of the time interval $[0,T]$ and approximation of $\{X(t)\}$ by $\{X_{t_i}^\tau\}$;*

2. *statistical approximation of random variables $X_{t_i}^\tau$ by their statistical samples $\{X_i^\tau(n)\}_{n=1}^N$.*

Approximation of univariate diffusions driven by Brownian motion.

According to the description of this class of stochastic processes in Section 4.10, we are interested in constructive computer methods of solution of the Itô–type stochastic differential equation of the following form

$$dX(t) = a(t, X(t)) \, dt + b(t, X(t)) \, dB(t) \quad \text{for} \quad t \in [0, T], \tag{6.2.16}$$

with $X(0) = X_0$ – given Gaussian random variable.

In order to describe the simplest possible method (Euler–like method) of approximation of the equation (6.2.16) it is enough to rewrite it in the following integral form

$$X(t) = X_0 + \int_0^t a(s, X(s)) \, ds + \int_0^t b(s, X(s)) \, dB(s), \tag{6.2.17}$$

and to modify slightly the method presented in Definition 6.2.1.

Definition 6.2.2 *With a fixed regular mesh on $[0,T]$ we approximate the process $\{X(t) : t \in [0,T]\}$ by a discrete time process $\{X_{t_i}^\tau\}_{i=0}^I$ defined in the following way:*

1. *$X_0^\tau \sim \mathcal{N}(\mu, \sigma^2)$;*

2. *for $i = 1, 2, ..., I$,*

$$X_{t_i}^\tau = X_{t_{i-1}}^\tau + a(t_{i-1}, X(t_{i-1})) \, \tau + b(t_{i-1}, X(t_{i-1}))\Delta B_i^\tau, \tag{6.2.18}$$

where, as before, ΔB_i^τ denotes the stochastic Gaussian measure of an interval (t_{i-1}, t_i) of the form

$$\Delta B_i^\tau = B((t_{i-1}, t_i)) \sim \mathcal{N}(0, \tau);$$

3. *the sequence $\{\Delta B_i^\tau; \ i = 1, ..., I\}$ being a given i.i.d. sequence.*

There are (as in the case of ordinary differential equations) some methods of higher order of approximation. We recall here two examples of such methods (see, e.g., Pardoux and Talay (1985) or Yamada (1976)).

Definition 6.2.3 *With a fixed regular mesh on $[0,T]$ we approximate the process $\{X(t) : t \in [0,T]\}$ by a discrete time process $\{X_{t_i}^\tau\}_{i=0}^I$ defined by*

1. $X_0^\tau \sim \mathcal{N}(\mu, \sigma^2)$;

2. *for $i = 1, 2, ..., I$,*

$$
\begin{aligned}
X_{t_i}^\tau &= X_{t_{i-1}}^\tau + \bar{a}(t_{i-1}, X(t_{i-1}))\,\tau + b(t_{i-1}, X(t_{i-1}))\Delta B_i^\tau \\
&+ \frac{1}{2}b'(t_{i-1}, X(t_{i-1}))b(t_{i-1}, X(t_{i-1}))(\Delta B_i^\tau)^2,
\end{aligned}
$$

where $\bar{a}(t,x) = a(t,x) - \frac{1}{2}b'(t,x)\,b(t,x)$ and, as before, ΔB_i^τ denotes the stochastic Gaussian measure of an interval (t_{i-1}, t_i) of the form

$$\Delta B_i^\tau = B((t_{i-1}, t_i)) \sim \mathcal{N}(0, \tau);$$

3. *the sequence $\{\Delta B_i^\tau;\ i = 1, ..., I\}$ being a given i.i.d. sequence.*

Definition 6.2.4 *With a fixed regular mesh on $[0,T]$ we approximate the process $\{X(t) : t \in [0,T]\}$ by a discrete time process $\{X_{t_i}^\tau\}_{i=0}^I$ defined in the following way:*

1. $X_0^\tau \sim \mathcal{N}(\mu, \sigma^2)$;

2. *for $i = 1, 2, ..., I$,*

$$
\begin{aligned}
X_{t_i}^\tau &= X_{t_{i-1}}^\tau + \bar{a}(t_{i-1}, X(t_{i-1}))\,\tau + b(t_{i-1}, X(t_{i-1}))\Delta B_i^\tau \\
&+ \frac{1}{2}b'(t_{i-1}, X(t_{i-1}))b(t_{i-1}, X(t_{i-1}))(\Delta B_i^\tau)^2 \\
&+ \{a'(t_{i-1}, X(t_{i-1}))b(t_{i-1}, X(t_{i-1})) + a(t_{i-1}, X(t_{i-1}))b'(t_{i-1}, X(t_{i-1})) \\
&\quad + \frac{1}{2}b^2(t_{i-1}, X(t_{i-1}))b''(t_{i-1}, X(t_{i-1}))\}\frac{\tau}{2}\Delta B_i^\tau \\
&+ \{a(t_{i-1}, X(t_{i-1}))a'(t_{i-1}, X(t_{i-1})) \\
&\quad + \frac{1}{2}b^2(t_{i-1}, X(t_{i-1}))a''(t_{i-1}, X(t_{i-1}))\}\frac{\tau^2}{2},
\end{aligned}
$$

where $\bar{a}(t,x) = a(t,x) - \frac{1}{2}b'(t,x)b(t,x)$ and ΔB_i^τ denotes the stochastic Gaussian measure of an interval (t_{i-1}, t_i) of the form

$$\Delta B_i^\tau = B((t_{i-1}, t_i)) \sim \mathcal{N}(0, \tau);$$

3. *the sequence $\{\Delta B_i^\tau;\ i = 1, ..., I\}$ being a given i.i.d. sequence.*

Remark 6.2.4 *In the above scheme one can replace the sequence of $\{\Delta B_i^\tau\}$ by any sequence of i.i.d. random variables $\{\sqrt{\tau}U_i\}$, where U_i satisfy*

$$E U_i = 0, \quad E(U_i)^2 = 1, \quad E(U_i)^3 = 0,$$

$$E(U_i)^4 = 3, \quad E(U_i)^5 = 0, \quad E(U_i)^6 < \infty.$$

One can choose, for example, discrete random variables distributed according to the following law

$$P\{U_i = \sqrt{3}\} = P\{U_i = -\sqrt{3}\} = \frac{1}{6}, \quad P\{U_i = 0\} = \frac{2}{3}.$$

Remark 6.2.5 *In all cases described above, having at our disposal* $\{X_{t_i}^\tau\}_{i=0}^{I-1}$*, we can construct a continuous time process*

$$X^\tau(t) = X_{t_{i-1}}^\tau \quad \text{for } t \in [t_{i-1}, t_i), \tag{6.2.19}$$

with $i = 1, 2, ..., I$ *and* $X^\tau(T) = X_{t_{I-1}}^\tau$.

Convergence of approximations of Gaussian diffusions.

Now we want to recall some theorems on convergence of approximate solutions of stochastic differential equations defined above in Definitions 6.2.2, 6.2.3, 6.2.4. The proofs can be found in Pardoux and Talay (1985) or in Kloeden and Platen (1992).

Theorem 6.2.1 *Suppose that coefficient functions a, b in (6.2.17) are uniformly Lipschitz continuous on* $[0, T]$.
   Then, for $\{X(t) : t \in [0, T]\}$ *satisfying (6.2.17) and* $\{X^\tau(t) : t \in [0, T]\}$ *described in Definition 6.2.2 and (6.2.19), we have*

$$\mathsf{E}|X(t) - X^\tau(t)|^2 = \mathcal{O}(\tau), \tag{6.2.20}$$

uniformly for $t \in [0, T]$.

Theorem 6.2.2 *Suppose that coefficient functions a, b in (6.2.17) are of the class* $C^2([0, t])$ *and* a, a', b, b' *are uniformly Lipschitz continuous on* $[0, T]$.
   Then, for $\{X(t) : t \in [0, T]\}$ *satisfying (6.2.17) and* $\{X^\tau(t) : t \in [0, T]\}$ *described in Definition 6.2.3 and (6.2.19), we have*

$$\mathsf{E}|X(t) - X^\tau(t)|^2 = \mathcal{O}(\tau^2), \tag{6.2.21}$$

uniformly for $t \in [0, T]$.

Theorem 6.2.3 *Suppose that coefficient functions a, b in (6.2.17) are of the class* $C^6([0, t])$ *and* b^2, b'', b^2, a'' *increase at infinity not faster then linear functions.*
   Then, for any given function f of class $C^6([0, T])$ *increasing at infinity not faster then some polynomial function and* $\{X(t) : t \in [0, T]\}$ *satisfying (6.2.17) and* $\{X^\tau(t) : t \in [0, T]\}$ *described in Definition 6.2.4 and (6.2.19), we have*

$$\mathsf{E}\, f(X(t)) - \mathsf{E}\, f(X^\tau(t)) = \mathcal{O}(\tau^2), \tag{6.2.22}$$

uniformly for $t \in [0, T]$.

Computer simulation and visualization of multivariate diffusions.

Dealing with an $\mathbb{R}^d$–valued diffusion processes $\{X(t) : t \in [0, T]\}$ driven by an $\mathbb{R}^p$–valued Brownian motion process $\{B(t) : t \geq 0\}$ we obtain an even simpler description of stochastic differential equations and their discrete–time approximations. In this case it is enough to write down the equation

$$X(t) = X_0 + \int_0^t a(X(s)) \, ds + \int_0^t b(X(s)) \, dB(s), \qquad (6.2.23)$$

keeping in mind that now $a : \mathbb{R}^d \to \mathbb{R}^d$ and $b : \mathbb{R}^d \to \mathbb{R}^d \times \mathbb{R}^p$, and, in an analogous way as before, to choose its (simplest) approximation

$$X_{t_i}^\tau = X_{t_{i-1}}^\tau + a(X(t_{i-1})) \, \tau + b(X(t_{i-1})) \Delta B_i^\tau. \qquad (6.2.24)$$

The theorems on convergence of approximate solutions can be rewritten as well.

Limited by computer capacity and stimulated by the most frequent examples in the physical and mathematical literature, we are mostly interested here in the case of a bivariate diffusion $\{(X(t), Y(t)) : t \in [0, T]\}$ that is described by the system of two equations

$$X(t) = X_0 + \int_0^t a(s, X(s), Y(s)) \, ds + \int_0^t b(s, X(s), Y(s)) \, dB^{(1)}(s), \quad (6.2.25)$$

$$Y(t) = Y_0 + \int_0^t c(s, X(s), Y(s)) \, ds + \int_0^t d(s, X(s), Y(s)) \, dB^{(2)}(s), \quad (6.2.26)$$

where a, b, c, d are smooth enough functions from $\mathbb{R}^3$ into $\mathbb{R}$, $(X(0), Y(0)) \sim \mathcal{N}(\mu_X, \sigma_X^2) \times \mathcal{N}(\mu_Y, \sigma_Y^2)$ and $\{B^{(1)}(t) : t \in [0, +\infty)\}$, $\{B^{(2)}(t) : t \in [0, +\infty)\}$ are two independent copies of an univariate Brownian motion process.

We decided to present results of computer simulations of bivariate diffusions by applying the technique developed in the context of scalar stochastic equations and repeating all presentation separately for $\{X(t)\}$ and for $\{Y(t)\}$. In this way we are able to expose more interesting details of obtained solutions. Unfortunately, graphs of trajectories of bivariate processes of the form $\{(X(t), Y(t)) : t \in [0, T]\}$ do not seem very impressive in the appropriate phase space.

We found it interesting, however, to construct for the process $\{(X(t), Y(t)) : t \in [0, T]\}$ approximate graphs of density kernel estimators for a few fixed values of t. It allows us to follow the transformations of diffusion densities in time. This technique proves sometimes advantageous in comparison with others, for example, it performs much better than computer methods based on an approximation of the Fokker–Planck equation in situations when initial values of processes are described by discrete distributions.

A demonstration of results of computer simulations of some interesting exemplary problems concerning univariate and bivariate diffusions can be found in Sections 6.4 and 7.3.

6.3   Approximation of Diffusions Driven by α–Stable Lévy Measure

Now we briefly describe the simplest constructive computer method of approximate solution of the stochastic differential equation with respect to the α–stable Lévy measure in the following form

$$dX(t) = a(t, X(t))\, dt + c(t)\, dM_\alpha(t) \quad \text{for} \quad t \in [0, T],$$

with $X(0) = X_0$ – given stable random variable.

In order to obtain the strict mathematical interpretation, we rewrite it in the following integral form

$$X(t) = X_0 + \int_0^t a(s, X(s-))\, ds + \int_0^t c(s)\, dM_\alpha(s). \qquad (6.3.1)$$

This stochastic differential equation involves only the stochastic integral of a deterministic function with respect to the α–stable Lévy measure, discussed in Section 4.3, and it is easier to handle it then the more general equation of the form

$$X(t) = X_0 + \int_0^t a(s, X(s-))\, ds + \int_0^t c(s, X(s-))\, dM_\alpha(s) \qquad (6.3.2)$$

involving much more general theory of stochastic integration (see Section 4.9. In turn, it also can be easily generalized by replacing the random α–stable Lévy measure with an ID stochastic measure.

But from the formal point of view of someone, who looks for the construction of appropriate computer approximate methods, these equations are very much the same. A method of approximating equations (6.3.1) and (6.3.2) discussed here consists of a slight modification of the method presented in the previous section.

Numerical methods of approximate solution of the stochastic differential equations involving an Itô integral with respect to Brownian motion, have existed for some time (see, e.g., Yamada (1976), Pardoux and Talay (1985) or Kloeden and Platen (1992)). Up to now these methods focused on such problems as mean–square approximation, pathwise approximation or approximation of expectations of the solution, etc.

Our aim has been to adapt some of these constructive computer techniques, based on discretization of the time parameter t, to the case of equation (6.3.1) or (6.3.2). So, looking for an approximation of the process $\{X(t) : t \in [0, T]\}$ solving such equations we have to approximate them by time discretized *explicit scheme* of the form

$$X_{t_i}^\tau = \mathcal{F}\left(X_{t_{i-1}}, \Delta M_{\alpha,i}^\tau\right), \qquad (6.3.3)$$

where the set $\{t_i = i\tau, \quad i = 0, 1, ..., I\}$, $\tau = T/I$, describes a fixed mesh on the interval $[0, T]$, $\Delta M_{\alpha,i}^\tau$ denotes the stochastic stable measure of the interval

$[t_{i-1}, t_i)$, i.e. an α–stable random variable defined by

$$\Delta M_{\alpha,i}^\tau = M_\alpha([t_{i-1}, t_i)) \sim S_\alpha(\tau^{1/\alpha}, \beta, 0), \qquad (6.3.4)$$

and where $\mathcal{F}$ stands for an appropriate operator defining the method.

Our idea consists in representing the discrete–time process $\{X_{t_i}\}$ solving this discrete system and approximating the solution of equation (6.3.1), by an appropriate sequence of random samples $\{X_{t_i}(n)\}_{n=1}^N$ calculated with the use of a computer generator of stable random variables. In this way we can obtain kernel estimators of densities of the discrete–time diffusions solving equation (6.3.3).

Now we describe briefly a constructive computer method providing approximate solutions to stochastic differential equations involving integrals with respect to the α–stable Lévy measure. An algorithm based on the Euler method consists of the following.

- With a fixed regular mesh on $[0, T]$ approximate the process $\{X(t) : t \in [0, T]\}$ which solves equation (6.3.1) by a discrete time process $\{X_{t_i}^\tau\}_{i=0}^I$ defined by

  1. $X_0^\tau = X_0$;
  2. for $i = 1, 2, ..., I$
  $$X_{t_i}^\tau = X_{t_{i-1}}^\tau + Y_{t_i}^\tau, \qquad (6.3.5)$$
  $$Y_{t_i}^\tau = a(t_{i-1}, X(t_{i-1}))\,\tau + c(t_{i-1})\Delta M_{\alpha,i}^\tau, \qquad (6.3.6)$$
  where $\Delta M_{\alpha,i}^\tau$ is defined by (6.3.4);
  3. the sequence $\{\Delta M_{\alpha,i}^\tau : i = 1, ..., I\}$ being a given i.i.d. sequence.

- In order to obtain an appropriate sequence of random samples $\{X_{t_i}(n)\}_{n=1}^N$ it is enough to replace random variables X_0^τ, $\Delta M_{\alpha,i}^\tau$, X_i^τ and Y_i^τ above by random samples $\{X_0^\tau(n)\}_{n=1}^N$, $\{\Delta M_{\alpha,i}^\tau(n)\}_{n=1}^N$, $\{X_i^\tau(n)\}_{n=1}^N$ and $\{Y_i^\tau(n)\}_{n=1}^N$, respectively, for $i = 1, 2, ..., I$.

- Random samples are simulated with the use of the direct method described by (3.5.1) or (3.5.2).

An approximate solution to the equation (6.3.2) can be obtained replacing (6.3.6) by

$$Y_{t_i}^\tau = a(t_{i-1}, X(t_{i-1}))\,\tau + c(t_{i-1}, X(t_{i-1}))\Delta M_{\alpha,i}^\tau. \qquad (6.3.7)$$

As in the Gaussian case a very important role is played by a linear stochastic equations

$$X(t) = X_0 + \int_0^t (a(s) + b(s)X(s))\,ds + \int_0^t c(s)\,dM_\alpha(s). \qquad (6.3.8)$$

In Section 8.4 we present some rather elementary results on convergence of approximate discrete numerical and statistical methods related to (6.3.8) and contained in Janicki, Podgórski and Weron (1992).

In order to obtain some information on convergence of numerical solutions defined by (6.3.5) and (6.3.7) and approximating equation (6.3.2) one can make use of some results concerning the question of *mathematical stability* of stochastic integrals and differential equations, and obtained by Jakubowski, Mémin and Pages (1989), Słomiński (1989), Kasahara and Yamada (1991), and Kurtz and Protter (1992).

6.4   Examples of Application in Mathematics

We find it interesting to begin the series of examples with the demonstration of a process that plays an important role in probability theory and statistics.

α–**Lévy Bridge.**   This process can be defined as a solution of the following linear stochastic equation

$$B_\alpha(t) = \int_0^t \frac{B_\alpha(s)}{s-1}\, ds + \int_0^t dL_\alpha(s). \qquad (6.4.1)$$

We believe that the series of figures presented here demonstrates in a surprisingly interesting manner how the behavior of the process $\{B_\alpha(t) : t \in [0,1]\}$ depends on α.

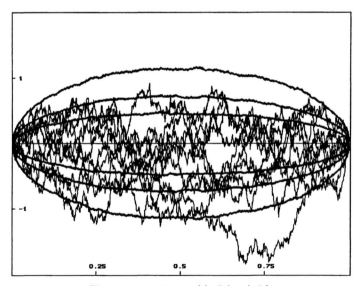

Figure 6.4.1. 2.0–stable Lévy bridge.

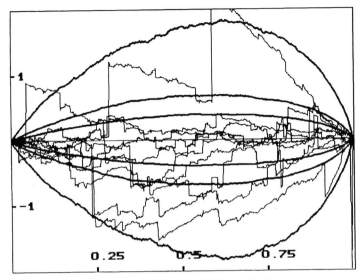

Figure 6.4.2. 1.2–stable Lévy bridge.

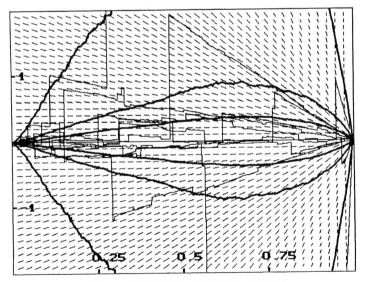

Figure 6.4.3. 0.8–stable Lévy bridge.

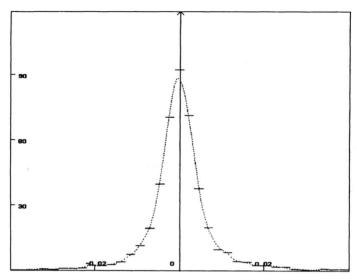

Figure 6.4.4. Density estimators for $B_{1.2}(1)$.

The role played by the drift coefficient is demonstrated in the Figure 6.4.3 (corresponding to the case of $\alpha = 0.8$), but it is similar for all other values of this parameter.

Figures 6.4.1–6.4.4 show 10 approximate trajectories of the process $B_\alpha(t)$ for three different values of α: $\alpha = 2$, $\alpha = 1.1$ and $\alpha = 0.7$. In all cases trajectories are included in the same rectangle $(t, B_\alpha(t)) \in [0,1] \times [-2,2]$. Trajectories are – as always – represented by thin lines. Vertical lines, in our convention, illustrate the effect of jumps of the process $B_\alpha(t)$.

Three pairs of thick lines represent our dynamical histogram, i.e. quantile lines, which at any fixed moment of time $t = t_i$ show the lengths of intervals including trajectories with probabilities 0.5, 0.7 and 0.9, respectively. Quantile lines were produced from statistical samples of $N = 2000$ realizations of $B_\alpha(t_i)$. In each case the time step τ was equal to 0.001.

Figure 6.4.3 contains a vector field corresponding to a deterministic part of equation (6.4.1), i.e.

$$\frac{dx}{dt}(t) = x(t)/(t-1).$$

We believe that this helps to figure out how the drift acts "against" dispersion in the process of forming a Lévy bridge, when t goes from 0 to 1.

In order to illustrate exactness of our computer simulations we present in Figure 6.4.4 the statistical estimators of $B_{1.2}(1)$. Observe that the error of this simulation can be derived as a difference between Dirac's δ function and its approximation given by the numerically constructed density estimators.

Examples of absorbing and overshooting processes.

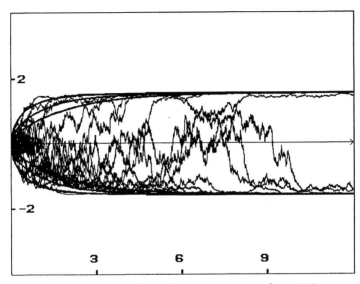

Figure 6.4.5. Absorbing process; case of $\alpha = 2.0$.

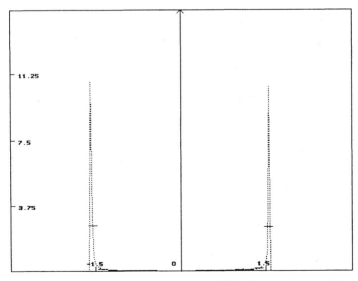

Figure 6.4.6. Density estimators of $X(12)$; case of $\alpha = 2.0$.

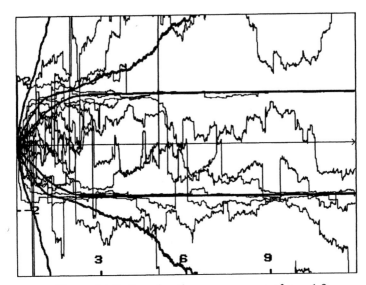

Figure 6.4.7. Overshooting process; case of $\alpha = 1.2$.

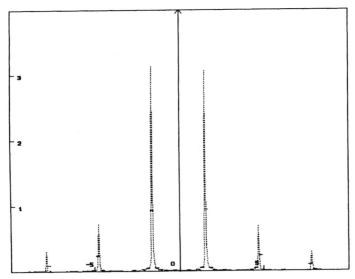

Figure 6.4.8. Density estimators of $X(12)$; case of $\alpha = 1.2$.

We find it interesting or even surprising to present the above figures. They present visualizations of two diffusion processes defined as follows

$$X(t) = \int_0^t \cos(X(s))\, dB(s), \quad t > 0; \tag{6.4.2}$$

$$X(t) = \int_0^t \cos(X(s))\, dL_{1.2}(s), \quad t > 0. \tag{6.4.3}$$

Supnorm densities. There is a quickly developing literature on this subject. It seems to us interesting to demonstrate how graphs of supnorm densities of typical α–stable processes responding, for instance, to changes of values of the index of stability α.

The problem is to construct the density of the random variable

$$Z_\alpha(\omega) = \sup_{t \in [0,T]} |L_\alpha(t,\omega)|, \tag{6.4.4}$$

with fixed index of stability $\alpha \in (0,2]$ and $T > 0$.

In Figures 6.4.9–6.4.10 we present a few estimators of such densities (unfortunately, they are a little bit distorted by some computational errors).

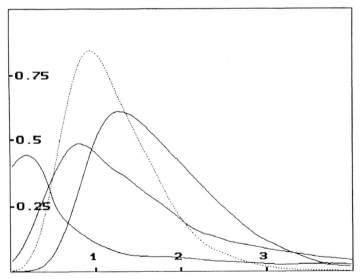

Figure 6.4.9. Supnorm density estimators for the processes $\{B(t)\}$ and $\{L_\alpha(t)\}$ with $\alpha \in \{0.5, 1.2, 2.0\}$.

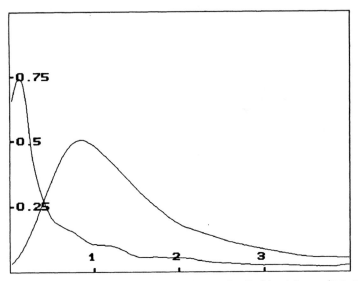

Figure 6.4.10. Supnorm density estimators for $L_\alpha(t)$ with $\alpha \in \{0.3, 1.2\}$.

Quadratic variation. We find it interesting to present visualizations of quadratic variation processes in the Gaussian and non–Gaussian cases. It is defined by the following formula (see e.g. Protter (1990))

$$[X, X]_t \stackrel{\mathrm{df}}{=} X^2(t) - 2 \int_0^t X(s-) \, dX(s).$$

Remark 6.4.1 *Notice that this definition can be applied in the case of an α–stable Lévy motion.*

For $\alpha = 2$ we have

$$[B, B]_t = B^2(t) - 2 \int_0^t B(s) \, dB(s) = B^2(t) - 2 \left\{ \frac{1}{2} B^2(t) - \frac{1}{2} t \right\} = t$$

and thus this example can serve as a test for correctness of computer simulations.

Remark 6.4.2 *The stochastic exponent is a process which solves the stochastic equation*

$$Z(t) = 1 + \int_0^t Z(s-) \, dX(s).$$

For an explicit formula for the solution $\{Z(t)\}$, given in terms of the quadratic variation process, see Protter (1990).

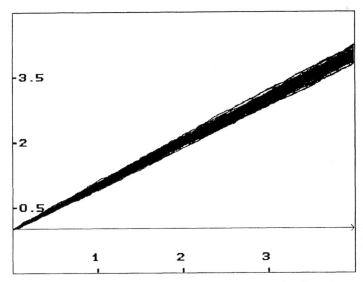

Figure 6.4.11. Visualization of the quadratic variation for Gaussian case.

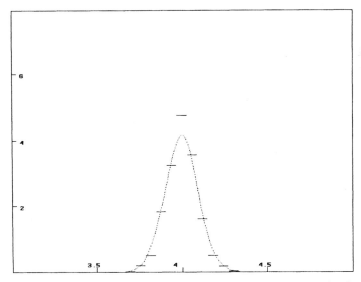

Figure 6.4.12. Density estimators of the quadratic variation at $t = 4$ for Gaussian case.

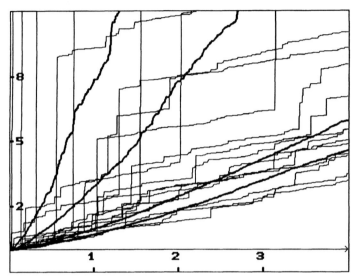

Figure 6.4.13. Visualization of the quadratic variation of the stable process for $\alpha = 1.5$.

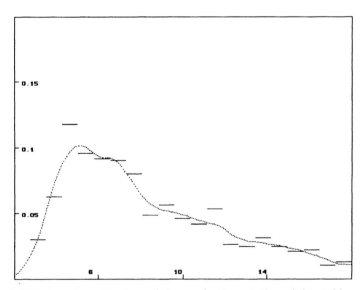

Figure 6.4.14. Density estimators of the quadratic variation of the stable process for $\alpha = 1.5$ at $t = 4$.

Stochastic exponent. We present the result of computer simulation of the so–called *stochastic exponent* process.

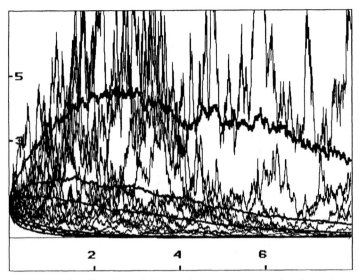

Figure 6.4.15. Visualization of the stochastic exponent for Gaussian case.

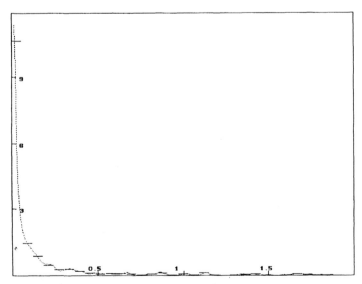

Figure 6.4.16. Density estimators of the stochastic exponent àt $t = 8$ for Gaussian case.

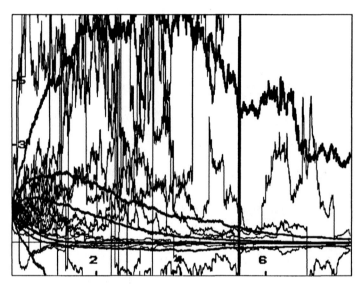

Figure 6.4.17. Visualization of the stochastic exponent for $\alpha = 1.5$.

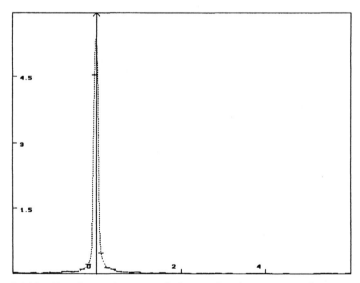

Figure 6.4.18. Density estimators of the stochastic exponent for $\alpha = 1.5$ at $t = 8$.

Totally skewed Lévy motion.   We find it interesting to present an example of α–stable subordinator (cf., Figures 6.4.13, 6.4.14).

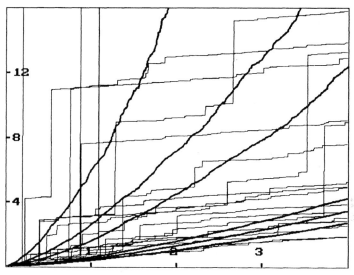

Figure 6.4.19. Visualization of the totally skewed Lévy motion for $\alpha = 0.7$.

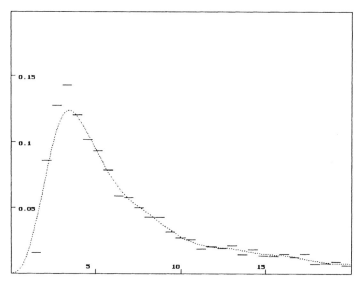

Figure 6.4.20. Density estimators of the totally skewed Lévy motion for $\alpha = 0.7$ at $t = 4$.

Chapter 7

Examples of α–Stable Stochastic Modeling

7.1 Survey of α–Stable Modeling

We believe that stable distributions and stable processes provide useful models for many phenomena observed in diverse fields. The CLT–type argument often used to justify the use of Gaussian models in applications may also be applied to support the choice of non–Gaussian stable models. That is, if the randomness observed is the result of summing many small effects, and those effects follow a heavy–tailed distribution then the non–Gaussian stable model may be appropriate. An important distinction between Gaussian and non–Gaussian stable distributions is that the latter are heavy–tailed, always with infinite variance, and in some cases with infinite first moment. Another distinction is that they admit asymmetry, or skewness, while a Gaussian distribution is necessarily symmetric about its mean. In certain applications, where an asymmetric or heavy–tailed model is called for, a stable model may be a viable candidate. In any case, the non–Gaussian stable distributions furnish tractable examples of non–Gaussian behavior and provide points of comparison with the Gaussian case, highlighting the special nature of Gaussian distributions and processes.

In order to gain some appreciation of the basic difference between a Gaussian distribution and a distribution with a long tail, Montroll and Shlesinger (1983b) proposed to compare the distribution of heights with the distribution of annual incomes for American adult males. An average individual who seeks a friend twice his height would fail. On the other hand, one who has an average income will have no trouble to discover a richer person, who, with a little diligence, may locate a third person with twice his income, etc. The income distribution in its upper range has a Pareto inverse power tail; however, most of the income distributions follow a log–normal curve, but the last few percent have a stable tail with exponent $\alpha = 1.6$, (see Badgar (1980)), i.e., the mean is finite but the variance of the corresponding 1.6–stable distribution diverges.

Failure of the least–squares method of forecasting in economic time series was first explained by Mandelbrot (1963). He introduced a radically new approach based on α–stable processes to the problem of speculative price variation.

Now it is commonly accepted that the distribution of returns on financial assets is non–Gaussian. Mandelbrot (1963) and Fama (1965) proposed the α–stable distribution for modeling stock returns. Mittnik and Rachev (1989) found that the geometric summation scheme provides a better model for describing the stability properties of stock returns computed from the Standard and Poor's ($S\&P$) 500 index. The problem of estimating multivariate α–stable distributions has received increasing attention in recent years in modeling portfolio of financial assets, see Mittnik and Rachev (1991) and references therein.

There are many physical phenomena which exhibit both space and time long tails and thus seem to violate the requirement of a Gaussian distribution as a limit in the traditional CLT, see Weron and Weron (1985). However, since these physical systems usually have nice scaling properties (self–similarity) one suspects the use of stable distributions which have long tails, infinite moments and elegant scaling properties to be relevant in the physics of these phenomena. Tunaley (1972) invoked physical arguments to suggest that if the frequency distributions in metallic films are stable then the observed noise characteristics in them may be understood. Based only on the experimental observation that near second order phase transition, where long tail spatial order develops, Jona–Lasinio (1975) considered stable distributions as a basic ingredient in understanding renormalization group notions in explaining such phenomena. See also a review article by Cassandro and Jona–Lasinio (1978). Scher and Montroll (1975) connected intermittent currents in certain xerographic films to a stable distribution of waiting times for the jumping of charges out of a distribution of deep traps. This was used to give the first explanation of experiments measuring transient electrical currents in amorphous semiconductors.

Stable distribution of first passage times appears both in the recombination reactions in amorphous materials, Montroll and Shlesinger (1983a), as well as in the dielectric relaxation phenomena described by the Williams–Watts formula: Montroll and Shlesinger (1984), Montroll and Bendler (1984), Bendler (1984) and Weron (1986). It turns out that the way stable distributions appear here is somewhat more refined and it has been a subject of extensive research in physics, see Scher, Shlesinger and Bendler (1991), Weron (1991), as well as in chemistry, see Płonka (1986, 1991) and Pittel, Woyczyński and Mann (1990).

As examples of the exploration of the stable process models in physical contexts we may cite a few very interesting papers. Doob (1942), West and Seshadri (1982) examined the response of a linear system driven by stable noise fluctuations and modeled by appropriately constructed stochastic differential equations. Takayasu (1984) demonstrated that the velocity of the fractal turbulence in $\mathbb{R}^3$ is the stable distribution with the index of stability $\alpha = D/2$, where D denotes the fractal dimension of the turbulence and that the diffusion process of particles in the turbulence and that of electrons in a uniformly magnetized plasma both can be approximated by the Lévy process.

Mandelbrot and Van Ness (1968) defined fractional Brownian motion. Hughes, Shlesinger and Montroll (1981) examined random walks with self–similar clusters leading to Lévy flights and 1/f–noise. Some connections between such clustered behavior in space or time of physical processes and fractal dimensionality of Lévy processes were studied by Seshadri and West (1982). Klafter, Blumen, Zumofen and Shlesinger (1990, 1992) described the Lévy walk scheme for diffusions in the framework of continuous time random walks with coupled memories. They concentrated on those Lévy walks, which lead to enhanced diffusions. Their approach was based on a modification of the Lévy flights. Schertzer and Love-joy (1990) made use of the self–similarity property of stable processes in order to make evident the multifractal behavior of some geophysical fields. For com-puter methods of construction of fractional Brownian motion and other processes mentioned above we refer the reader to Barnsley, Devaney, Mandelbrot, Peitgen, Saupe and Voss (1992), pp. 42 - 132.

Stable and infinitely divisible (or Lévy) processes are beginning to attract the interest of mathematicians working in the field of applied probability. Let us mention, among others, Hardin, Samorodnitsky and Taqqu (1991), Kasahara and Yamada (1991), Kella and Whitt (1991), or McGill (1989).

In this context let us remark that, as far as stable distributions are concerned, in commonly known probability textbooks only a reference to Holtsmark's work from 1915 on the gravitational field of stars (3/2–stable distribution) is made. For example, Feller devotes considerable space to stable distributions in volume II of his probability textbook, but he admits that their role in applied sciences seems to be almost nonexistent. The above–mentioned and related findings should be viewed as a step forward toward fulfilling the prophecy of Gnedenko and Kolmogorov (1954): "It is probable that the scope of applied problems in which stable distributions will play an essential role will become in due course rather wide".

7.2   Chaos, Lévy Flight, and Lévy Walk

It is now widely appreciated that complex, seemingly random behaviors can be governed by deterministic nonlinear equations. Such complex deterministic be-havior has been termed chaotic, see Devaney (1989). The stochastic properties of dynamical systems exhibiting deterministic chaos have attracted considerable interest over the last few years. The rise of nonlinear dynamics has opened up a host of new challenges for stochastic processes, see Berliner (1992), Chatterjee and Yilnez (1992), and Nicolis, Piasecki and McKernan (1992). These challenges focus on characterizing, predicting, and controlling the spatial–temporal evolu-tion of complex nonlinear physical processes. In most studies emphasis is placed on ergodic properties, in particular, on the existence and main features of an in-variant probability density. An ultimate goal is to obtain a kinetic description of chaotic dynamics which will provide the starting point for different applications. One expects a possibility to map, in an exact manner, deterministic chaos onto a stochastic process governed by a "master equation" describing the evolution of an initial nonequilibrium distribution toward the invariant equilibrium form. For example, in a paper by Nicolis et al. (1992) a systematic method for casting chaos into a stochastic processes via the Chapman–Kolmogorov equation is developed.

Real orbits in dynamical systems are always theoretically predictable since they represent solutions of a rather simple system of equations, namely Newton's equations. However, under conditions that guarantee dynamical chaos, these orbits are highly unstable. Generally, for chaotic motion, the distance between two initially close orbits grows exponentially with time as

$$d(t) = d(0) \exp(\sigma t)$$

and the rate σ is called the *Lyapunov exponent*. This dependence holds for sufficiently long times. Local instabilities, described by the above equation, lead to a rapid mixing of orbits within the time interval $\tau_\sigma = \frac{1}{\sigma}$. Nevertheless, some properties of the system remain unchanged and their evolution occurs at a significantly longer time $\tau_D \gg \tau_\sigma$ as a result of averaging over the fast process of mixing, caused by instability of the above equation, see Shlesinger, Zaslavsky and Klafter (1993). Kinetic equations arise as a consequence of such averaging. The Gaussian and Poissonian distributions can, under certain conditions, give an approximate description of the apparent randomness of chaotic orbits.

It has been realized, however, that behaviors much more complex than standard diffusion can occur in dynamical Hamiltonian chaos. The α–stable distributions can appear both in space and time, and the fractal processes they describe have been found to lie at the heart of important complex processes such as turbulent diffusion, chaotic phase diffusion in Josephson junctions, and slow relaxation in glassy materials, see Takayasu (1990), Klafter et al. (1990), Scher, Shlesinger and Bendler (1991), or Weron and Jurlewicz (1993).

The α–stable distributions can be generated by stochastic processes which are scale invariant. This means that a trajectory will possess many scales, but no scale will be charecteristic and dominate the process. Geometrically, this implies the fractal property that a trajectory, viewed at different resolutions, will look self–similar. One example of a scale invariant stochastic process is a random walk with infinite mean–squared jump distances.

Random walks have been shown to be a powerful tool in investigations of the transport properties of ordered and disordered systems, see Weiss and Rubin (1983), Montroll and Shlesinger (1984), and Klafter et al. (1990). The simplest type of random walk, Brownian motion, which was introduced into physics by Einstein and which is common to a broad spectrum of systems, is characterized by linear in time mean square displacement and a Gaussian propagator.

For random walks known as *Lévy flights* each step in the process is chosen from an α–stable distribution

$$p(r) \approx \mid r \mid^{-\alpha-1}, \quad 0 < \alpha < 2.$$

There is no characteristic size in this process in contradistinction to the Gaussian distribution for $\alpha = 2$. Steps of all sizes occur and it can be shown that a self–similar set of points of fractal dimension α is visited by the walker.

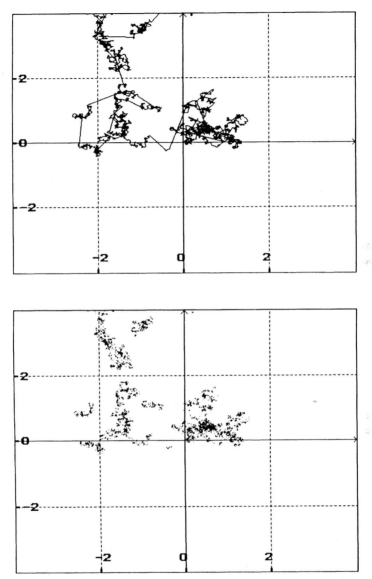

Figure 7.2.1. One path of the 2–dimensional α–stable Lévy walk and, below, the same realization of the Lévy flight for $\alpha = 1.7$.

Consider a random walk on a plane with the distribution of jump lengths following the above power law. The direction at each jump is chosen completely at random. Plot a point at each place the walker jumps to. The resulting set of points is called a Lévy flight or a Lévy dust, see Mandelbrot (1983). The lower part of Figure 7.2.1 shows an example of a Lévy flight drawn by a computer for $\alpha = 1.7$. We emphasize here the crucial point that the power–law behavior of the tail of the α–stable distribution for $\alpha < 2$ defines the non–small probability of large values of displacement r. It explains the reason for the notion of "Lévy flights". Lévy flights jump between successively visited sites, however distant, which leads to a divergence of the mean square displacement. One is interested in modifying the Lévy flight law so that the motion still follows an α–stable distribution but with a finite mean square displacement. We are therefore interested in a stochastic process which visits the same sites as in the Lévy flight, but with a time "cost" which depends on the distance.

For the flight we only need to specify $p(r)$, the probability density that a jump of distance r occurs. In the framework of the continuous time random walk (CTRW) theory the nature of the walks is entirely specified by the probability distribution $\Psi(r, t)$ of making a step of length r in the time interval t to $t + \delta t$. This probability density has the form

$$\Psi(r, t) = p(r)\Phi(t|r),$$

where $p(r)$ is defined as for the flight and $\Phi(t|r)$ is the conditional probability density that the transition takes time t, given that it was of displacement r. Following Klafter et al. (1992), we shall choose

$$\Phi(t|r) = A \mid r \mid^{-\alpha-1} \delta(\mid r \mid -t^{\nu}),$$

where r and t are coupled through the δ–function. These processes are called Lévy walks and the above equation allows steps of arbitrary lenghts as for Lévy flights, but long steps are penalized by requiring longer time to be performed. Stated differently, in a given time window, only a finite shell of points may be reached. Hierarchically, nearer points occur no more and further points are not yet accessible.

To visualize a realization of such a Lévy walk we present in the upper part of Figure 7.2.1 the situation for a two–dimensional geometry where we choose $\alpha = 1.7$ and $\nu = 1$. The connected lines show the trajectory of the 1.7–stable Lévy walk. The isolated dots presented in the lower part show the points visited by the corresponding 1.7–stable Lévy flight which represents the turning points of the Lévy walk. One should notice the self–similar aspect of the picture: a series of small steps is followed by larger steps which are, after a while, followed by still larger ones. Furthermore, no particular length scale dominates. Thus Lévy flights and Lévy walks can be applied to dynamical systems whose orbits possess fractal properties. Also, it is clear that strange kinetics for chaotic Hamiltonian systems falls outside the domain of Brownian motion processes, where the cluster structure does not exist as a consequence of the continuity of trajectories.

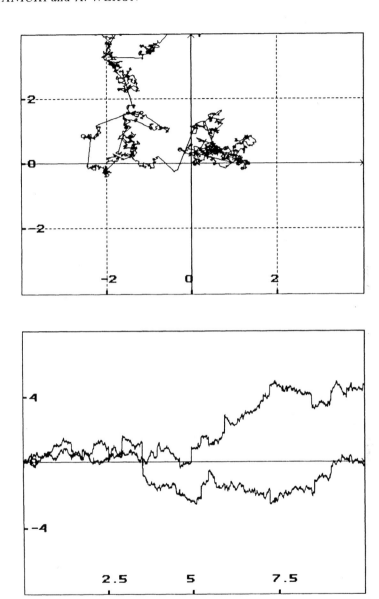

Figure 7.2.2. One path of the 2–dimensional 1.7–stable Lévy walk and, below, its both 1–dimensional projections on the axes, presented as functions of time.

Figure 7.2.2 gives a general impression about a relationship between 2-dimensional Lévy walks and their 1-dimensional projections on the axes. It turns out that these projections can be identified as α-stable processes, which are the main subject of this book. Note that the upper trajectory represents the projection on the vertical axis and the lower trajectory on the horizontal axis, respectively. Figure 7.2.3 contains the visualization of the horizontal projection of the 2-dimensional 1.7-stable Lévy walk. According to the method described in the previous chapter, the process is represented by ten trajectories and three pairs of quantile lines.

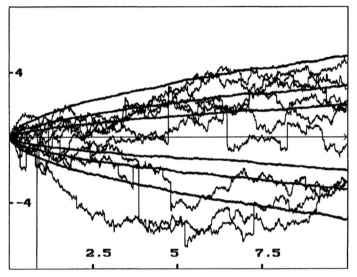

Figure 7.2.3. The visualization of the stochastic process obtained as the horizontal projection of the Lévy walk from Figure 7.2.1 and Figure 7.2.2.

In a recent paper by Shlesinger, Zaslavsky and Klafter (1993) several examples of dynamical systems with "symptoms" of α-stable processes are considered. These include turbulent diffusion, the Arnold–Beltrami–Childress flow, and phase rotation in the Josephson's junctions.

7.3   Examples of Diffusions in Physics

In this section we provide a few examples of diffusion processes constructed and visualized with the use of our method described in previous chapters.

Resistive–inductive electrical circuit.   Here we present an example of a linear stable stochastic equation involving this kind of stochastic measure (see West and Seshadri (1982)), that has a nice physical interpretation, emphasizing the role of the parameter α.

The deterministic part of the stochastic differential equation

$$dX(t) = (4\sin(t) - X(t))\, dt + \frac{1}{2}dL_\alpha(t) \qquad (7.3.1)$$

can be interpreted as a particular case of the ordinary differential equation

$$\frac{di}{ds} + \frac{R}{L}i = \frac{E}{L}\sin(\gamma s),$$

which describes the resistive–inductive electrical circuit, where i, R, L, E and γ denote, respectively, electric force, resistance, induction, electric power and pulsation. (Similar examples can be found in Gardner (1986).) In order to obtain a realistic model it is enough to choose, for example, $R = 2.5[k\Omega]$, $L = 0.005[H]$, $E = 10[V]$, $\gamma = 500[1/s]$ and to rescale real time s using the relation $t = \gamma s$.

The results of computer simulation and visualization of the equation (7.3.1) with the initial random variable $X(0)$ chosen as an α–stable variable from $S_\alpha(2,0,1)$ for $t \in [0,4]$ and three different values of the parameter α ($\alpha \in \{2.0, 1.3, 0.7\}$) are included in the following two series of figures. The first series of figures shows the behavior of trajectories in the same way and with the same values of technical parameters as before in the case of α–stable Lévy motion. They contain also a field of directions corresponding to the deterministic part of (7.3.1), i.e., the equation

$$\frac{dx}{dt}(t) = -x(t) + 4\sin(t).$$

This helps us to figure out how the drift acts "against" the diffusion, when t tends to infinity.

The second series shows density estimators of $X(4.0)$ for these three values of α.

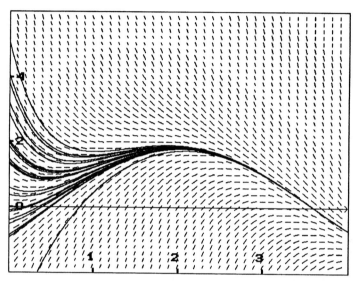

Figure 7.3.1. Deterministic electric circuit equation; a few solutions with different values at a starting point.

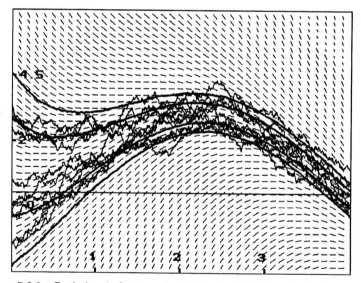

Figure 7.3.2. Resistive–inductive electrical circuit driven by Lévy motion for $\alpha = 2.0$.

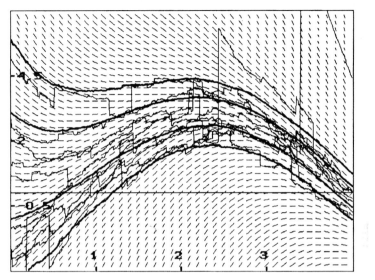

Figure 7.3.3. Resistive–inductive electrical circuit driven by Lévy motion for $\alpha = 1.3$.

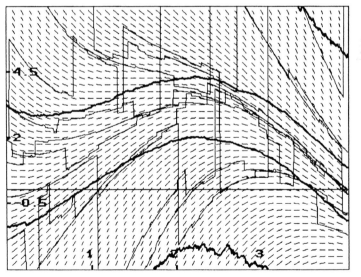

Figure 7.3.4. Resistive–inductive electrical circuit driven by Lévy motion for $\alpha = 0.7$.

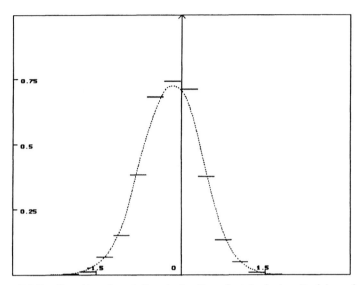

Figure 7.3.5. Density of resistive–inductive electrical circuit driven by Lévy motion for $\alpha = 2.0$ at time $t = 4.0$.

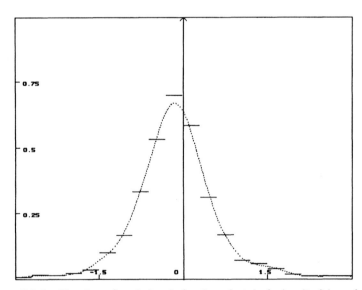

Figure 7.3.6. Density of resistive–inductive electrical circuit driven by Lévy motion for $\alpha = 1.3$ at time $t = 4.0$.

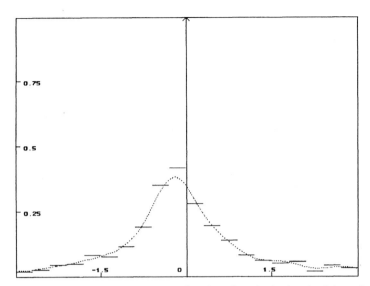

Figure 7.3.7. Density of resistive–inductive electrical circuit driven by Lévy motion for $\alpha = 0.7$ at time $t = 4.0$.

Ornstein–Uhlenbeck process versus Brownian motion.

For a long time (see Doob (1942)) there has been two commonly used models describing the movement of a Brownian particle. In the first model, the position is described by a Brownian motion process, in the second – by a stationary Ornstein–Uhlenbeck process. Each of them can be described by a very simple system of 2 stochastic differential equations.

Example 7.3.1 *(Brownian motion process.)*

A Brownian motion process and its Lebesgue integral can be regarded as a solution to the following problem

$$dX(t) = Y(t)\,dt, \qquad dY(t) = dB(t), \tag{7.3.2}$$

with $X(0) = 0$ a.s., $Y(0) \sim \mathcal{N}(0,1)$.

Example 7.3.2 *(Ornstein–Uhlenbeck process.)*

An Ornstein–Uhlenbeck process and its Lebesgue integral can be regarded as a solution to the following problem

$$dX(t) = Y(t)\,dt, \qquad dY(t) = -2\,Y(t)\,dt + 2\,dB(t), \tag{7.3.3}$$

with $X(0) = 0$ a.s., $Y(0) \sim \mathcal{N}(0,1)$.

The solutions obtained from computer simulations are presented in Figures 7.3.8 – 7.3.11. It seems a little bit surprising that for times t small enough both models give very similar results.

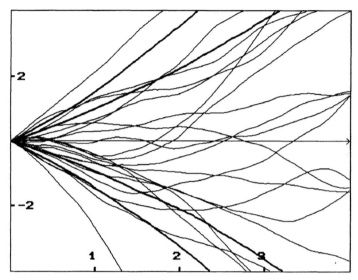

Figure 7.3.8. Lebesgue integral of the Brownian motion.

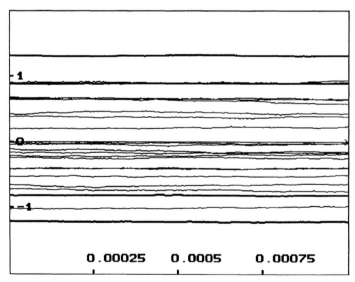

Figure 7.3.9. Brownian motion starting from $B(0) \sim \mathcal{N}(0,1)$.

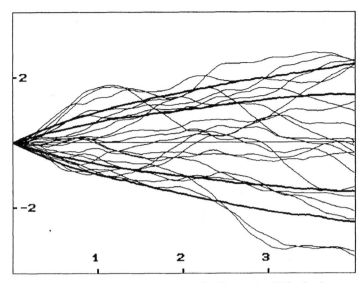

Figure 7.3.10. Lebesgue integral of the Ornstein–Uhlenbeck process.

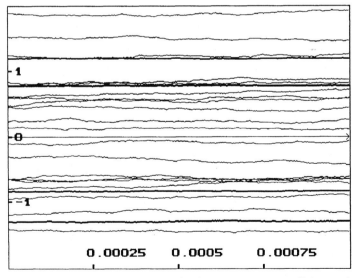

Figure 7.3.11. Ornstein–Uhlenbeck process on a small interval.

There are a lot of well–known models described by means of stochastic differential equations driven by Brownian motion (see, e.g., Gardiner (1983) or Gardner (1986)). They still play very important role in applications.

Nonlinear harmonic oscillator.
Here we propse two probabilistic models of the mathematical pendulum. First we propose an example where randomness appears in the initial condition.

Example 7.3.3 *Nonlinear Brownian oscilator; deterministic equation.*

$$d\,X(t) = Y(t)\,dt, \quad d\,Y(t) = \{-\sin(X(t)) - \frac{1}{2}Y(t)\}\,dt, \qquad (7.3.4)$$

where $X(0) = 0$ a.s. and $Y(0) \sim S_{1.7}(1,0,1)$.

Example 7.3.4 *Nonlinear α-stable oscilator.*

As an example of a nonlinear stochastic physical model submitted to random external forces described by α-stable "colored noise" we consider a system of two equations describing a harmonic oscillator. We look for a solution $\{(X(t), Y(t)); \; t \in [0, 16]\}$ of the following system of stochastic differential equations

$$d\,X(t) = Y(t)\,dt, \quad d\,Y(t) = \{-\sin(X(t)) - \frac{1}{2}Y(t)\}\,dt + d\,L_\alpha(t), \qquad (7.3.5)$$

where $X(0) = 0$ a.s. and $Y(0) = 1$ a.s.

Notice that, of course, $X(t)$ and $Y(t)$ are not independent and the joint distributions of $(X(t), Y(t))$ are not α-stable.

The dependence of solutions of (7.3.5) on the parameter α is similar to that discussed in connection with the resistive–inductive electrical circuit equation, see Figures 7.3.2 – 7.3.7. Therefore we restrict ourselves to the presentation of the Gaussian case only.

This example demonstrates that a wide class of non–linear multidimensional problems can be successfully solved with the use of computer simulation and visualization techniques when analytical calculations are inaccessible.

The solution of equation (7.3.5) with ($\alpha = 2.0$) is represented here by eight consecutive figures (Fig. 7.3.14 – 7.3.21). The last four of them describe the evolution of a bivariate distribution function of vectors $(X(t), Y(t))$ for four fixed values of time t, i.e. for $t \in \{0.2, 2.0, 4.0, 16.0\}$. All of these figures were obtained with the use of the same technical parameters (e.g., defining kernel function) and their graphs are included in the same part of $\mathbb{R}^3$; their domains are cut to the rectangle $[-8, 8] \times [-8, 8]$. It is impossible to construct the density of $(X(0), Y(0))$, which is of the form of a product of two Dirac's delta functions: $\delta(0) \times \delta(1)$ or even for values of t close to 0, so the first value for which we decided to present the density estimator in a chosen frame in $\mathbb{R}^3$ was $t = 0.2$. To give an idea of the scaling of the vertical axis, let us mention that the maximum of a density for $t = 16$ is about 0.090.

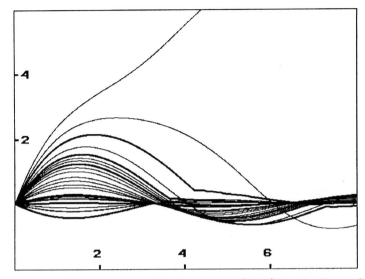

Figure 7.3.12. Representation of displacements $\{X(t)\}$ for the Example 7.2.3.

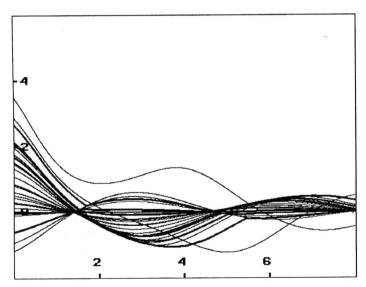

Figure 7.3.13. Representation of velocities $\{Y(t)\}$ for the Example 7.2.3.

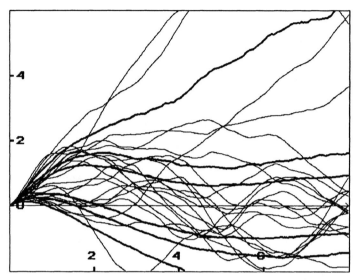

Figure 7.3.14. Representation of displacements $\{X(t)\}$ for nonlinear stochastic oscillator.

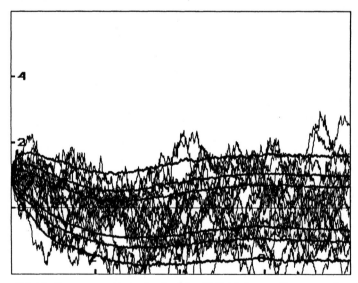

Figure 7.3.15. Representation of velocities $\{Y(t)\}$ for nonlinear stochastic oscillator.

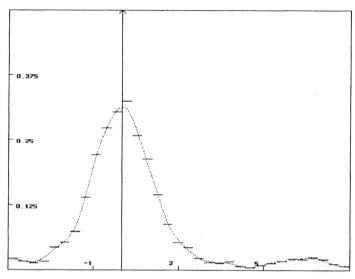

Figure 7.3.16. Density estimators of X(8).

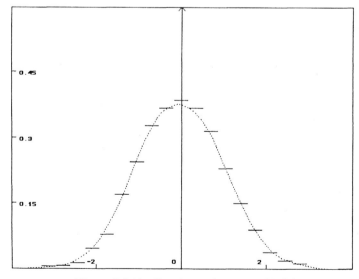

Figure 7.3.17. Density estimators of Y(8).

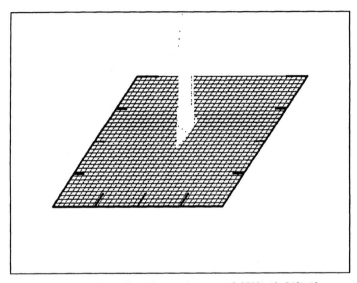

Figure 7.3.18. Density estimator of $(X(0.2), Y(0.2))$.

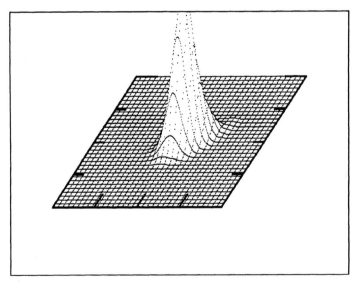

Figure 7.3.19. Density estimator of $(X(2), Y(2))$.

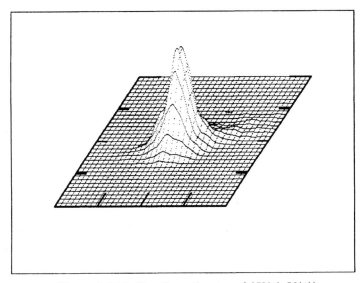

Figure 7.3.20. Density estimator of $(X(4), Y(4))$.

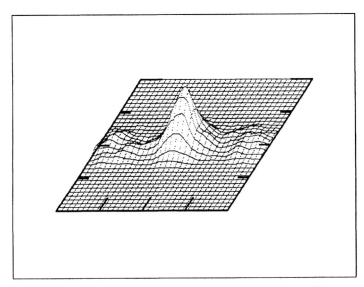

Figure 7.3.21. Density estimator of $(X(16), Y(16))$.

7.4   Logistic Model of Population Growth

We present here some computer visualizations of different possible models of
a population growth. The logistic model, developed by Verhulst in 1844 to
describe the growth of a population subject to a fixed food supply, has the basic
form

$$\frac{dX}{dt} = X\,(K - X),\tag{7.4.1}$$

where K essentially represents the limit, in population units, of the food supply.

Assuming that the limit K is subject to the random disturbances modeled by
the Gaussian white noise of the form $a_K\xi(t)$, we obtain the following stochastic
differential equation

$$dX(t) = X\,(K - X)\,dt + a_K X\,dB(t).\tag{7.4.2}$$

We propose the following generalization of this equation

$$dX(t) = X\,(K - X)\,dt + a_K X\,dM_\alpha(t),\tag{7.4.3}$$

where $\mathbf{M}_\alpha$ denotes α–stable Lévy measure.

In Figures 7.4.2, 7.4.3 we plotted visualizations of solutions to (7.4.3) for two
measures $\mathbf{M}_\alpha$: $S1.3S$ measure and 0.7-stable totally skewed measure. Figures
7.4.6, 7.4.7 contain estimators of densities of obtained solutions at time $t = 1$.
One can compare these solutions with solutions to (7.4.1), (7.4.2). In all cases
$K = 4$, $a_K = 0.5$, and the initial condition is $X(0) \equiv 2$.

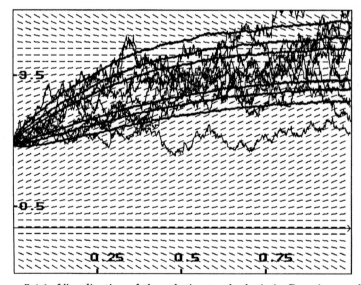

Figure 7.4.1. Visualization of the solution to the logistic Gaussian model.

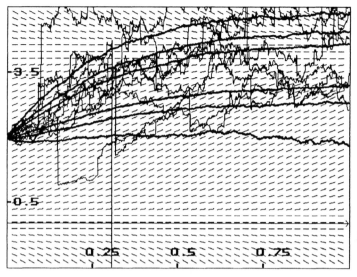

Figure 7.4.2. Visualization of the solution to the logistic model with the $S1.3S$ Lévy measure.

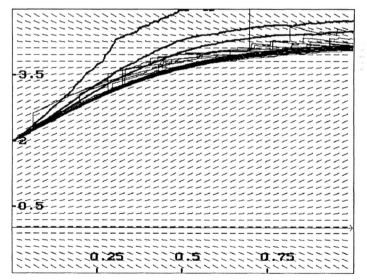

Figure 7.4.3. Visualization of the solution to the logistic model with the totally skewed 0.7–stable Lévy measure.

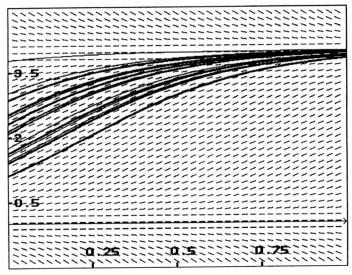

Figure 7.4.4. Visualization of the solution to the deterministic equation of a population growth.

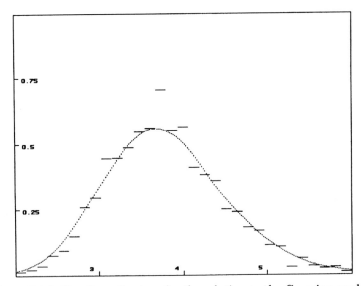

Figure 7.4.5. Density estimators for the solution to the Gaussian model.

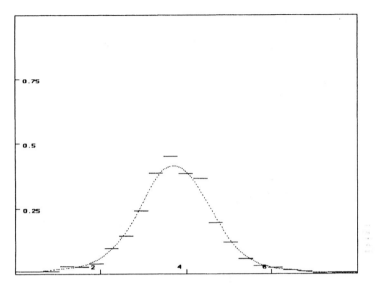

Figure 7.4.6. Density estimators for the solution to the logistic model with the $S1.3S$ Lévy random measure.

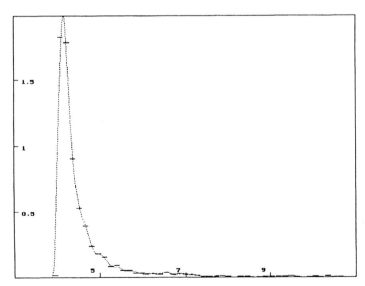

Figure 7.4.7. Density estimators for the solution to the logistic model with the totally skewed 0.7–stable Lévy measure.

7.5   Option Pricing Model in Financial Economics

A special application of Gaussian and stable processes in financial economics is *option pricing*.

Any financial contract has option features or can be decomposed in options. Determining the economic value of the contract obligations is in many cases a matter of valuing the underlying option. A great many options models have been developed relative to very different assets. An important problem is to evaluate the options on shares, bonds, foreign currency, futures, commodities, etc. Since the first paper of Black and Scholes (1973) much progress has been made in applying option techniques and results in valuation of real assets and contracts, for example in international trade and investments, bank loan commitment, deposit insurance, portfolio insurance, agency problems. Thus, precise knowledge about evaluating the options seems to be of importance.

Let us start with the basic results under the Gaussian hypothesis of the market behavior, which is based on a pioneer article of Black–Scholes (1973). It completely revolutionized the approach to the subject. Next, we briefly present attempts to price the option, when the returns on asset are assumed to follow a stable law.

Definitions.   An *option* is the right to buy or sell shares or property within a stated period (or at the end of stated period) at a predetermined price. The right to buy is known as a *call option*, the right to sell as a *put option*. Individuals may write (sell) as well as purchase options, and are obligated to deliver or to buy the stock at the specified price. One can distinguish four components of the option

1. *Expire date* T – the date on which all unexercised options in the particular series expire,

2. *Exercise (striking) price* K – the price at which the buyer can purchase or sell the shares during the currency of the option,

3. *Number of shares N,*

4. *Premium* – final element of the option and the only variable; it is the price paid to acquire the right to buy or sell, but not the obligation to exercise.

Black–Scholes model.   Let the banksavings $X(t)$ and the stock price $Y(t)$ satisfy the stochastic differential equation

$$dX(t) = rX(t)\, dt, \tag{7.5.1}$$
$$dY(t) = \rho Y(t)\, dt + \sigma Y(t)\, dW(t), \tag{7.5.2}$$

respectively, where $W(\cdot)$ is the standard Wiener process on a probability space $(\Omega, \mathcal{F}, P)$, r denotes riskless interest rate and $\rho > r$ is the expected rate of return on risky asset Y. The above model is commonly used in the description of the stock market. One assumes that relative increments of the price are independent (the market is perfect), stationary, and follow the Gaussian law. In such a model the value of an option at the expire time T is

$$C_T = \max\{Y(T) - K, 0\}.$$

Since money today is more worthy than money tomorrow, to know the present value of an option, we have to discount a future payment and project the future value of an option on the information which we have today, i.e.,

$$C_t = e^{-\rho'(T-t)} E[C_T \mid F_t], \qquad (7.5.3)$$

where $\rho' > \rho$ reflects the higher variability in C_t than in $Y(t)$, σ–field $\mathcal{F}_t$ includes the whole information about the behavior of the process $\mathbf{Y}$ up to time t. Black and Scholes put $\rho = \rho' = r$ and discover that *having a call is equivalent to the portfolio of shares and banksavings* in proper proportions. With continuously changing amounts of shares and written calls, where the payments for calls are used to buy shares and vice versa, it will turn out to be possible to guard against all possible losses due to decreasing stock prices. For constant proportions in stocks and calls, the market value of such portfolio V_t is

$$V_t = aY(t) - bC_t,$$

and then

$$dV_t = adY(t) - bdC_t.$$

On the other hand, the present value of the option C_t is a function of $Y(t)$ and t, so using the Itô Formula we get

$$dV_t = \left(a - b\frac{\partial C}{\partial y}\right) dy - b\left(\frac{1}{2}\frac{\partial^2 C}{\partial y^2}\sigma^2 y^2 + \frac{\partial C}{\partial t}\right) dt$$

and

$$\frac{\partial C}{\partial t} = rC - ry\frac{\partial C}{\partial y} - \frac{1}{2}\frac{\partial^2 C}{\partial y^2}\sigma^2 y^2,$$

with boundary condition $C(y, T) = \max\{y - K, 0\}$. Black and Scholes transformed the upper equation into the heat equation, for which an explicit solution is well known, and finally got

$$
\begin{aligned}
C_t &= Y(t)\Phi\left(\frac{1}{\sigma\sqrt{T-t}}\ln\frac{Y(t)}{K\exp(-r(T-t))} + \frac{1}{2}\sigma\sqrt{T-t}\right) \qquad (7.5.4) \\
&\quad - Ke^{-r(T-t)}\Phi\left(\frac{1}{\sigma\sqrt{T-t}}\ln\frac{Y(t)}{K\exp(-r(T-t))} - \frac{1}{2}\sigma\sqrt{T-t}\right).
\end{aligned}
$$

α–stable model.   The first idea is to mimic the previous method, presented before. However, one can notice that the language of stochastic integrals, used in the Black and Scholes model, creates many difficulties under hypothesis of α-stable returns. Therefore, we have to follow an alternative method. We shall present the results of Rachev and Samorodnitsky (1992). Let S be the current stock price. In order to model a stock price whose logarithm is α-stable Levy motion, assume that the stock price S_1 at the end of an unit period is described by

$$S_1 = \begin{cases} u_1 S & \text{with probability } \frac{1}{2} \\ d_1 S & \text{with probability } \frac{1}{2} \end{cases}$$

In order to obtain a stable distribution as a limit, one have to establish that u and d are random, and $\ln u$ and $\ln d$ have heavy tailed distribution.

Consequently, after n periods the stock price is

$$S_n = S \prod_{k=1}^{n} u_k^{\delta_k} d_k^{(1-\delta_k)},$$

or

$$\ln(S_n/S) = \sum_{k=1}^{n} \left(U_k \delta_k + D_k (1 - \delta_k) \right),$$

where $U_k = \ln u_k$, $D_k = \ln d_k$ and δ_k are i.i.d. Bernoulli's random variables independent of u_k and d_k. Rachev and Samorodnitsky assume that

$$U_k = \sigma |X_k^{(n)}|, \quad D_k = -U_k,$$

where n represents the number of movements up to time T and $X_k^{(n)}$ are i.i.d. symmetric Pareto r.v. with

$$P[|X_k^{(n)}| > x] = \frac{1}{nx^\alpha}, \quad x \geq \frac{1}{n^{1/\alpha}}, \, 1 < \alpha < 2.$$

Thus, the changes of log price are symmetric. Finally, the process

$$\xi_n(t) = \ln(S_n/S) = \sigma \sum_{k=1}^{n} X_k^{(n)}, \quad T\frac{k-1}{n} < t \leq T\frac{k}{n},$$

converges weakly to a symmetric α-stable Lévy motion. Let r_k denote "riskless interest rate" at kth period and let it be connected with u_k and d_k in the following manner

$$r_k = \frac{1}{2}(u_k + d_k).$$

Selecting an "equivalent portfolio" one can easily show that the only rational value of the call with expiration date T which is n periods away and the striking price K equals

$$c^{(n)} = \frac{2^{-n}}{r_1 \dots r_n} \quad \{ (u_1 \dots u_n S - K)_+ + (u_1 \dots u_{n-1} d_n S - K)_+ \qquad (7.5.5)$$
$$+ \dots + (d_1 u_2 \dots u_n S - K)_+ + (d_1 \dots d_n S - K)_+ \},$$

where $(x)_+ = \max(0, x)$. Further calculation consists in averaging (7.5.5), i.e., taking $C^{(n)} = E[c^{(n)}]$ and seeking the limit of $C^{(n)}$ as n tends to infinity.

Let Z_k be i.i.d. uniforms on $(0, 1)$ and ϵ_k be Rademacher random signs independent of Z_k. Then one can write down $X_k^{(n)}$ as

$$X_k^{(n)} = \epsilon_k n^{-1/\alpha} Z_k^{-1/\alpha},$$

and show that the order statistics of the "sample" $(n^{-1/\alpha} Z_1^{-1/\alpha}, \ldots, n^{-1/\alpha} Z_n^{-1/\alpha})$, say $(X_{1,n}^{(n)}, \ldots, X_{n,n}^{(n)})$, have the same joint distribution as the vector

$$\left(\frac{\tau_{n+1}}{n}\right)^{1/\alpha} \left(\tau_1^{-1/\alpha}, \ldots, \tau_n^{-1/\alpha}\right),$$

where $\tau_1, \tau_2, \ldots$ are Poisson arrivals times with intensity 1, independent of ϵ_k. Averaging the equation (7.5.5), Rachev and Samorodnitsky obtain the *binomial option pricing formula* for heavy tailed distributed stock returns:

$$C^{(n)} = E \frac{\left[S \exp\left(\sigma(\frac{\tau_{n+1}}{n})^{1/\alpha} \sum_{i=k}^{n} \epsilon_k \tau_k^{-1/\alpha}\right) - K\right]_+}{E_{(\epsilon_1, \ldots, \epsilon_n)} \exp\left[\sigma \left(\frac{\tau_{n+1}}{n}\right)^{1/\alpha} \sum_{k=1}^{n} \epsilon_k \tau_k^{-1/\alpha}\right]} \tag{7.5.6}$$

and $E_{(\epsilon_1, \ldots, \epsilon_n)}$ denotes that the expectation is taken with respect to $\epsilon_1, \ldots, \epsilon_n$.

In the last step, letting $n \to \infty$, the authors find option pricing formula

$$C = \lim_{n \to \infty} C^{(n)} = E \frac{\left[S \exp\left(\sigma \sum_{k=1}^{\infty} \epsilon_k \tau_k^{-1/\alpha}\right) - K\right]_+}{E_{(\epsilon_1, \ldots, \epsilon_n)} \exp\left[\sigma \sum_{k=1}^{\infty} \epsilon_k \tau_k^{-1/\alpha}\right]}. \tag{7.5.7}$$

Unfortunately, the above formula (7.5.7) is difficult to analyze from the practical point of view, see Sections 3.3 and 3.4. Thus other approaches to the problem would be of interest. By the numerical method providing approximate solutions to stochastic differential equations driven by the α–stable Lévy motion we can construct the process $\{Y_\alpha(t) : t \in [0, T]\}$ solving the equation

$$dY_\alpha(t) = \rho Y_\alpha(t) \, dt + \sigma Y_\alpha(t) \, dL_\alpha(t), \tag{7.5.8}$$

with the initial condition $Y(0) = Y_o$ a.s., where Y_o is a given constant. The value of the option is a discounted mean of payments given by the following formula

$$C_T^\alpha = \max\{Y_\alpha(T) - K, 0)\}. \tag{7.5.9}$$

Making use of the statistical estimation methods we can construct approximate density of the random variable C_T^α.

The results of computer experiments with different values of the index of stability α are included in Figures 7.5.1 – 7.5.3.

The fixed values of parameters are: $\rho = 0.1$, $\sigma = 0.01$, $T = 1$, $K = 110$, $Y_o = 100$.

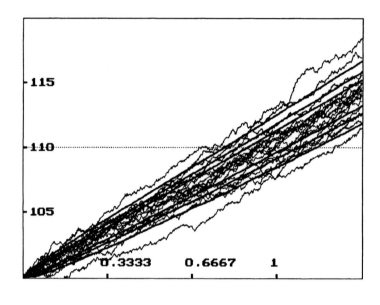

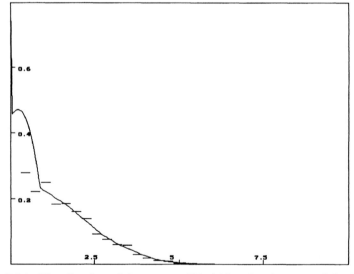

Figure 7.5.1. Visualization of the process $\{Y_{2.0}(t)\}$ and estimators of the density of the corresponding random variable $C_1^{2.0}$.

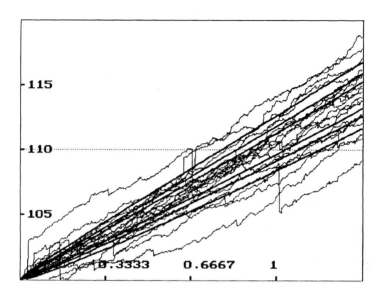

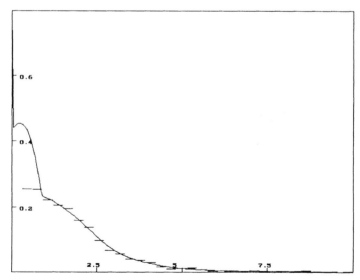

Figure 7.5.2. Visualization of the process $\{Y_{1.7}(t)\}$ and estimators of the density of the corresponding random variable $C_1^{1.7}$.

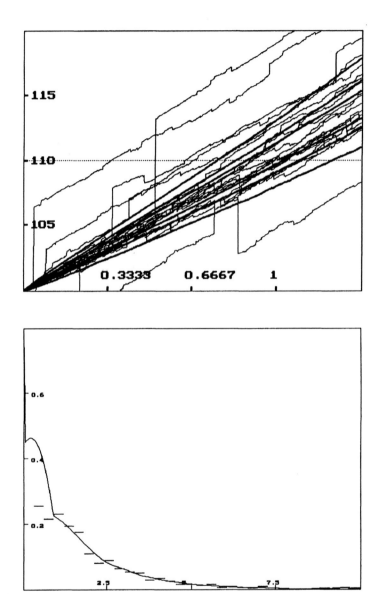

Figure 7.5.3. Visualization of the process $\{Y_{1.3}(t)\}$ and estimators of the density of the corresponding random variable $C_1^{1.3}$.

Chapter 8

Convergence of Approximate Methods

8.1 Introduction

In this section we collect some basic facts concerning probability measures on spaces of trajectories of continuous and cadlag processes. For omitted proofs we refer the reader to Parthasaraty (1969) and to Jacod and Shiryaev (1987).

Theorem 8.1.1 *The class $\mathcal{B}_C$ of the Borel subsets of $C[0,1]$ coincides with the smallest σ-algebra of subsets of $C[0,1]$ with respect to which the maps π^t : $x \to x(t)$ are measurable for all $t \in [0,1]$. If μ and ν are two measures on $C[0,1]$ then a necessary and sufficient condition that $\mu = \nu$ is that $\mu^{t_1,\ldots,t_k} = \nu^{t_1,\ldots,t_k}$ for all k and $t_1, t_2, \ldots, t_k$ from $[0,1]$, where $\mu^{t_1,\ldots,t_k}$ and $\nu^{t_1,\ldots,t_k}$ are measures in the k-dimensional vector space $\mathbb{R}^k$ induced by μ and ν, respectively, through the map $\pi^{t_1,\ldots,t_k} : x \to (x(t_1), \ldots, x(t_k))$.*

Let us recall that the Wiener measure W on $C[0,1]$ was defined in Section 2.2.

Theorem 8.1.2 *Let $a_1 < 0 < a_2$ and $[c,d] \subset [a_1, a_2]$. Then*

$$W\{x : \min_{0 \le t \le T} x(t) > a_1, \max_{0 \le t \le T} x(t) < a_2, x(T) \in [c,d]\}$$

$$= \frac{1}{\sqrt{2\pi T}} \sum_{k=-\infty}^{\infty} \int_c^d \left\{ e^{[-\frac{1}{2T}(u+2k(a_2-a_1))^2]} - e^{[-\frac{1}{2T}(u-2a_2+2k(a_2-a_1))^2]} \right\} \, du. \quad (8.1.1)$$

For any function $f : [0,1] \to \mathbb{R}$ and any $\delta > 0$ let us define

$$w_f(\delta) \stackrel{\mathrm{df}}{=} \sup_{|t'-t''| \le \delta} |f(t') - f(t'')|, \quad (8.1.2)$$

$$\tilde{w}_f(\delta) \stackrel{\mathrm{df}}{=} \max\{s_1, s_2, s_3\}, \quad (8.1.3)$$

where

$$s_1 \overset{\mathrm{df}}{=} \sup_{0 \le t \le \delta} |f(t) - f(0)|,$$

$$s_2 \overset{\mathrm{df}}{=} \sup_{t-\delta \le t' \le t \le t'' \le t+\delta} \min\{|f(t') - f(t)|, |f(t'') - f(t)|\},$$

$$s_3 \overset{\mathrm{df}}{=} \sup_{1-\delta \le t \le 1} |f(t) - f(1)|.$$

Note that $\tilde{w}_f(\delta)$ and $w_f(\delta)$ monotonically decrease if δ decreases to 0 for any fixed $f \in D[0,1]$.

Theorem 8.1.3 *Let Γ be a set of probability measures on $D[0,1]$. In order that $\overline{\Gamma}$ be compact it is necessary and sufficient that the following conditions be satisfied:*

(i)

$$\forall_{\epsilon>0} \exists_{M_\epsilon} \forall_{\mu \in \Gamma} \mu(\{x : \sup_t |x(t)| \le M_\epsilon\}) > 1 - \epsilon/2;$$

(ii)

$$\forall_{\delta>0} \forall_{\epsilon>0} \exists_{\eta=\eta(\delta,\epsilon)>0} \forall_{\mu \in \Gamma} \mu(\{x : \tilde{w}_x(\eta) \le \delta\})1 - \epsilon/2.$$

Proposition 8.1.1 *Let $\{\mu_n\}$ be a sequence of probability measures on the space $D[0,1]$, such that*

$$\lim_{\delta \to 0} \limsup_{n \to \infty} \mu_n(\{x : \tilde{w}_x(\delta) > \epsilon\}) = 0 \quad \text{for any } \epsilon > 0.$$

Let further the sequence $\{\mu_n^{t_1,\dots,t_k}\}$ be conditionally compact in the k–dimensional vector space $\mathbb{R}^k$ for each fixed k and $t_1, \dots, t_k$ from $[0,1]$, where $\{\mu_n^{t_1,\dots,t_k}\}$ is the measure in $\mathbb{R}^k$ induced by μ_n through the map

$$\pi^{t_1,\dots,t_k} : x \to (x(t_1), \dots, x(t_n)), \quad \text{for } x \in D[0,1].$$

Then the set $\{\mu_n\}$ is conditionally compact.

Let us recall that the definition of Lévy processes was already introduced (see Definition 4.4.1). It is known that $X(s)$ have to have infinitely divisible distribution and its characterisic function has the form (known as the Lévy-Khintchine Formula).

$$E e^{i\theta X(s)} = \exp(s\psi(\theta)), \tag{8.1.4}$$

where

$$\psi(\theta) = i\theta b - \frac{1}{2} c\theta^2 + \int_{-\infty}^{\infty} (e^{i\theta x} - 1 - \frac{i\theta x}{1+x^2}) \, d\nu(x) \tag{8.1.5}$$

and ν is called the Lévy measure. This measure has the following properties: $\nu(\{0\}) = 0$, $\nu \ge 0$, $\int_{|x| \le 1} x^2 \, d\nu < \infty$ and $\nu\{x : |x| > \delta\} < \infty$ for all $\delta > 0$. Every measure with above properties is a Lévy measure of some infinitely divisible distribution.

8.2 Error of Approximation of Itô Integrals

We assume that $\mathbf{B} = \{B(t) : t \in [0,1]\}$ is the standard Brownian motion on a probability space $(\Omega, \mathcal{F}, P)$ with the usual increasing family $\{\mathcal{F}(t) : t \in [0,1]\}$ of σ–algebras such that $B(t) \in \mathcal{F}(t)$ and independent of σ–algebras generated by $\{B(s) - B(t) : t < s \leq 1\}$.

It is natural to make use of the definition of the Itô stochastic integral $\int_0^t \Phi(s)\, dB(s)$ of an adapted integrand Φ with respect to a Brownian motion process $\mathbf{B} = \{B(t) : t \in [0,1]\}$ understood as a limit of integrals of approximating simple integrands $\{\Phi_n\}$ in order to obtain a numerical method of computation of such stochastic integrals, which converges in the space $C[0,1]$. We are interested in a general theorem on convergence of stochastic integrals $\int_0^t \Phi_n(s)\, dB(s)$ to $\int_0^t \Phi(s)\, dB(s)$.

The error of approximation is expressed in terms of stochastic integrals $\int_0^t \Psi(s)\, dB(s)$, where the integrands $\{\Psi_n(t) : t \in [0,1]\}_{n=1}^\infty$ are supposed to be measurable, adapted to $\{\mathcal{F}(t)\}$, and satisfy

$$\int_0^1 \Psi_n^2(s)\, ds < \infty \quad \text{a.s.}, \quad n = 1, 2, \ldots$$

One of the Rootzén's theorems can be stated as follows.

Theorem 8.2.1 *Let us define*

$$\tau_n(t) \overset{\mathrm{df}}{=} \int_0^t \Psi_n^2(s)\, ds.$$

Suppose that

$$\sup_{0 \leq t \leq 1} \left| \int_0^t \Psi_n(s)\, ds \right| \overset{p}{\to} 0, \quad as \quad n \to \infty \tag{8.2.1}$$

and that

$$\tau_n(t) = \int_0^t \Psi_n^2\, ds \overset{p}{\to} \tau(t), \quad as \quad n \to \infty, \tag{8.2.2}$$

for some continuous stochastic process $\{\tau(t) : t \in [0,1]\}$ with values in $C[0,1]$. Then

$$\int_0^{(\cdot)} \Psi_n(s)\, dB(s) \overset{d}{\to} W \circ \tau, \quad as \quad n \to \infty, \tag{8.2.3}$$

where the Brownian motion W is independent of τ.

In order to facilitate checking conditions (8.2.1) and (8.2.2) of Theorem 8.2.1, let us formulate a useful lemma.

Lemma 8.2.1 *Suppose that the stochastic process $\{\Upsilon(t) : t \in [0,1]\}$ is a.s. Riemann integrable over $[0,1]$. Let*

$$\Psi(t) = n^{k/2} \sum_{i=0}^{n-1} \Upsilon(i/n)(B(t) - B(i/n))^k I_{A_{n,i}}(t)$$

where $A_{n,i} = [i/n, (i+1)/n)$, *and let* $E_k = E \int_0^1 B(s)^k \, ds$.
Then

$$\sup_{0 \le t \le 1} \left| \int_0^t \Psi(s) \, ds - E_k \int_0^t \Upsilon(s) \, ds \right| \xrightarrow{p} 0, \quad \text{as } n \to \infty, \qquad (8.2.4)$$

for $k > 0$.

Now we are in a position to discuss the problem of the numerical approximation of the following stochastic integral

$$I(t) = \int_0^t f(B(sw), s) \, dB(s).$$

In the simplest possible case we can approximate $I(t)$ by

$$I_n(t) = \sum_{i=0}^{[nt]} f(B(i/n), i/n) \, \Delta_n^i.$$

where $\Delta_n^i = B((i+1)/n) - B(i/n)$ if $i < [nt]$ and $\Delta_n^i = B(t) - B(i/n)$ if $i = [nt]$. We look for the error of approximation, i.e. for the difference $I(t) - I_n(t)$.

Rootzén (1980) obtained the following theorem.

Theorem 8.2.2 *Suppose that $f(x,t) - f(x,s) = o(|t - s|^{1/2})$ uniformly on compacts, and that the derivative $\frac{\partial}{\partial x} f(x,t) = f_x(x,t)$ is continuous. Then*

$$\left\{ n^{\frac{1}{2}} (I(t) - I_n(t)) : t \in [0,1] \right\} \xrightarrow{d} \{ W \circ \tau(t) : t \in [0,1] \}, \qquad (8.2.5)$$

where the Brownian motion W is independent of τ and where

$$\tau(t) = \frac{1}{2} \int_0^t f_x(B(s), s)^2 \, ds.$$

PROOF. Put

$$\Phi(t) = f(B(t), t), \quad \text{and} \quad \Phi_n(t) = \sum_{i=0}^{n-1} f(B(i/n), i/n) I_{A_{n,i}}(t)$$

and let

$$\Psi_n(t) = n^{\frac{1}{2}} \sum_{i=0}^{n-1} f_x(B(i/n), i/n)(B(t) - B(i/n)) I_{A_{n,i}}(t).$$

The first step of the proof is to approximate $n^{1/2} \int_0^t (\Phi - \Phi_n) dB$ by $\int_0^t \Psi_n dB$. We have

$$n^{\frac{1}{2}} \int_0^t (\Phi(s) - \Phi_n(s)) \, dB(s) = n^{\frac{1}{2}} \sum_{i=0}^{n-1} \int_{A_{n,i} \cap [0,t]} (f(B(s), s) - f(B(s), i/n)) \, dB(s)$$

$$(8.2.6)$$

$$+ \int_0^t \Psi_n(s) \, dB(s) + n^{\frac{1}{2}} \sum_{i=0}^{n-1} \int_{A_{n,i} \cap [0,t]} o(|B(s) - B(i/n)|) \, dB(s)$$

$$= R_n^1(t) + \int_0^t \Psi(s) \, dB(s) + R_n^2(t).$$

Definitions of R_n^1 and R_n^2 are obvious. To prove that they converge uniformly to 0 in probability, one can use the fact that, if $\int_0^1 \Theta_n^2(s) \, ds \xrightarrow{P} 0$ for a given sequence of integrands $\{\Theta_n(t) : t \in [0,1]\}$, then $\sup_{0 \le t \le 1} |\int_0^t \Theta_n(s) \, dB(s)| \xrightarrow{P} 0$. Since almost every sample path of $\{B(t) : t \in [0,1]\}$ is bounded, we have from the first condition of the theorem that

$$n \sum_{i=0}^{n-1} \int_{A_{n,i}} (f(B(s),s) - f(B(s),i/n))^2 \, ds$$

$$= o(1)n \sum_{i=0}^{n-1} \int_{A_{n,i}} (B(s) - B(i/n))^2 \, ds \xrightarrow{P} 0, \quad \text{as} \quad n \to \infty,$$

since $o(1) \xrightarrow{a.s.}$ and $\int_{A_{n,i}} (B(s) - B(i/n))^2 ds$ has the same distribution as the expression $n^{-2} \int_0^1 B(s)^2 ds$, and hence $\{n \sum_{i=0}^{n-1} \int_{A_{n,i}} (B(s) - B(i/n))^2 ds\}$ is tight. Thus, we have proved that

$$\sup_{0 \le t \le 1} |R_n^i(t)| \xrightarrow{P} 0 \quad \text{as} \quad n \to \infty \quad \text{for} \quad i = 1, 2. \tag{8.2.7}$$

Next, by Lemma 8.2.1 with $\Upsilon(t)$ replaced by $f_x(B(t),t)$ and $f_x(B(t),t)^2$, respectively, which are continuous and hence Riemann integrable,

$$\sup_{0 \le t \le 1} |\int_0^t \Psi(s) \, ds| \xrightarrow{P}, \quad \text{as} \quad n \to \infty,$$

and

$$\int_0^t \Psi(s)^2 \, ds \xrightarrow{P} \frac{1}{2} \int_0^t f_x(B(s),s)^2 \, ds, \quad t \in [0,1], \quad \text{as} \quad n \to \infty,$$

since, clearly, $E_1 = 0$ and $E_2 = \frac{1}{2}$. Thus, $\{\Psi_n\}$ satisfies the hypothesis of Theorem 8.2.1, and hence

$$\int_0^{(\cdot)} \Psi_n(s) \, dB(s) \xrightarrow{d} W \circ \tau, \quad \text{as} \quad n \to \infty,$$

and by (8.2.6) and (8.2.7) this proves the theorem. $\qquad\square$

In addition, let us notice that it is possible to treat other approximation schemes in a quite similar way.

8.3   The Rate of Convergence of LePage Type Series

In this section (based on Janicki and Kokoszka (1991)) we study a representation of an α–stable random variable X as an $a.s.$–limit of a series $X_n \stackrel{\mathrm{df}}{=} \sum_{j=1}^n \xi_j \tau_j^{-1/\alpha}$, where the τ_j's are the arrival times of a Poisson process and the ξ_j's – appropriately chosen random variables. This allows us to evaluate the expectation $E|X_{n+m} - X_n|^2$ for $n > 0$, $m > 0$ and apply our estimations to establish an upper bound for

$$P\left\{ \max_{1 \le k \le 2^w} \left| X_n(2^{-w}k) - X(2^{-w}k) \right| > d \right\},$$

where the distribution of the vector $\{X(2^{-w}k)\}_{k=1}^{2^w}$ coincides with the appropriate finite dimensional distribution of the Lévy motion and the variables $X_n(2^{-w}k)$, constructed by means of LePage–type sums, converge to the $X(2^{-w}k)$'s.

This section contains also some new results pertinent to series representations of skewed stable random variables.

Theoretical Preliminaries.   Let us recall that Lévy measures are concentrated on $\mathbb{R}\backslash\{0\}$. Though our first lemma seems to be elementary, we outline its proof for completeness.

Lemma 8.3.1 *Let $\alpha \in (0,2)$ and let $c \in \mathbb{R}$. Define*

$$F_C(A) = \int_0^\infty I_A(cu^{-1/\alpha})\, du, \quad A \in \mathcal{B}(\mathbb{R}\backslash\{0\}).$$

Then F_C is the Lévy measure of a stable law on $\mathbb{R}$ and

$$\int_{-\infty}^\infty \left(e^{i\theta x} - 1 - i\theta x\right) F_C(dx) = \alpha\Gamma(-\alpha)|c\theta|^\alpha e^{-\mathrm{isgn}(c\theta)\alpha\pi/2}.$$

PROOF. By elementary arguments the proof can be reduced to the case of $c > 0$, $\theta > 0$. Using approximation by simple functions one arrives at the formula

$$\int_{-\infty}^\infty f(x)\, F_C(dx) = \alpha c^\alpha \int_0^\infty f(x) x^{-\alpha-1}\, dx, \qquad (8.3.1)$$

which holds whenever the left or right–hand side of (8.3.1) exists. Elementary calculations show that

$$\int_0^\infty \left(e^{it} - 1 - it\right) t^{-\alpha-1}\, dt = \Gamma(-\alpha)e^{i\alpha\pi/2}. \qquad (8.3.2)$$

Combine (8.3.1) and (8.3.2) to complete the proof.   □

In the sequel $\{\tau_1, \tau_2, ...\}$ denotes the sequence of arrival times (jump points) of the right continuous Poisson process with unit rate. Thus, the random variable τ_j has the density

$$f_j(x) = x^{j-1}\, e^{-x} I_{[0,\infty)}(x)/\Gamma(j). \qquad (8.3.3)$$

Further on, $\{\xi_1, \xi_2, ...\}$ will stand for a sequence of i.i.d. random variables which will be assumed to be independent of the sequence $\{\tau_j\}$.

Proposition 8.3.1 *Let $\alpha \in (1, 2)$ and $\beta \in (-1, 1)$. Let $P\{\xi_1 = t_1\} = p_1$ and $P\{\xi_1 = t_2\} = p_2$, where*

$$
\begin{aligned}
t_1 &= (1 + \beta)^{1/(\alpha-1)}, \quad t_2 = -(1 - \beta)^{1/(\alpha-1)}, \\
p_1 &= -t_2/(t_1 - t_2), \quad p_2 = 1 - p_1.
\end{aligned}
$$

Then, for any $s > 0$

$$
s \sum_{j=1}^{\infty} \xi_j \, \tau_j^{-1/\alpha} \sim S_\alpha(\sigma, \beta, 0), \tag{8.3.4}
$$

i.e. the series on the left-hand side converges a.s. to a stable random variable with characteristic function of the form

$$
\phi(\theta) = \exp\left\{-\sigma^\alpha |\theta|^\alpha \left(1 - i\beta \operatorname{sgn}(\theta) \tan(\alpha\pi/2)\right)\right\}, \tag{8.3.5}
$$

where

$$
\sigma^\alpha = s^\alpha \, L(\alpha, \beta),
$$
$$
L(\alpha, \beta) = 2K(\alpha) p_1 t_1,
$$
$$
K(\alpha) = -\alpha \Gamma(-\alpha) \cos(\alpha\pi/2).
$$

PROOF. Let λ be the law of ξ_1. Then F defined by

$$
F(A) = \int_0^\infty \int_{\{t_2, t_1\}} \mathbf{I}_A(svu^{-1/\alpha}) \, \lambda(dv) \, du = p_1 \, F_{st_1}(A) + p_2 \, F_{st_2}(A),
$$

for A such that $0 \notin A$, is the Lévy measure of a stable law. Therefore,

$$
\int_{\{|x|>1\}} |x|^p \, F(dx) < \infty, \quad \text{whenever } p \in (1, \alpha),
$$

and we are in a position to apply Theorem 3.1 of Rosinski (1990). First of all note that

$$
p_1 t_1 + p_2 t_2 = 0, \tag{8.3.6}
$$
$$
p_1 t_1^{<\alpha>} + p_2 t_2^{<\alpha>} = \beta \left(p_1 |t_1|^\alpha + p_2 |t_2|^\alpha\right), \tag{8.3.7}
$$

where $t^{<\alpha>} = |t|^\alpha \operatorname{sgn}(t)$. It follows from (8.3.6) that the centering constant $C(t)$ appearing in Theorem 3.1 of Rosinski (1990) vanishes. Consequently, series (8.3.4) converges a.s. to a random variable X with

$$
\widehat{\mathcal{L}}(X)(\theta) = \exp\left\{\int_{-\infty}^{\infty} \left(e^{i\theta x} - 1 - i\theta x\right) \, F(dx)\right\}.
$$

Now Lemma 8.3.1 yields

$$
\int_{-\infty}^{\infty} \left(e^{i\theta x} - 1 - i\theta x\right) \, F(dx)
$$

$$
\begin{aligned}
&= \; p_1 \alpha \Gamma(-\alpha) s^\alpha |\theta|^\alpha |t_1|^\alpha \left(\cos\left(\frac{\alpha\pi}{2}\right) - i \sin\left(\frac{\alpha\pi}{2}\right) \mathrm{sgn}(\theta) \right) \\
&+ \; p_2 \alpha \Gamma(-\alpha) s^\alpha |\theta|^\alpha |t_2|^\alpha \left(\cos\left(\frac{\alpha\pi}{2}\right) + i \sin\left(\frac{\alpha\pi}{2}\right) \mathrm{sgn}(\theta) \right) \\
&= \; -K(\alpha) s^\alpha |\theta|^\alpha \left(p_1 |t_1|^\alpha + p_2 |t_2|^\alpha \right) \left(1 - i\beta \mathrm{sgn}(\theta)\tan\left(\frac{\alpha\pi}{2}\right) \right)
\end{aligned}
$$

and (8.3.5) follows.   □

A notable shortcoming of the above result is the exclusion of the case $|\beta| = 1$. However, as Proposition 8.3.2 demonstrates, this cannot be remedied as long as we insist that $E\xi_1 = 0$, a requirement which proves decisive in Section 2. Before formulating the next proposition we set forth a simple lemma in which the τ_j's and ξ_j's denote the same variables as above.

Lemma 8.3.2 *For an $\alpha > 0$ set $T_n = \sum_{j=1}^n \xi_j \tau_j^{-1/\alpha}$. If $E|\xi_1|^\alpha = \infty$, then the sequence $\{T_1, T_2, ...\}$ diverges a.s..*

PROOF. The event $\Omega_0 = \{\omega : \lim_{n\to\infty} \tau_n(\omega)/n = 1\}$ has probability one. Therefore, to prove that $\{T_n\}$ diverges a.s., it suffices to show that it diverges a.s. for each fixed sequence $\{\tau_n\}$ belonging to Ω_0. Fix such a sequence. Then, the summands of T_n are independent and $\tau_j^{-1/\alpha} > 2^{-1/\alpha} j^{-1/\alpha}$ eventually. Consequently

$$
P\left\{ \left| \xi_j \tau_j^{-1/\alpha} \right| > 2^{-1/\alpha} \right\} \geq P\left\{ |\xi_j|^\alpha > j \right\}.
$$

Thus, the assertion follows from the three series theorem and the fact that for any positive random variable ζ we have $E\zeta < \infty$ if and only if $\sum_{j=1}^\infty P\{\zeta > j\} < \infty$.
□

Proposition 8.3.2 *Let $\alpha \in (1,2)$ and $E\xi_1 = 0$. Set $T_n = \sum_{j=1}^n \xi_j \tau_j^{-1/\alpha}$.*

(i)) *If $\{T_n\}$ converges a.s., then its limit T_∞ is a strictly stable random variable.*

(ii) *If T_∞ is nondegenerated, then its skewness parameter satisfies the condition $|\beta| \neq 1$.*

PROOF. Let λ denote the law of ξ_1 and $D = supp(\lambda)$.

(i) Since, by Lemma 8.3.2, $E|\xi_1|^\alpha < \infty$, the series $\sum_{j=1}^\infty \epsilon_j \xi_j \tau_j^{-1/\alpha}$ converges a.s. to a symmetric stable random variable (Theorem 1.5.1 of Samorodnitsy and Taqqu (1993)). Here $\{\epsilon_1, \epsilon_2, ...\}$ is a sequence of i.i.d. Rademacher random variables, i.e. $P\{\epsilon = 1\} = 1/2 = P\{\epsilon_1 = -1\}$, independent of all other sequences introduced so far. By Corollary 3.6 of Rosinski (1990), G defined by

$$
G(A) = \int_0^\infty \int_D I_A(vu^{-1/\alpha})\lambda(dv)du, \quad 0 \notin A,
$$

is a Lévy measure and thus the symmetrization of G is the Lévy measure of a stable law. Consequently, G is the Lévy measure of a stable law. (See Propositions 6.3.1 and 6.3.2 of Linde (1983).) Thus,

$$\int_{\{|x|>1\}} |x|^p \, G(dx) < \infty \quad \text{for all } p \in (1, \alpha),$$

and it remains to apply Theorem 3.1 of Rosinski (1990).

(ii) By applying Lemma 8.3.1 one gets, similarly as in the proof of Proposition 8.3.1,

$$\log \widehat{\mathcal{L}}(T_\infty)(\theta) = \int_{-\infty}^{\infty} \left(e^{i\theta x} - 1 - i\theta x \right) \, G(dx)$$

$$= \int_D \int_{-\infty}^{\infty} \left(e^{i\theta x} - 1 - i\theta x \right) \, G_v(dx) \, \lambda(dv)$$

$$= \int_D \alpha \Gamma(-\alpha) |v\theta|^\alpha e^{i \operatorname{sgn}(v\theta) \alpha \pi / 2} \, \lambda(dv)$$

$$= -K(\alpha) |\theta|^\alpha \left\{ \int_D |v|^\alpha \lambda(dv) - i \operatorname{sgn}(\theta) \tan(\frac{\alpha\pi}{2}) \int_D v^{<\alpha>} \lambda(dv) \right\}.$$

Recall that G_v is the Lévy measure defined by the following relation: $G_v(A) = \int_0^\infty I_A(vu^{-1/\alpha}) \, du$. If T_∞ is nondegenerated, then $\int_D |v|^\alpha \lambda(dv) > 0$ and T_∞ has the skewness parameter

$$\beta = \frac{\int_D v^{<\alpha>} \lambda(dv)}{\int_D |v|^\alpha \lambda(dv)}.$$

Thus, requirements $|\beta| = 1$ and $\int_D v \, \lambda(dv) = 0$ are incompatible.   □

Now some comment seems in place. If $\alpha < 1$, then there is a clear difference between totally skewed ($|\beta| = 1$) and remaining stable random variables. Suppose $X \sim S_\alpha(\sigma, 1, 0)$ and $Y \sim S_\alpha(\sigma, \beta, 0)$ with $0 < \alpha < 1$ and $|\beta| < 1$, then $\operatorname{supp}(\mathcal{L}(X)) = [0, \infty)$ and $\operatorname{supp}(\mathcal{L}(Y)) = (-\infty, \infty)$. By contrast, if $\alpha > 1$, each α-stable random variable has positive density on the whole line. Looking at numerically obtained graphs of stable densities, one can see only a quantitative difference between, say, cases $\beta = 1$ and $\beta = 0.75$, provided $\alpha > 1$. However, as remarked by LePage (1980), the series representations of the kind discussed here provide a fine insight into the structure of stable distributions. In this light, Propositions 8.3.1 and 8.3.2 exhibit a qualitative distinction between totally skewed and remaining stable random variables in the case of $\alpha > 1$. Proposition 8.3.1 also raises the question whether the LePage representation of stable vectors taking values in Banach spaces can be so modified that no centering is needed.

Case of $\alpha \in (1, 2)$, $|\beta| \neq 1$. First we prove some technical results.

Lemma 8.3.3 *Let $\alpha \in (1,2)$ and $|\beta| \neq 1$. For $L(\alpha, \beta)$ and the ξ_j's, τ_j's as in Proposition 8.3.1 and $h > 0$ define*

$$X_n(h) = h^{1/\alpha} \, L(\alpha, \beta)^{-1/\alpha} \sum_{j=1}^{n} \xi_j \tau_j^{-1/\alpha}.$$

Then, for any $n > 2/\alpha$ and $m > 0$,

$$E|X_{n+m}(h) - X_n(h)|^2 < h^{2/\alpha} \, R_n(\alpha, \beta), \tag{8.3.8}$$

where

$$R_n(\alpha, \beta) = L(\alpha, \beta)^{-2/\alpha}(1 - \beta)^{1/(\alpha-1)} \sum_{j=n+1}^{\infty} (j - 2/\alpha)^{-2/\alpha}. \tag{8.3.9}$$

PROOF. Write

$$E|X_{n+m}(h) - X_n(h)|^2 = E\left| h^{1/\alpha} L(\alpha, \beta)^{-1/\alpha} \sum_{j=n+1}^{n+m} \xi_j \tau_j^{-1/\alpha} \right|^2$$

$$= h^{2/\alpha} L(\alpha, \beta)^{-2/\alpha} \, E|\xi_j|^2 \, E\tau_j^{-2/\alpha}.$$

The last equality being justified by the fact that for $j \neq k$ we have

$$E\left(\xi_j \xi_k (\tau_j \tau_k)^{-1/\alpha}\right) = E\left(\xi_j \xi_k\right) E\left((\tau_j \tau_k)^{-1/\alpha}\right)$$

$$= E\xi_j \, E\xi_k \, E\left((\tau_j \tau_k)^{-1/\alpha}\right) = 0.$$

Since $E|\xi_j|^2 = E|\xi_1|^2 = t_1^2 p_1 + t_2^2 p_2 = (1 - \beta^2)^{1/(\alpha-1)}$, we get

$$E|X_{n+m}(h) - X_n(h)|^2 = h^{2/\alpha} \, L(\alpha, \beta)^{-2/\alpha}(1 - \beta^2)^{1/(\alpha-1)} \sum_{j=n+1}^{\infty} E\tau_j^{-2/\alpha}.$$
$$\tag{8.3.10}$$

By (8.3.3) $E\tau_j^{-2/\alpha} = \Gamma(j - 2/\alpha)/\Gamma(j)$ whenever $j > 2/\alpha$. It is obvious that $\Gamma(j - 2/\alpha)/\Gamma(j) = (j - 2/\alpha)^{-2/\alpha}$ for $\alpha = 2$, so, after some calculations, we derive the inequality

$$\Gamma(j - 2/\alpha)/\Gamma(j) < (j - 2/\alpha)^{-2/\alpha} \quad \text{for all} \ \ \alpha \in (1,2)$$

and thus for $j > 2/\alpha$ we have

$$E\tau_j^{-2/\alpha} \leq (j - 2/\alpha)^{-2/\alpha}. \tag{8.3.11}$$

Combine (8.3.10) and (8.3.11) to get (8.3.8). $\square$

Now set

$$X(h) = h^{1/\alpha} \, L(\alpha, \beta) \sum_{j=1}^{\infty} \xi_j \tau_j^{-1/\alpha}.$$

The *a.s.*–convergence of $\{X_n(h)\}$ to $X(h)$ follows from Proposition 8.3.1. Using Theorem 3.1 of Rosinski (1990), one can easily check that the convergence is also in L^p for each $p \in (1, \alpha)$.

Proposition 8.3.3 *For any $n > 2/\alpha$, $h > 0$ and $p \in (1, \alpha)$,*

$$P\{|X(h) - X_n(h)| > h\} < \eta^{-2}h^{2/\alpha}R_n(\alpha, \beta), \qquad (8.3.12)$$

$$E\left\{(|X(h) - X_n(h)|^p)^{1/p}\right\} \leq h^{1/\alpha}(R_n(\alpha, \beta))^{1/2}. \qquad (8.3.13)$$

PROOF. To get (8.3.12) note that for any $\delta \in (0, \eta)$ and $m > 0$

$$P\{|X(h) - X_n(h)| > \eta\} \leq P\{|X(h) - X_{n+m}(h)| > \delta\} + (\eta - \delta)^{-2}h^{2/\alpha}R_n(\alpha, \beta).$$

First let $m \to \infty$ and then $\delta \to 0$. $\qquad\qquad\qquad\qquad\qquad\qquad\qquad\qquad$ □

Remark 8.3.1 *It follows from the proof of Lemma 8.3.3 that for $n > 2/\alpha$ and $m > 0$*

$$E|X(h) - X_n(h)|^2 = h^{2/\alpha}\, L(\alpha, \beta)^{-2/\alpha}(1 - \beta^2)^{1/(\alpha-1)} \sum_{j=n+1}^{\infty} \Gamma(j)^{-1}\Gamma(j - 2/\alpha).$$

Now let us return to the question how the trajectories of the Lévy motion can be generated on a computer (see Section 2.5). Our approach, based on the fact that $\{X(t)\}$ has stationary independent increments, is standard. Without loss of generality we can restrict ourselves to the simulation of the Lévy motion on the unit interval. Recall that throughout this section we assume that $\alpha \in (1, 2)$, $\beta \in (-1, 1)$.

Let us partition $(0, 1]$ into 2^w subintervals $(t_{k-1}, t_k]$, $k = 1, 2, ..., 2^w$, of equal length $h = 2^{-w}$ and set

$$X_{i,n} = h^{1/\alpha}\, L(\alpha, \beta)^{-1/\alpha} \sum_{j=1}^{n} \xi_{i,j}\tau_{i,j}^{-1/\alpha}, \quad i = 1, 2, ..., 2^w, \qquad (8.3.14)$$

$$X_n(t_k) = \sum_{i=1}^{k} X_{i,n}, \quad k = 1, 2, ..., 2^w, \qquad (8.3.15)$$

the families $\{\xi_{i,1}, \xi_{i,2}, ...; \tau_{i,1}, \tau_{i,2}, ...\}$ being independent (independently generated), and for each $i \in \{1, 2, ..., 2^w\}$ the families $\{\xi_{i,1}, \xi_{i,2}, ...\}$ and $\{\tau_{i,1}, \tau_{i,2}, ...\}$ being independent and distributed in the same way as $\{\xi_1, \xi_2, ...\}$ and $\{\tau_1, \tau_2, ...\}$, respectively, in Proposition 8.3.1.

To obtain the value $X(t_k)$ of the Lévy motion at point t_k we have to generate and sum up k increments $X(t_i) - X(t_{i-1})$ for $i = 1, 2, ..., k$. Now, it follows from Proposition 8.3.1 that each of these increments can be obtained as a sum of an infinite series, i.e., $X(t_i) - X(t_{i-1}) = X_i = \lim_{n\to\infty} X_{i,n}$. Therefore, $X_n(t_k)$ is an approximation of $X(t_k)$. By increasing n we make our approximation more accurate. However, what is needed is an estimation of the probability of generating an approximate trajectory which deviates too far away from a real trajectory. Let d and ϵ be arbitrary, small enough, positive numbers. It seems reasonable to require that

$$P\left\{\exists_{k \,\in\, \{1,2,...,2^w\}} : |X_n(t_k) - X(t_k)| > d\right\} < \epsilon. \qquad (8.3.16)$$

Thus, we require the probability that the approximate value $X_n(t_k)$ differs from the exact value $X(t_k)$ more than d at any of the points t_k to be less than ϵ. It will follow from Theorem 8.3.1 that, given d, ϵ and w, (8.3.16) holds if n satisfies

$$R_n(\alpha, \beta) < d^2 (2^w)^{2/\alpha-1} \,\epsilon.$$

Theorem 8.3.1 *Let $\alpha \in (1,2)$ and $|\beta| \neq 1$. Let random variables $X_n(t_k)$ be defined by (8.3.15) and (8.3.14). Then*

$$X(t_k) = \lim_{n \to \infty} X_n(t_k)$$

is the value at point t_k of Lévy motion $\{X(t) : t_k \in [0,1]\}$ satisfying the definition from Section 2.5. Moreover, for any positive d and $n > 2/\alpha$ we have

$$P\left\{\exists_{k \,\in\, \{1,2,...,2^w\}} : |X_n(t_k) - X(t_k)| > d\right\} < d^{-2}(2^w)^{(1-2/\alpha)} R_n(\alpha, \beta), \qquad (8.3.17)$$

where $R_n(\alpha, \beta)$ is given by (8.3.9).

PROOF. By the Kolmogorov Inequality and Lemma 8.3.3 we can write for any $n > 2/\alpha$ and $m > 0$ the following relations

$$P\left\{\exists_{k \,\in\, \{1,2,...,2^w\}} : |X_n(t_k) - X(t_k)| > d\right\}$$

$$= P\left\{\max_{1 \leq k \leq 2^w} : \left|\sum_{i=1}^{k}(X_{n,i} - X_{n+m,i})\right| \geq d\right\}$$

$$\leq d^{-2}\sum_{k=1}^{2^w}\sum\sum E|X_{n,k} - X_{n+m,k}|^2$$

$$\leq d^{-2}(2^w)^{(1-2/\alpha)} R_n(\alpha, \beta).$$

Since $\max_{1 \leq k \leq 2^w} |X_{n+m}(t_k) - X(t_k)| \to 0$ a.s. as $m \to \infty$, the argument used in the proof of Proposition 8.3.3 completes the proof. $\square$

Note that with d and n fixed the right–hand side of (8.3.17) tends to zero as $w \to \infty$. This means that the tails $|X_n(t_k) - X(t_k)|$ are so small that even their maximum over the points $2^{-w}k$ for $k \in \{1,2,...,2^w\}$ tends to zero in probability as $w \to \infty$. To avoid confusion, note that neither the $X_n(t_k)$'s nor $X(t_k)$''s are elements of any prior processes $\{X_n(t)\}$ or $\{X(t)\}$ but simply appropriately constructed random variables.

Case of $\alpha \in (0,1)$.   Now we deal with the Lévy motion $\{X(t) : t \in [0,1]\}$ for $\alpha(0,1)$ and β arbitrary from $[-1,1]$. This means that, in contrast to Section 2, we admit here totally skewed processes.

Let $\{\gamma_1, \gamma_2, ...\}$ be a sequence of i.i.d. random variables distributed as follows

$$P\{\gamma_1 = 1\} = \frac{1 + \beta}{2}, \quad P\{\gamma_1 = -1\} = \frac{1 - \beta}{2}.$$

The sequence $\{\gamma_1, \gamma_2, ...\}$ is again assumed to be independent of the sequence $\{\tau_1, \tau_2, ...\}$. Define

$$X_n(h) = (C_\alpha h)^{1/\alpha} \sum_{j=1}^{n} \gamma_j \tau_j^{-1/\alpha},$$

where

$$C_\alpha = \frac{1}{L(\alpha, 0)} = \frac{1 - \alpha}{\Gamma(2 - \alpha) \cos(\alpha \pi / 2)}.$$

The sequence $\{X_n(h)\}$ is known to converge $a.s.$ to random variable $X(h)$ with characteristic function

$$\widehat{\mathcal{L}}(X(h))(\theta) = \exp\left\{ -h|\theta|^\alpha \left(1 - i\beta \mathrm{sgn}(\theta) \tan\left(\frac{\alpha\pi}{2}\right) \right) \right\}$$

(cf. Rosinski (1990) or Samorodnitsky and Taqqu (1992)). Since $\mathbf{E}\gamma_1 = \beta$, the procedure presented in Section 2 cannot be repeated. However, in the present case we can proceed as follows. First note that

$$\mathbf{E}|X_{n+m}(h) - X_n(h)| \le h^{1/\alpha} Q_n(\alpha)$$

for $n > 1/\alpha$, where

$$Q_n(\alpha) = C_\alpha \sum_{j=n+1}^{\infty} (j - 1/\alpha)^{-1/\alpha}.$$

Our objective consists in determining the values of n for which

$$P\left\{ \exists_{k \,\in\, \{1,2,...,2^w\}} : |X_n(t_k) - X(t_k)| > d \right\} < \epsilon. \tag{8.3.18}$$

The above inequality will be satisfied if

$$P\left\{ \forall_{i \,\in\, \{1,2,...,2^w\}} : |X_{i,n}(t_k) - X_i(t_k)| \le d2^{-w} \right\} > 1 - \epsilon, \tag{8.3.19}$$

where the $X_{i,n}$'s are defined by (8.3.14) and (8.3.15), with $\xi_{i,j}$ replaced by appropriately defined $\gamma_{i,j}$. Since random variables $X_{i,n} - X_i$ are i.i.d., (8.3.19) is equivalent to

$$\left(1 - P\left\{ |X_n(h) - X(h)| \ge d2^{-w} \right\} \right)^{2^w} > 1 - \epsilon.$$

Similarly as in the proof of Proposition 8.3.3 we see that

$$P\{|X_n(h) - X(h)| > \eta\} \le \eta^{-1} h^{1/\alpha} Q_n(\alpha).$$

Consequently, all n satisfying the condition

$$Q_n(\alpha) > d \left(2^{-w} \right)^{(1/\alpha - 1)} \left(1 - (1 - \epsilon)^2 \right)$$

also satisfy (8.3.18).

Symmetric Case. In spite of some success in applying our method to the skewed Lévy motion, we have the impression that this is essentially a "symmetric" method, mainly because the important totally skewed processes are out of its reach (see Proposition 8.3.2). Moreover, this approach applies only to the symmetric 1-stable Lévy motion. On the other hand, the method can be extended to wider classes of symmetric infinitely divisible processes with homogeneous independent increments, as it is exemplified by semistable processes discussed in this section.

Let $\{\epsilon_1, \epsilon_2, ...\}$ be the Rademacher sequence defined in the proof of Proposition 8.3.2. Let $C_\alpha = 1/L(\alpha, 0)$ if $\alpha \neq 1$, and $C_\alpha = 2/\pi$ if $\alpha = 1$. It is known (see, e.g., Rosinski (1990)) that for each $\alpha \in (0, 2)$ we have

$$C_\alpha^{1/\alpha} \sum_{j=1}^n \epsilon_j \tau_j^{-1/\alpha} \to S(h), \quad \text{as } n \to \infty, \tag{8.3.20}$$

the convergence being a.s. and in L^p for $p \in (0, \alpha)$. The $S(h)$ is a symmetric random variable with scale parameter $\sigma = h^{1/\alpha}$. Proceeding as in Section 2 and using (8.3.20) instead of Proposition 8.3.1 one can easily formulate and prove results analogous to Proposition 8.3.3 and Theorem 8.3.1.

We devote the rest of this section to symmetric semistable processes (see Rajput and Rama–Murthy (1987)).

Definition 8.3.1 *A stochastic process* $\{Y(t) : t \in [0, \infty)\}$ *is called a symmetric semistable Lévy process if*

(i) $Y(0) = 0$ *a.s.,*

(ii) Y *has stationary and independent increments,*

(iii) $\widehat{\mathcal{L}}(Y(h))(\theta) = \exp\left\{h \sum_{n=-\infty}^{\infty} r^{-n} \int_\Delta \left(\cos(r^{n/\alpha}\theta x) - 1\right) \sigma(dx)\right\}$,

where $r \in (0, 1)$, $\alpha \in (0, 2)$, $\Delta = \{x \in \mathbb{R} : r^{1/\alpha} < |x| \leq 1\}$ *and* σ *is a finite symmetric Borel measure on* Δ.

If $\eta_1, \eta_2,...$ are i.i.d. with $\mathcal{L}(\eta_1) = \sigma(\Delta)\sigma$, then it follows from Example 4.11 of Rosinski (1990) that $Y_n(h) \to Y(h)$ as $n \to \infty$, where

$$Y_n(h) = \sum_{j=1}^n \epsilon_j \eta_j \left[(\frac{1}{r} - 1)h^{-1}\sigma^{-1}(\Delta)\tau_j\right]_r^{-1/\alpha}, \tag{8.3.21}$$

and $[t]_r = r^k$ if $r^k \leq t < r^{k-1}$. Using (8.3.21), one easily gets

$$E|Y_{n+m}(h) - Y_n(h)| \leq h^{2/\alpha} R_n(r, \alpha, \sigma), \tag{8.3.22}$$

where

$$R_n(r, \alpha, \sigma) = (1 - r)^{-2/\alpha}\sigma(\Delta)^{2/\alpha} \sum_{j=n+1}^{\infty} (j - 2/\alpha)^{-2/\alpha}.$$

By (8.3.22) one can, in a familiar way, obtain an estimation similar to (8.3.17).

8.4   Approximation of Lévy α–Stable Diffusions

Let us present now (after Janicki, Podgórski and Weron (1992)) the simple case of the linear stochastic differential equation involving only the stochastic integral of a deterministic function. (In this case the solution is an α–stable process.)

Theorem 8.4.1 *The family* $\{X^\tau(t) : t \in [0,T]\}$ *of approximate solutions of the linear stochastic differential equation (6.3.8), defined by a linear analogue of (6.3.5) and (6.3.6), uniformly converges in probability to the exact solution* $\{X(t) : t \in [0,T]\}$ *of (6.3.8) on* $[0,T]$ *when* $\tau \to 0$.

PROOF. In order to shorten our notation we define

$$\sigma(t, X(t)) = f(t) + g(t)X(t),$$

$$\sigma_\tau(t, X^\tau(t)) = \sum_{i=0}^{I-1} \sigma(t_i, X^\tau(t_i)) I_{[t_i,t_{i+1})}(t),$$

$$h_\tau(t) = \sum_{i=0}^{I-1} h(t_i) I_{[t_i,t_{i+1})}(t).$$

Then the formulas defining $\{X(t)\}$ and $\{X^\tau(t)\}$ can be rewritten as follows

$$X(t) = X_0 + \int_0^t \sigma(s, X(s))ds + \int_0^t h(s)dL_\alpha(s),$$

$$X^\tau(t) = X_0 + \int_0^t \sigma_\tau(s, X_\tau(s))ds + \int_0^t h_\tau(s)dL_\alpha(s).$$

By the triangle inequality we have

$$|X(t) - X^\tau(t)| \le \int_0^t |\sigma(s, X(s)) - \sigma_\tau(s, X^\tau(s))|ds + \int_0^T |h(s) - h_\tau(s)|dL_\alpha(s).$$

Observe that $h_\tau(s)$ uniformly approximates $h(s)$, so the second term converges in probability to 0.

From the continuity of $f(\cdot)$ and $g(\cdot)$ it follows that

$$\int_0^t |\sigma(s, X(s)) - \sigma_\tau(s, X^\tau(s))|ds \le \int_0^t |\sigma(s, X(s)) - \sigma_(s, X^\tau(s))|ds$$

$$+ \int_0^t |\sigma(s, X^\tau(s)) - \sigma_\tau(s, X^\tau(s))|ds$$

$$\le K \int_0^t |X(s) - X^\tau(s)|dt$$

$$+ T \sup_{0 \le s \le T} |\sigma(s, X^\tau(s)) - \sigma_\tau(s, X^\tau(s))|.$$

Since $X^\tau(t)$ is a continuous process and $\sigma_\tau(\cdot, \cdot)$ uniformly approximates $\sigma(\cdot, \cdot)$, thus the last term converges to 0 with probability 1.

Hence we obtain the following estimate

$$|X(t) - X^\tau(t)| \le K \int_0^t |X(s) - X^\tau(s)| ds + \epsilon_\tau,$$

where $\epsilon_\tau \to 0$ in probability. Thus from the Gronwall Lemma we obtain

$$|X(t) - X^\tau(t)| \le K \int_0^t e^{K(t-s)} \epsilon_\tau ds + \epsilon_\tau \le \epsilon_\tau e^{KT},$$

so we have the uniform convergence to 0 in probability for $t \in [0, T]$.   □

8.5   Applications to Statistical Tests of Hypotheses

One of the most important problems in statistics is to test the hypothesis based on observations that a population is distributed according to a given distribution function.

Suppose we make n independent observations $\xi_1, \xi_2, ..., \xi_n$ on a statistical population, which is distributed according to a continuous distribution function $F(x)$. The most natural way to estimate $F(x)$ on the basis of these observations is to construct the sample distribution $F_n(x)$, where

$$F_n(x) \overset{\mathrm{df}}{=} \nu_n(x)/n; \qquad \nu_n(x) \overset{\mathrm{df}}{=} \sum_{i=1}^n I_{(-\infty, x]}(\xi_i);$$

i.e. $\nu_n(x)$ is the number of ξ_i's in the interval $(-\infty, x]$. From the classical theorem of Glivienko–Cantelli we have

$$\lim_{n \to \infty} \sup\{|F_n(x) - F(x)| : x \in \mathbb{R}\} = 0 \quad a.e.$$

For testing the hypothesis mentioned above it would be useful to find the asymptotic distribution of $\sup\{|F_n(x) - F(x)| : x \in \mathbb{R}\}$. It is possible to solve this problem by applying some results on convergence of sequences of measures on the space $D[0, 1]$.

We consider

$$\eta_n(x) = \sqrt{n}(F_n(x) - F(x)), \qquad x \in \mathbb{R},$$

as a random process by taking x as time.

Lemma 8.5.1 *The finite dimensional distributions of the process $\eta_n(x)$ converge weakly to the corresponding finite dimensional distributions of the Gaussian process $\eta(x)$, for which*

$$\mathbb{E}\eta(x) = 0, \quad for \ -\infty < x < \infty,$$

$$\mathbb{E}\eta(x)\eta(y) = F(x)(1 - F(y)), \quad for \ -\infty < x \le y < \infty.$$

PROOF. We have

$$\eta_n(x) = \frac{1}{\sqrt{n}} \sum_{k=1}^{n} [\boldsymbol{I}_{[0,\infty)}(x - \xi_k) - F(x)].$$

It is clear that

$$\boldsymbol{E}\,\boldsymbol{I}_{[0,\infty)}(x - \xi_k) = F(x),$$

$$\boldsymbol{E}[\boldsymbol{I}_{[0,\infty)}(x - \xi_k)\boldsymbol{I}_{[0,\infty)}(y - \xi_k)] = F(x), \quad \text{if} \quad x < y.$$

Since $\boldsymbol{I}_{[0,\infty)}(x - \xi_k) - F(x)$ are independent for different k, the lemma follows from the classical central limit theorem in finite–dimensional vector spaces. $\quad\square$

Let now the function $F^{-1}(t)$ be defined as

$$F^{-1}(t) = \inf\{x : F(x) = t\}.$$

Since $F(x)$ is continuous, it takes every value between 0 and 1 as x varies in $(-\infty, \infty)$. Thus $F^{-1}(t)$ is well defined and monotonic in the interval $[0, 1]$. We introduce

$$\vartheta_n(t) = \eta_n(F^{-1}(t)), \qquad \vartheta(t) = \eta(F^{-1}(t)),$$

where η_n and η are as in the preceding Lemma 8.5.1. It is then clear that

$$\vartheta_n(t) = \frac{1}{\sqrt{n}} \sum_{k=1}^{n} \{\boldsymbol{I}_{[0,\infty)}(t - \zeta_k) - t\}, \quad t \in [0, 1],$$

where $\zeta_k = F(\xi_k)$ are independently distributed according to the uniform distribution in the interval $[0, 1]$.

Lemma 8.5.2 *Almost all trajectories of the process $\{\vartheta_n(t) : t \in [0, 1]\}$ belong to $D[0, 1]$. Further, the finite dimensional distributions of the process $\{\vartheta_n(t) : t \in [0, 1]\}$ converge weakly to the corresponding finite dimensional distributions of the process $\{\vartheta(t) : t \in [0, 1]\}$ as $n \to \infty$. The process $\{\vartheta(t) : t \in [0, 1]\}$ is Gaussian with $\boldsymbol{E}\vartheta(t) = 0$ and $\boldsymbol{E}\vartheta(t)\vartheta(s) = t(1 - s)$ for $0 \le t \le s \le 1$.*

PROOF. The proof immediately follows from the previous lemma.

Remark 8.5.1 *One can notice easily that the process $\{\vartheta(t) : t \in [0, 1]\}$ is just the Brownian bridge process, i.e. the diffusion process solving the following stochastic differential equation*

$$\vartheta(t) = \int_0^t \frac{\vartheta(s)}{s - 1}\, dB(s), \quad t \in [0, 1),$$

where $\{B(t) : t \ge 0\}$ denotes the Brownian motion process.

(See e.g. Revuz and Yor (1991).)

Theorem 8.5.1 *The sequence of distributions $\{\mu_n\}$ of processes $\{\vartheta_n(t)\}$ converges weakly to the distribution μ of the process $\{\vartheta(t)\}$ in the space $D[0,1]$.*

PROOF. It follows from Remark 8.5.1 that the process $\{\vartheta(t)\}$ is continuous with probability 1, and hence its distribution can be considered as a measure in $D[0,1]$.

It is clear that for any $\delta > 0$ and any $x \in D[0,1]$,

$$\tilde{w}_x(\delta) \leq \sup_{|t'-t''|\leq\delta} |x(t') - x(t'')|,$$

where $\tilde{w}_x(\delta)$ is defined by (8.1.3). Thus, in order to establish the conditional compactness of the sequence of distributions μ_n of ϑ_n it is enough to prove that for any $\epsilon > 0$ we have

$$\lim_{\delta\to 0}\limsup_{n\to\infty} P\{\sup_{|t'-t''|\leq\delta} |\vartheta_n(t') - \vartheta_n(t'')| > \epsilon\} = 0 \qquad (8.5.1)$$

and to make use of Propositon 8.1.1 and Lemma 8.5.2.

Since the process $\vartheta_n(t) + \sqrt{n}t$ is monotonically increasing in t, we have

$$-\sqrt{n}(t_3 - t_2) \leq \vartheta_n(t_3) - \vartheta_n(t_2) \leq \vartheta_n(t_4) - \vartheta_n(t_1) + \sqrt{n}(t_2 - t_1 + t_4 - t_3),$$

for any sequence $0 \leq t_1 < t_2 < t_3 < t_4 \leq 1$. By an easy calculation we obtain

$$\sup_{|t'-t''|\leq\delta} |\vartheta(t') - \vartheta(t'')|$$

$$\leq 3 \sup_{|k_1/2^m-k_2/2^m|\leq\delta+1/2^{m-1}} \left|\vartheta_n\left(\frac{k_1}{2^m}\right) - \vartheta_n\left(\frac{k_2}{2^m}\right)\right| + 4\frac{\sqrt{n}}{2^m}$$

for any positive integer m. Let m_n be so chosen that $\sqrt{n}/2^{m_n} \to 0$ as $n \to \infty$ and $n/2^{m_n} \geq 1$. Then in order to prove (8.5.1) it is sufficient to establish that for any $\epsilon > 0$ we have

$$\lim_{\delta\to 0}\limsup_{n\to\infty} P\left\{\sup_{|k_1/2^{m_n}-k_2/2^{m_n}|\leq\delta} \left|\vartheta_n\left(\frac{k_1}{2^{m_n}}\right) - \vartheta_n\left(\frac{k_2}{2^{m_n}}\right)\right| > \epsilon\right\} = 0. \qquad (8.5.2)$$

Let $m(\delta)$ be the largest positive integer such that $\delta 2^{m(\delta)} \leq 1$. Then

$$\sup_{|k_1/2^{m_n}-k_2/2^{m_n}|\leq\delta} \left|\vartheta_n\left(\frac{k_1}{2^{m_n}}\right) - \vartheta_n\left(\frac{k_2}{2^{m_n}}\right)\right|$$

$$\leq \sup_{|k_1/2^{m_n}-k_2/2^{m_n}|\leq 1/2^{m(\delta)}} \left|\vartheta_n\left(\frac{k_1}{2^{m_n}}\right) - \vartheta_n\left(\frac{k_2}{2^{m_n}}\right)\right|. \qquad (8.5.3)$$

If $|k_1/2^{m_n} - k_2/2^{m_n}| \leq 1/2^{m(\delta)}$ we can find an integer j_n such that

$$\frac{j_n}{2^{m(\delta)}} \leq \frac{k_1}{2^{m(\delta)}} \leq \frac{k_2}{2^{m(\delta)}} \leq \frac{j_n + 1}{2^{m(\delta)}}.$$

Hence

$$\frac{k_1}{2^{m_n}} = \frac{j_n}{2^{m(\delta)}} + \frac{1}{2^{\tau_{1,1}}} + \ldots + \frac{1}{2^{\tau_{1,q}}},$$

where $m(\delta) \leq \tau_{1,1} < \tau_{1,2} < \ldots < \tau_{1,q} \leq m_n$ are positive integers. Similarly,

$$\frac{k_2}{2^{m_n}} = \frac{j_n}{2^{m(\delta)}} + \frac{1}{2^{\tau_{2,1}}} + \ldots + \frac{1}{2^{\tau_{1,r}}},$$

where $m(\delta) \leq \tau_{2,1} < \tau_{2,2} < \ldots < \tau_{2,r} \leq m_n$ are positive integers. Hence,

$$\left| \vartheta_n\left(\frac{k_1}{2^{m_n}}\right) - \vartheta_n\left(\frac{k_2}{2^{m_n}}\right) \right|$$

$$\leq \left| \vartheta_n\left(\frac{k_1}{2^{m_n}}\right) - \vartheta_n\left(\frac{k_2}{2^{m(\delta)}}\right) \right| + \left| \vartheta_n\left(\frac{k_1}{2^{m(\delta)}}\right) - \vartheta_n\left(\frac{k_2}{2^{m_n}}\right) \right|$$

$$\leq \sum_{i=1}^{q} \left| \vartheta_n\left(\frac{j_n}{2^{m(\delta)}} + \sum_{k=1}^{i} \frac{1}{2^{\tau_{1,k}}}\right) - \vartheta_n\left(\frac{j_n}{2^{m(\delta)}} + \sum_{k=1}^{i-1} \frac{1}{2^{\tau_{1,k}}}\right) \right|$$

$$+ \sum_{j=1}^{r} \left| \vartheta_n\left(\frac{j_n}{2^{m(\delta)}} + \sum_{k=1}^{j} \frac{1}{2^{\tau_{2,k}}}\right) - \vartheta_n\left(\frac{j_n}{2^{m(\delta)}} + \sum_{k=1}^{j-1} \frac{1}{2^{\tau_{2,k}}}\right) \right|$$

$$\leq 2 \sum_{j=m(\delta)}^{m_n} \sup_{i} \left| \vartheta_n\left(\frac{i+1}{2^j}\right) - \vartheta_n\left(\frac{i}{2^j}\right) \right|.$$

Thus from (8.5.3) and the above inequality we obtain

$$\sup_{|k_1/2^{m_n} - k_2/2^{m_n}| \leq \delta} \left| \vartheta_n\left(\frac{k_1}{2^{m_n}}\right) - \vartheta_n\left(\frac{k_2}{2^{m_n}}\right) \right|$$

$$\leq 2 \sum_{j=m(\delta)}^{m_n} \sup_{i} \left| \vartheta_n\left(\frac{i+1}{2^j}\right) - \vartheta_n\left(\frac{i}{2^j}\right) \right|.$$

Let now $0 < a < 1$ be such that $2a^4 > 1$. If the left–hand side of the above inequality exceeds ϵ, then the jth term within the sum must exceed the value of $(1-a)a^{s-m(\delta)}\epsilon/2$ for at least one j. Otherwise, the left–hand side will be less than or equal to ϵ. Thus, by the Chebyshev Inequality,

$$P\left\{ \sup_{|k_1/2^{m_n} - k_2/2^{m_n}| \leq \delta} \left| \vartheta_n\left(\frac{k_1}{2^{m_n}}\right) - \vartheta_n\left(\frac{k_2}{2^{m_n}}\right) \right| > \epsilon \right\}$$

$$\leq \sum_{j=m(\delta)}^{m_n} \sum_{i=0}^{2^s-1} P\left\{ \left| \vartheta_n\left(\frac{i+1}{2^j}\right) - \vartheta_n\left(\frac{i}{2^j}\right) \right| > (1-a)a^{j-m(\delta)}\frac{\epsilon}{2} \right\}$$

$$\leq \sum_{j=m(\delta)}^{m_n} \sum_{i=0}^{2^j-1} \frac{2^4}{(1-a)^4 \epsilon^4 a^{4(j-m(\delta))}} \, E \left| \vartheta_n\left(\frac{i+1}{2^j}\right) - \vartheta_n\left(\frac{i}{2^j}\right) \right|^4. \qquad (8.5.4)$$

But, for any $h > 0$,

$$\vartheta_n(t + h) - \vartheta_n(t) = \sqrt{n}\left(\frac{z_n}{n} - h\right),$$

where z_n is a random variable taking values in the set $\{1, 2, ..., n\}$ with probabilities

$$P\{z_n = k\} = \binom{n}{k} h^k (1 - h)^{n-k}.$$

The standard calculation of moments of a binomial distribution yields

$$
\begin{aligned}
E|\vartheta_n(t + h) - \vartheta_n(t)|^4 &= \frac{1}{n}h(1 - h)(h^3 + (1 - h)^3) + 3\frac{n-1}{n}h^2(1 - h)^2 \\
&\le \frac{2h}{n} + 3h^2 \le \frac{2h}{2^{mn}} + 3h^2,
\end{aligned}
$$

since $n/2^{mn} \ge 1$. Thus, when $h \ge 1/2^{mn}$, we have

$$E|\vartheta_n(t + h) - \vartheta_n(t)|^4 \le 5h^2.$$

Substituting this estimate in (8.5.4), we get

$$P\left\{\sup_{|k_1/2^{mn} - k_2/2^{mn}| \le \delta} \left|\vartheta_n\left(\frac{k_1}{2^{mn}}\right) - \vartheta_n\left(\frac{k_2}{2^{mn}}\right)\right| > \epsilon\right\}$$

$$\le \sum_{j=m(\delta)}^{mn} \frac{2^4}{(1 - a)^4 \epsilon^4} \cdot \frac{5}{2^j a^{4(j-m(\delta))}}$$

$$\le \frac{80}{(1 - a)^4 \epsilon^4}\left(\sum_{i=0}^{\infty} \frac{1}{(2a^4)^i}\right) \cdot \frac{1}{2^{m(\delta)}} = c_\epsilon \cdot \frac{1}{2^{m(\delta)}}.$$

Since $m(\delta) \to \infty$, thus (8.5.2) is proven. This shows that the sequence of distributions μ_n is conditionally compact in $D[0, 1]$.

We have to show that any limit measure μ of some subsequence $\{\mu_{n_k}\}$ must be concentrated on the subset $C[0, 1]$ of continuous functions in $D[0, 1]$. For any $x \in D[0, 1]$, let $j_\lambda(x)$ be the number of jumps whose absolute value is greater than λ. It is clear that the function $x \to j_\lambda(x)$ is continuous in the Skorohod topology for every $\lambda > 0$. Hence, if any subsequence $\{\mu_{n_k}\}$ of $\{\mu_n\}$ converges weakly to a measure μ, then the distribution of j_λ according to μ_{n_k} converges weakly to the distribution of j_λ according to μ. But from (8.5.1) we see that for all sufficiently large n the distribution of j_λ according to μ_n is degenerate at the origin. This shows that $\mu(C[0, 1]) = 1$. Hence, for every fixed $t_1, t_2, ..., t_k \in [0, 1]$, the map

$$x \to (x(t_1), ..., x(t_k))$$

is μ–almost everywhere continuous in $D[0, 1]$. Therefore, the finite–dimensional distributions of μ_{n_k} converge weakly to the finite–dimensional distributions of μ. Now Remark 8.5.1 and Theorem 8.1.1 imply that μ is the distribution of the process $\{\vartheta(t)\}$. This completes the proof.   □

Theorem 8.5.2 *Let* $\xi_1, \xi_2, ..., \xi_n$ *be independent and identically distributed random variables with a continuous distribution function* $F(x)$. *Let* $F_n(x)$ *be the sample distribution function based on* $\xi_1, \xi_2, ..., \xi_n$. *Then for all* $\delta > 0$ *we have*

$$\lim_{n\to\infty} P\{\sqrt{n} \sup_{x \in \mathbb{R}} |F(x) - F_n(x)| < \delta\} = \sum_{k=-\infty}^{\infty} (-1)^k \exp(-2k^2\delta^2),$$

$$\lim_{n\to\infty} P\{\sqrt{n} \sup_{x \in \mathbb{R}} [F(x) - F_n(x)] < \delta\} = 1 - \exp(-2\delta^2).$$

PROOF. It is clear that the functionals

$$f_1(x) = \sup_t |x(t)|, \qquad x \in D[0,1],$$

$$f_2(x) = \sup_t x(t), \qquad x \in D[0,1]$$

are continuous in the Skorohod topology. Further,

$$\sqrt{n} \sup_{x \in \mathbb{R}} |F_n(x) - F(x)| = f_1(\vartheta), \qquad (8.5.5)$$

$$\sqrt{n} \sup_{x \in \mathbb{R}} [F_n(x) - F(x)] = f_2(\vartheta), \qquad (8.5.6)$$

where $\{\vartheta_n(t)\}$ is the process defined in Theorem 8.5.1. Since by this theorem the distributions of processes $\{\vartheta_n(t)\}$ converge weakly to the distribution of $\{\vartheta(t)\}$ in the space $D[0,1]$, it follows that the distributions of (8.5.5) and (8.5.6) converge weakly to the distributions of $f_1(\vartheta)$ and $f_2(\vartheta)$, respectively. By Remark 8.5.1 the distribution of $\vartheta(t)$ is the same as that of the Brownian bridge, i.e. the same as that of the Brownian motion $B(t)$ under the condition that $B(1) = 0$.

But, by Theorem 8.1.2 we have for all $\delta > 0$, $x < \delta$,

$$P\{\sup_{0\leq t\leq 1} |B(t)| < \delta, B(1) < x\}$$

$$= \frac{1}{\sqrt{2\pi}} \sum_{k=-\infty}^{\infty} \int_{-\delta}^{x} \left\{ e^{[-\frac{1}{2}(u+4k\delta)^2]} - e^{[-\frac{1}{2}(u+4k\delta-2\delta)^2]} \right\} \, du,$$

$$P\{\sup_{0\leq t\leq 1} B(t) < \delta, B(1) < x\}$$

$$= \frac{1}{\sqrt{2\pi}} \int_{-\infty}^{x} \left\{ e^{[-u^2/2]} - e^{[-(u-2\delta)^2/2]} \right\} \, du.$$

Differentiating these two expressions with respect to x, putting $x = 0$, and dividing by $1/\sqrt{2\pi}$, we obtain

$$P\{\sup_{0\leq t\leq 1} |B(t)| < \delta|\, B(1) = 0\} \qquad \sum_{k=-\infty}^{\infty} (-1)^k e^{(-2k^2\delta^2)},$$

$$P\{\sup_{0\leq t\leq 1} B(t) < \delta|\, B(1) = 0\} \qquad = 1 - e^{(-2\delta^2)}.$$

This completes the proof. $\qquad\qquad\qquad\qquad\qquad\qquad\qquad\qquad\qquad\qquad$ □

8.6   Lévy Processes and Poisson Random Measures

We start with a formal definition of a Poisson random measure. Let $(S, \mathcal{S}, n)$ be a measure space and $\mathcal{S}_O = \{A \in \mathcal{S} : n(A) < \infty\}$

Definition 8.6.1 *A **Poisson random measure** N on (S, S, n) is an independently scattered σ-additive set function $N \colon \mathcal{S}_O \longrightarrow L^O(\Omega)$ such that for each $A \in \mathcal{S}_O$, $N(A)$ has the Poisson distribution with mean $n(A)$, that is*

$$P(N(A) = k) = e^{-n(A)} \frac{(n(A))^k}{k!}$$

$k = 0, 1, 2 \ldots$ *and n is called the control measure of N (mean).*

We know that the trajectories of any Lévy process are cadlag functions (right continuous with left limits). Thus we can define the *counting measure*

Definition 8.6.2 *Let $x \in D[0, \infty)$ and $0 \notin \bar{A}$. Then we have*

$$N([t_1, t_2], A, x) = card\{t_1 \leq s \leq t_2 : \Delta x(s) \in A\}$$

Let $N([t_1, t_2], A, X) \stackrel{\mathrm{df}}{=} N([t_1, t_2], A)$, where $\{X(t)\}$ denotes a Lévy process. Then

Theorem 8.6.1 *N generates a measure on $\mathbb{R}_+ \times \mathbb{R}_O$ ($\mathbb{R}_O = \mathbb{R} \setminus 0$) which is a Poisson measure with mean*

$$E N(ds, dx) = ds \, d\nu(x)$$

Now we can construct an integral with respect to N. Let $f : \mathbb{R}_O \longrightarrow \mathbb{R}$ be a measurable function.

Theorem 8.6.2   *(i) Let $\int_{\mathbb{R}} |f(x)| \, d\nu(x) < \infty$. Then*

$$\int_0^{t+} \int_{\mathbb{R}_O} f(x) \, N(ds, dx) \stackrel{\mathrm{df}}{=} \int_{[0,t] \times \mathbb{R}_O} f(x) \, N(ds, dx)$$

is well–defined and

$$E\left(\int_0^{t+} \int_{\mathbb{R}_O} N(ds, dx)\right) = \int_0^{t+} \int_{\mathbb{R}_O} f(x) \, ds \, d\nu(x)$$

$$= t \int_{\mathbb{R}_O} f(x) \, d\nu(x)$$

(ii) Let $\int_{\mathrm{IR}_o} f^2(x) \, d\nu(x) < \infty$, then

$$\int_0^{t+} \int_{\mathrm{IR}_o} f(x) \, \tilde{N}(ds, dx)$$

is well defined and

$$E\left(\int_0^{t+} \int_{\mathrm{IR}_o} f(x) \, \tilde{N}(ds, dx)\right)^2 = \int_0^t \int_{\mathrm{IR}_o} f^2(x) \, ds \, d\nu(x)$$

$$= t \int_{\mathrm{IR}_o} f^2(x) \, ds \, d\nu(x),$$

where $\tilde{N} = N - EN$.

The next theorem describes a Lévy process $\{X(t)\}$ in terms of Poisson random measures constructed above.

Theorem 8.6.3 *Let $\{X(t)\}$ be a Lévy process and N the Poisson measure constructed by $\{X(t)\}$. Then*

$$X(t) = B(t) + \int_0^{t+} \int_{|x| \leq 1} x \, N(ds, dx) + \int_0^{t+} \int_{|x| > 1} x \, N(ds, dx) + t\gamma$$

where $B(t)$ is the Brownian motion.

We defined in Section 4.3 the stable integral

$$\int_0^t f(s) \, dZ(s),$$

where $Z(s)$ is α-stable Lévy motion and $\int_0^t |f(s)|^\alpha < \infty$. It is known that its Lévy measure has the form

$$\nu(dx) = \begin{cases} \alpha\{C_+ I[x > 0] + C_- I[x < 0]\}|x|^{-\alpha-1} \, dx & \text{if } 0 < \alpha < 2 \\ 0 & \text{if } \alpha = 2 \end{cases} \qquad (8.6.1)$$

where C_+, $C_- \geq 0$ and $C_+ + C_- > 0$. We can write the above integral for $0 < \alpha < 2$ using the Poisson random measure (see Samorodnitsky and Taqqu (1993)).

$$\int_0^t f(s) \, dZ(s) = \begin{cases} \int_0^{t+} \int_{\mathrm{IR}_o} f(s)x \, N(ds, dx) & \text{if } 0 < \alpha < 1 \\ \int_0^{t+} \int_{|x| > \delta} f(s)x \, N(ds, dx) & \text{if } \alpha = 1 \\ \quad + \int_0^{t+} \int_{|x| \leq \delta} f(s)x \, \tilde{N}(ds, dx) & \\ \int_0^{t+} \int_{\mathrm{IR}_o} f(s)x \, \tilde{N}(ds, dx) & \text{if } 1 < \alpha < 2 \end{cases} \qquad (8.6.2)$$

8.7 Limit Theorems for Sums of i.i.d. Random Variables

Let $\{X_j\}_{j=1}^{\infty}$ be a sequence of i.i.d. random variables such that

$$\frac{1}{\varphi(n)} \sum_{j=1}^{n} \xrightarrow{\mathcal{L}} Z_\alpha(1) \tag{8.7.1}$$

as $n \to \infty$ for some $\varphi(n)$, where $Z_\alpha(t)$ is a strictly stable Lévy motion. In order that $F(x)$ (F is common distribution of $\{X_j\}$) belong to the domain of attraction of a stable law with index α, $0 < \alpha \leq 2$, it is necessary and sufficient that there exists a slowly varying function $\varphi(n)$ with index $1/\alpha$ such that as $n \to \infty$, when $\alpha \neq 2$

$$\begin{aligned} n\{1 - F(\varphi(n)x)\} &\longrightarrow C_+ x^{-\alpha} & x > 0, \\ nF(\varphi(n)x) &\longrightarrow C_- |x|^{-\alpha} & x < 0, \end{aligned} \tag{8.7.2}$$

with C_+ and C_- as defined in Section 8.6. When $\alpha = 2$,

$$n\{1 - F(\varphi(n)x)\} \longrightarrow 0 \qquad\qquad\qquad \text{if } x > 0,$$

$$nF(\varphi(n)x) \longrightarrow 0 \qquad\qquad\qquad \text{if } x < 0, \tag{8.7.3}$$

$$n\{\textstyle\int_{|x|\leq 1} x^2 \, dF(\varphi(n)x) - (\int_{|x|\leq 1} x \, dF(\varphi(n)x))^2\} \longrightarrow \sigma^2$$

We note that (8.7.2) is written as

$$\nu_n(dx) \xrightarrow{n\to\infty} \nu(dx) \tag{8.7.4}$$

in the vague topology in $\mathbb{R}_O$, and

$$\int_{|x|\leq a} x^2 \, \nu_n(dx) - \frac{1}{n}\left(\int_{|x|\leq a} x \, \nu(dx)\right)^2 \to \int_{|x|\leq a} x^2 \, \nu(dx) \tag{8.7.5}$$

for any $a > 0$, where $\nu_n(dx) = n \, dF(\varphi(n)x)$ and $\nu(dx)$ were defined in (8.6.1) and (8.7.3) is also written as

$$\begin{aligned} &\nu_n(dx) \to 0 \qquad \text{vaguely in } \mathbb{R}_O \\ &\textstyle\int_{|x|\leq 1} x^2 \, \nu_n(dx) - \frac{1}{n}(\int_{|x|\leq 1} x \, \nu_n(dx))^2 \to \sigma^2. \end{aligned} \tag{8.7.6}$$

For later use we also note that from (8.7.2) it follows that

$$\lim_{n\to\infty} \textstyle\int_{|x|\leq a} |x|^\beta \, \nu_n(dx) = \int_{|x|\leq a} |x|^\beta \, \nu(dx) \quad \text{if } \beta > \alpha \tag{8.7.7}$$

and

$$\lim_{n\to\infty} \textstyle\int_{|x|\geq a} |x|^\beta \, \nu(dx) = \int_{|x|\geq a} |x|^\beta \, \nu(dx) \quad \text{if } \beta < \alpha \tag{8.7.8}$$

Above properties 8.7.7 and 8.7.8 can be shown by the definition of ν_n, the property of regular variation of F and theorem VIII.9 in Feller (1971). Now we give a special case of a result in a paper by Kasahara and Watanabe (1986).

Let $\{X_{n,i}\}_{i=1}^{\infty}$, $n = 1, 2 \ldots$ be a collection of random variables which are independent for each n, and let $\nu(dx)$ be a Borel measure such that $\nu(x : |x| > \epsilon) < \infty$ for any $\epsilon > 0$. Also let $\{g_n(u, x)\}$, $g(u, x)$, $h_n(u, x)$ and $h(u, x)$ be measurable functions defined on $\mathbb{R}_O \times \mathbb{R}$ such that $h_n(u, 0) = 0$ for all $u \in \mathbb{R}_O$. Suppose the conditions (i)–(v) to be satisfied:

(i)

$$\sum_{i \le nt} P\{X_{n,i} > x\} \longrightarrow t\nu((x, \infty))$$

as $n \to \infty$ at continuity points $x > 0$ of ν and

$$\sum_{i \le nt} P\{X_{n,i} < x\} \longrightarrow t\nu((-\infty, x))$$

as $n \to \infty$ at continuity points $x < 0$ of ν;

(ii) $g_n(u, x)$ and $h_n(u, x)$ coverge continuosly to $g(u, x)$ and $h(u, x)$ almost surely with respect to $du\, \nu(dx)$, respectively;

(iii)

$$\lim_{\epsilon \downarrow 0} \limsup_{n \to \infty} \sum_{i \le nt} E[|g_n(i/n, X_{n,i})| I(|X_{n,i}| < \epsilon)] = 0$$

for any $t > 0$;

(iv)

$$\lim_{\epsilon \downarrow 0} \limsup_{n \to \infty} \left| \sum_{i \le nt} \left\{ E[h_n^2(i/n, X_{n,i}) I(|X_{n,i}| < \epsilon)] \right. \right.$$
$$\left. \left. [E[h_n(i/n, X_{n,i}) I(|X_{n,i}| < \epsilon)]]^2 \right\} - m(t) \right| = 0$$

for some continuous function $m(t)$;

(v) for any $T > 0$

$$\int_0^T \int_{\mathbf{R}_O} \{|g(u, x)| I(|x| \le 1) + [h(u, x)]^2\}\, du\, \nu(dx) < \infty$$

$$\sup_{0 \le u \le T} |h_n(u, x)| < \infty$$

and

$$\lim_{\epsilon \downarrow 0} \sup_n \sup_{\substack{0 \le u \le T \\ 0 \le |x| \le \epsilon}} |h_n(u, x)| = 0.$$

Then the following theorem holds (for details see Kasahara and Watanabe (1986)).

Theorem 8.7.1

$$\left(\sum_{i \le nt} g_n(i/n, X_{n,i}); \sum_{i \le nt} \{h_n(i/n, X_{n,i}) - E[h_n(i/n, X_{n,i})]\} \right) \xrightarrow{\mathcal{L}}$$

$$\left(\int_0^{t+} \int_{\mathbf{R}_O} g(u, x) \, N(du, dx); \int_0^{t+} \int_{\mathbf{R}_O} h(u, x) \, \tilde{N}(du, dx) + M(t) \right)$$

as $n \to \infty$ *over the space* $(D([0, \infty)) : \mathbb{R}^2)$, *where* $N(du, dx)$ *is the Poisson random measure with intensity* $du \, \nu(dx)$ *and* $M(t)$ *is a continuous martingale independent of* $N(du, dx)$ *such that the quadratic variation* $< M(t) >$ *is* $m(t)$.

Let us return to a sequence $\{X_j\}_{j=1}^\infty$ of i.i.d. random variables with common distribution $F(x)$ belonging to the domain of attraction of a stable law with index α, $0 < \alpha \le 2$.

In order to proof the next theorem we need three following assumptions.

(A) When $1 < \alpha \le 2$, $\int_{-\infty}^\infty x \, dF(x) = 0$. When $\alpha = 1$, $C_+ = C_-$ and $n \int_{|x| \le a} x \, dF(\varphi(n)x) \to 0$ as $n \to \infty$.

(B) Let $\{f_n(u)\}_{n=1}^\infty$ and $f(u)$ be measurable functions defined on $(0, \infty)$, $\{f_n(n)\}$ being uniformly bounded on finite intervals.

(C) $\{f_n(u)\}$ converges continuously to $f(u)$ on $(0, \infty)$ almost surely, namely, it is for almost all u that for any u_n tending to u, $f_n(u_n)$ coverges to $f(u)$.

So we obtain the following result (see Kasahara and Maejima (1986)).

Theorem 8.7.2 *Under the assumptions stated above*

$$Z_n(t) = \frac{1}{\varphi(n)} \sum_{i \le nt} f_n\left(\frac{i}{n}\right) X_i \xrightarrow{\mathcal{L}} \int_0^t f(s) \, dZ(s)$$

over the space $D([0, \infty) : \mathbb{R})$.

PROOF. To apply Theorem 8.7.1, put

$$g_n(u, x) = f_n(u) x \, \boldsymbol{I}(|x| > \delta)$$

$$g(u, x) = f(u) x \, \boldsymbol{I}(|x| > \delta)$$

$$h_n(u, x) = f_n(u) x \, \boldsymbol{I}(|x| \le \delta)$$

$$h(u, x) = f(u) x \, \boldsymbol{I}(|x| \le \delta)$$

and

$$X_{n,i} = \frac{1}{\varphi(n)} X_i.$$

We have

$$Z_n(t) = \sum_{i \le nt} f_n(i/n)\{X_i/\varphi(n) - \int_{|x| \le \delta} x \, dF(\varphi(n)x)\} +$$

$$+ \int_{|x| \leq \delta} x \, dF(\varphi(n)x) \sum_{i \leq nt} f_n(i/n) = Z_n^\delta(t) + A_n^\delta(t).$$

Let us check the conditions of the previous theorem.

(i) Let $x > 0$

$$\sum_{i \leq nt} P(X_{n,i} > x) = [nt]P\left(\frac{1}{\varphi(n)}X_1 > x\right)$$

$$= \frac{[nt]}{n} n \int_{|y| > x} dF(\varphi(n)y) \to t \int_{|y| > x} \nu(dy).$$

We proceed similarly for $x < 0$, (see (8.7.8))

(ii) It follows from assumption (C).

(iii) From assumption (B) we have $|f_n(x)| < M$ for all x and n, so that

$$\sum_{i \leq nt} E\left[|g_n\left(i/n, X_{n,i}\right)| \, I\{|X_{n,i}| < \epsilon\}\right]$$

$$= \sum_{i \leq nt} E\left[\left|f_n\left(\frac{i}{n}\right)\frac{X_i}{\varphi(n)}\right| I\left\{\delta < \frac{|X_i|}{\varphi(n)} < \epsilon\right\}\right]$$

$$\leq \sum_{i \leq nt} E\left[\epsilon M I\left\{\frac{X_i}{\varphi(n)} > \delta\right\}\right] \xrightarrow{n \to \infty} \epsilon M \, t\nu(x : |x| > \delta)$$

(see equation (8.7.8).

(iv) If $\alpha \in (0,2)$ then taking $\epsilon < \delta$ and $m(t) = 0$ we obtain

$$\left|\sum_{i \leq nt}\left\{E\left[\left(f_n\left(\frac{i}{n}\right)\frac{|X_i|}{\varphi(n)}\right)^2 I\left\{\frac{X_i}{\varphi(n)} < \epsilon\right\}\right]\right.\right.$$

$$\left.\left. - \left[E\left[f_n\left(\frac{i}{n}\right)\frac{X_i}{\varphi(n)} I\left\{\frac{|X_i|}{\varphi(n)} < \epsilon\right\}\right]\right]^2\right\} - m(t)\right|$$

$$\leq M^2[nt]\left|E\left[\left(\frac{X_1}{\varphi(n)}\right)^2 I\left\{\frac{|X_1|}{\varphi(n)} < \epsilon\right\}\right]\right.$$

$$\left. - \left[E\left[\frac{X_1}{\varphi(n)} I\left\{\frac{X_1}{\varphi(n)} < \epsilon\right\}\right]\right]^2\right| \xrightarrow{n \to \infty} tM^2 \int_{|x| \leq \epsilon} x^2 \, \nu(dx).$$

If $\alpha = 2$ then taking $m(t) = \sigma^2 \int_0^t f^2(s) \, ds$ we obtain 0 in the limit, since

$$n\left\{E\left[\left(\frac{X_1}{\varphi(n)}\right)^2 I\left\{\frac{|X_1|}{\varphi(n)} < \epsilon\right\}\right] - \left[E\left[\frac{X_1}{\varphi(n)} I\left\{\frac{X_1}{\varphi(n)} < \epsilon\right\}\right]\right]^2\right\} \xrightarrow{n \to \infty} \sigma^2$$

and

$$\frac{1}{n} \sum_{i \leq nt} f_n^2(i/n) \xrightarrow{n \to \infty} \int_0^t f^2(s) \, ds$$

(using equation (8.7.6) and assumptions (B) and (C), respectively).

So, for $\alpha = 2$, $M(t) = \sigma B(t)$, where $B(t)$ is the standard Brownian motion.

(v) It follows from assumption (B) and the properties of measure ν

From Theorem 8.7.1 we obtain, if $0 < \alpha < 2$,

$$Z_n^\delta(t) \xrightarrow{\mathcal{L}} \int_0^{t+} \int_{|x|>\delta} f(u)x \; \tilde{N}(dx, du) + \int_0^t \int_{|x|\leq\delta} f(u)x \; N(du, dx)$$

and, if $\alpha = 2$,

$$Z_n^\delta(t) \xrightarrow{\mathcal{L}} \sigma \int_0^t f(u) \, dB(u)$$

over the space $D([0, \infty) : \mathbb{R})$.

Now it is enough to calculate the limit of $A_n^\delta(t)$. From assumption (A) and (8.7.2) and (8.7.3) we have

$$n \int_{|x|\leq\delta} x \; dF(\varphi(n)x) \longrightarrow \begin{cases} \int_{|x|\leq\delta} x \; \nu(dx) & 0 < \alpha < 1 \\ -\int_{|x|>\delta} x \; \nu(dx) & 1 < \alpha < 2 \\ 0 & \alpha = 1, \quad \alpha = 2 \end{cases}$$

It remains to calculate the limit of $\frac{1}{n} \sum_{i \leq nt} f_n(\frac{i}{n})$. It follows from assumptions (B) and (C) that

$$\frac{1}{n} \sum_{i \leq nt} f_n\left(\frac{i}{n}\right) = \int_0^{([nt]+1)/n} f_n\left(\frac{[nt]}{n}\right) du \longrightarrow \int_0^t f(u) \, du$$

and covergence is uniform on any bounded interval of t. Hence,

$$\lim_{n \to \infty} A_n^\delta(t) = \begin{cases} \int_0^t \int_{|x|\leq\delta} f(u)x \; du \; \nu(dx) & 0 < \alpha < 1, \\ -\int_0^t \int_{|x|>\delta} f(u)x \; du \; \nu(dx) & 1 < \alpha < 2, \\ 0 & \alpha = 1, \quad \alpha = 2. \end{cases}$$

This ends the proof. $\square$

Chapter 9

Chaotic Behavior of Stationary Processes

9.1 Examples of Chaotic Behavior

Let us consider three examples motivating the concept of ergodicity, as well as a hierarchy of chaotic behavior.

Example 9.1.1 *Measurement of average rainfall.*

Suppose that rainfall data are collected at a very large number of observation points $p_0, p_1, \ldots$ at times $t = 0, 1, \ldots$. Records and analysis of the rainfall pattern are indispensable to agriculturists and builders of reservoirs, dams, flood control works, irrigation systems, power plants, water works, airports and urban storm sewers, etc. Assume that the statistical character of the observations at p_i is the same for all i, and is represented by a stationary sequence of random variables $X_0, X_1, \ldots$, with X_n – the amount of rainfall at time n. Assume also that the p_i correspond to a sequence of independent performances of a random experiment, where a performance means an observation of the entire random sequence $X_0, X_1, \ldots$.

Suppose that the problem is to measure the average rainfall. Analyst A_1 might take the following approach. He might take measurements at each observation point at a given time, say $t = 0$, and average the results. Analyst A_2 might reason as follows. Since all observation points have the same statistical character, we can simply go to one observation point, take a large number of observations, say, at $t = 0, 1, \ldots, n$, and average the results. Analyst A_1 is using what might be called a vertical measuring scheme, and analyst A_2 a horizontal scheme, as illustrated below.

231

Table 9.1.1. Scheme of measurements.

observation points	$t = 0$	$t = 1$	$t = 2$	$\ldots$
p_0	X_{00}	X_{01}	X_{02}	$\ldots$
p_1	X_{10}	X_{11}	X_{12}	$\ldots$
p_2	X_{20}	X_{21}	X_{23}	$\ldots$
$\vdots$	$\vdots$			

A_1's observations correspond to the first column, A_2's to the first row. It is clear that A_1 and A_2 will not necessarily obtain the same result. What A_2 is computing is the time average, namely

$$\frac{1}{n} \sum_{k=0}^{n-1} X_k(\omega)$$

for a particular ω. But A_1 is observing the ensemble average at a particular time, namely,

$$n^{-1}(Y_0 + Y_1 + Y_2 + \ldots + Y_{n-1}),$$

where the Y_j's are independent random variables, all having the same distribution as X_0. Thus A_1's result would approximate $E(X_0) = \int_\Omega X_0 dP$. Hence, for A_1 and A_2 to get approximately the same answer, we must have

$$\frac{1}{n} \sum_{k=0}^{n-1} X_k(\omega) \to \int_\Omega X_0 \, dP.$$

If we observe that here $X_k(\omega) = X_0(S^k\omega)$, where S is the one–sided shift on $\Omega = \mathbb{R}^\infty$, i.e.,

$$S(\omega_0, \omega_1, \ldots) = (\omega_1, \omega_2, \ldots),$$

then, more generally, we might ask when it will be true that for each integrable function f on Ω we get

$$\frac{1}{n} \sum_{k=0}^{n-1} f(S^k\omega) \to \int_\Omega f \, dP$$

at least for almost every ω. In particular, if f is an indicator function I_B, the property to be verified is simply the convergence of the relative frequency of visits to B in the first n steps to the probability of B. Now suppose that B is an "almost invariant" set, i.e., B and $S^{-1}B$ differ only by a set of measure 0. Then the LHS of the above formula is almost everywhere equal to

$$\frac{1}{n} \sum_{k=0}^{n-1} I_B(S^k\omega) = I_B(\omega)$$

for all n. Thus the relative frequency of visits to B cannot converge to $P(B)$, except when $P(B) = 0$ or 1. Conversely, the Birkhoff Ergodic Theorem (see Section 9.3) implies that if every almost invariant set has probability 0 or 1, the above convergence result holds.

Example 9.1.2 *Hierarchy of chaos in statistical mechanics.*

It was believed that integrable systems reducible to free particles, for which a single function, namely the Hamiltonian $H(p, q)$, describes the dynamics completely, were the prototype of dynamical systems. Generations of physicists and mathematicians tried hard to find for each kind of systems the "right" variables that would eliminate the interactions. This program failed. At the end of the nineteenth century, Bruns and Poincaré demonstrated that most of the dynamical systems, starting with the famous "three body" problem (the moon's motion, influenced by both the earth and the sun), were not integrable. On the other hand, the idea of approaching equilibrium in terms of the theory of ensembles introduced by Gibbs and Einstein requires that we go beyond the idealization of integrable systems. According to the theory of ensembles, an isolated system is in equilibrium if and only if it is represented by a "microcanonical ensemble", i.e., when all points on the surface of a given energy have the same probability. This means that, for a system to evolve to equilibrium, energy must be the only quantity conserved during its evolution. It must be the only invariant. But for an integrable system energy is far from being the only invariant. In fact, there are as many invariants as degrees of freedom, since each generalized momentum remains constant. Therefore, we have to expect that such a system is "imprisoned" in a very small fraction of the constant–energy surface formed by the intersection of all these invariant surfaces. To avoid these difficulties, Maxwell and Boltzmann introduced a new, quite different type of dynamical systems. For these systems energy would be the only invariant. Such systems are called *ergodic*.

If we consider the temporal evolution of a cell in the (p, q)–phase space then the "volume" of the cell and its form are maintained in time; moreover, most of the phase space is inaccessible to the system. In contrast, typical evolution of a cell corresponding to an ergodic system is quite different. When time increases, the "volume" and the form are conserved but the cell now spirals through the whole space. Today we know that there are large classes of dynamical (though non–Hamiltonian) systems that are ergodic. We also know that even relatively simple systems may have properties stronger than ergodicity: weak mixing, strong mixing, Kolmogorov property, exactness. This hierarchy exhibits gradually stronger chaotic properties. Kolmogorov systems, which are invertible and therefore cannot be exact are stronger than mixing. To some extent they are parallel to exact systems. For the significance of these properties for studying chaotic behavior of physical systems we refer the reader to Lasota and Mackey (1985). For these systems the motion in phase space becomes highly chaotic, while always preserving "volume" in agreement with the Liouville Theorem.

Suppose that our knowledge of initial conditions permits us to localize a system in a small cell of the phase space. During its evolution, we shall see this initial cell twist and turn, and, like an amoeba, send out "pseudopods" in all directions, spreading out in increasingly thinner and ever more twisted filaments until it finally invades the whole space. Schematic representations of temporal evolutions for such systems are presented in physical textbooks, however no sketch can do justice to the complexity of the actual situation.

To illustrate this situation we present below some simple mathematical models of dynamical systems with the discrete time. Let D be the unit square in a plane, i.e., $D = [0,1] \times [0,1]$. The Borel σ–algebra is now generated by all possible rectangles of the form $[0,a] \times [0,b]$, and the Borel measure μ is the unique measure on the Borel σ–algebra such that $\mu([0,a] \times [0,b]) = ab$. Define the following three transformations $S_i : D \to D$ from the family of the *Anosov transformations*

$$\begin{aligned} S_1(x,y) &= (\sqrt{2}+x, \sqrt{3}+y) \quad (mod \quad 1), \\ S_2(x,y) &= (x+y, x+2y) \quad (mod \quad 1), \\ S_3(x,y) &= (3x+y, x+3y) \quad (mod \quad 1), \end{aligned}$$

(cf. Lasota and Mackey (1985)). To see, for example, the effect of S_2 transformation, observe that in the first step we divide the unit square D into four triangular areas. In the second step the transformation

$$(x,y) \to (x+y, x+2y)$$

maps $[0,1] \times [0,1]$ into $[0,3] \times [0,3]$. Finally, in the third step, applying the "modulo one" operation, we get the result of this transformation in the unit square D. It is clear that the effect of transformation S_2 will be to very quickly scramble, or mix, various regions of D. To illustrate this, we present the first six iterates of a cell formed by a random distribution of 1000 points chosen at the uper left corner of the unit square D by four transformations, see Figure 9.1.1.

Successive iterates of this cell by transformations S_1, S_2, and S_3 are shown in the first, second, and third column, respectively. Note how differently the same initial distribution of points in the given cell moves in the phase space D. It turns out that the above Anosov transformations illustrate three levels of chaotic behavior that a dynamical system can display. These three levels are known as ergodicity, mixing, and exactness.

The fourth column in Figures 9.1.1 – 9.1.3 presents the iterates of the same cell by the *baker transformation*

$$S_b(x,y) = \begin{cases} (2x, \frac{1}{2}) & \text{if } x \in [0, \frac{1}{2}), \\ (2x-1, \frac{1}{2}y + \frac{1}{2}) & \text{if } x \in [\frac{1}{2}, 1]. \end{cases}$$

This is an example of an automorphism, which is a Kolmogorov system (see Cornfeld, Fomin and Sinai (1982)).

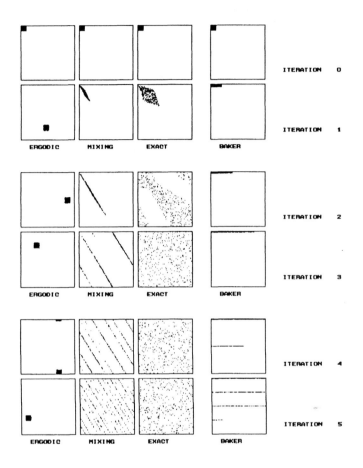

Figure 9.1.1. Graphical representation of the first six iterations of three Anosov transformations versus the baker transformation.

In oder to show what happens to these systems when the number of iterates increases, we refer to Figure 9.1.2 and Figure 9.1.3. The application of these four transformations are presented in the ranges $8 - 25$ and $40 - 121$, respectively. In Figure 9.1.2 we observe a rapid spread of the initial distribution of points throughout the phase space except for the ergodic transformation. Further application of the Anosov transformations presented in Figure 9.1.3 demonstrates their different behavior. It is also clear that the baker transformation, which is invertible and therefore cannot be exact, is stronger than mixing. To some extent its behavior is parallel to exact transformation S_3. The interested reader can put in the fourth column any transformation $S : D \to D$ and try to test its level of chaotic behavior.

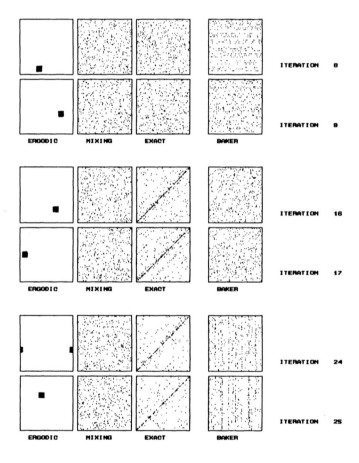

Figure 9.1.2. Graphical representation of three Anosov transformations versus the baker transformation for six iterations in the range 8–25.

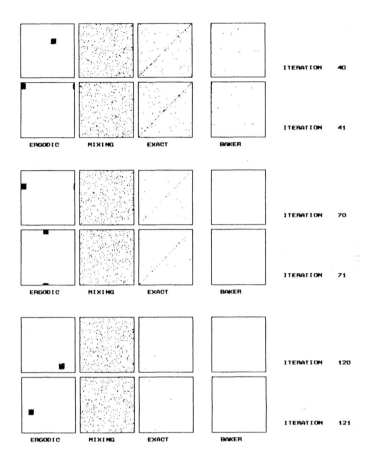

Figure 9.1.3. Graphical representation of three Anosov transformations versus the baker transformation for further six iterations in the range 40–121.

Example 9.1.3 *Consistency question in statistics.*

Consistency in classical statistics means that the estimate θ_n^* tends to the true value θ in probability as the sample size tends to infinity. This is related to perpendicularity if we consider infinite, rather than potentially infinite, samples with $\mathbb{T} = \mathbb{N}$.

A better way of expressing this is to say that $\theta^* = \theta^*(X)$ is a *consistent estimator* of θ, with the observed stochastic element $X = X(t)$, $t \in \mathbb{T}$, if $P_\theta[\theta^*(X) = \theta] = 1$ for all $\theta \in A$. When $\mathbb{T} = \mathbb{N}$, this statement can of course be expressed by the asymptotic relation as above, but in the present context the latter way of formulating the definition is more concise.

For stochastic processes with continuous time we encounter the same problem. Suppose that we observe a weakly stationary process $X(t)$, continuous in the mean, and with mean θ, $t \in \mathbf{T} = (0, L)$. If we form the *time average*

$$\overline{X}_L = \frac{1}{L} \int_0^L X(t) \, dt$$

we know that it is an unbiased estimate of θ with variance

$$\text{Var}(\overline{X}_L) = \frac{1}{L^2} \int_0^L \int_0^L r(s - t) \, ds \, dt = \frac{1}{L} \int_{-L}^L r(h)(1 - \frac{|h|}{L}) \, dh \qquad (9.1.1)$$

Is $\overline{X}_L$ consistent? This can be answered by studying the limit of (9.1.1) as $L \to \infty$. Another way is by appealing to the von Neumann Mean Ergodic Theorem (see Section 9.3), which tells us that $\overline{X}_L$ converges in the mean. We can be more specific. Expressing the process in the Cramér Representation and integrating over $(0, L)$, we get

$$\overline{X}_L = \int_{-\infty}^{\infty} K_L(\lambda) Z(d\lambda),$$

where

$$K_L(\lambda) = \frac{e^{iL\lambda} - 1}{iL\lambda}.$$

This function is bounded uniformly, $|K_L(\lambda)| \leq 1$, and $K_L(0) = 1$. When L tends to infinity, it tends to zero everywhere except at $\lambda = 0$. Hence $K_L \to K_\infty$ in the mean with respect to the F–measure associated with the process $\mathbf{Z}$, where

$$K_\infty(\lambda) = \begin{cases} 1 & \text{for } \lambda = 0, \\ 0 & \text{otherwise.} \end{cases}$$

The Isometry Theorem then tells us that the limit in the mean of $\overline{X}_L$, as L tends to infinity, is the jump of the process $\mathbf{Z}$ at frequency zero

$$\lim_{L \to \infty} \|\overline{X}_L - [Z(0+) - Z(0-)]\| = 0.$$

We also know that the variance of $\Delta Z(0)$ is equal to $\Delta F(0)$, the jump of the spectral measure at 0. In other words, the time average is a consistent estimate of the mean if and only if the spectrum is continuous at 0 and contains no discontinuous components.

While this is useful to know, it is not general enough since it only tells us something about a particular parameter, the mean, and its particular estimate $\overline{X}$. To probe deeper into the consistency question, we must consider more general parameters. Let us suppose that our parameter θ can be expressed as a continuous function $\theta = h(g_1, g_2, ..., g_p)$ of quantities g_κ that themselves are expected values of functions of a stochastic process at certain time points. For example, we may have:

$$
\begin{aligned}
g_1 &= E_\theta[X^2(t)], \\
g_2 &= E_\theta[X(t)X(t+h)], & h > 0, \\
g_3 &= E_\theta[|X(t+h) - X(t)|], & h > 0, & \qquad (9.1.2)
\end{aligned}
$$

.
.
.

If we can find consistent estimators $g_1^*, ..., g_p^*$ of $g_1, ..., g_p$, then $h(g_1^*, g_2^*, ..., g_p^*)$ is a consistent estimator of θ. We cannot deal with this problem using second–order properties only since we have allowed nonlinear functions in the definition of the g_j's (see (9.1.2)). Instead we now assume that $X(t)$ is strictly stationary, so that we can appeal to the Individual Ergodic Theorem (see Section 9.3). It is then natural to estimate the g_j's by time averages approximating ensemble averages:

$$
\begin{aligned}
g_1^* &= \tfrac{1}{L} \int_0^L X^2(t)\, dt, \\
g_2^* &= \tfrac{1}{L} \int_0^{L-h} X(t)X(t+h)\, dt, \\
g_3^* &= \tfrac{1}{L} \int_0^{L-h} |X(t+h) - X(t)|\, dt,
\end{aligned}
$$

.
.
.

Of course we must assume that $X(t)$ is measurable and has sufficiently high moments so that the integrals defining g_j^*'s exist. The Individual Ergodic Theorem then guarantees that the averages converge with probability 1 to some limiting stochastic variables. If, moreover, $X(t)$ is ergodic, then all these stochastic variables are equal to the expected values in (9.1.2) a.e. so that we can estimate θ consistently.

9.2   Ergodic Property of Stationary Gaussian Processes

As we have seen in the previous section the question of *consistent estimation* leads us to study *ergodicity*, and we will discuss this problem in detail now in the Gaussian case. A beautiful and complete answer to this question is given by the following theorem (Maruyama (1949), Grenander (1950) and Fomin (1950)).

Theorem 9.2.1 *In order that the real and stationary Gaussian process with continuous covariance function be ergodic, it is necessary and sufficient that its spectrum be continuous.*

PROOF. Writing the covariance function in its Bochner representation

$$
r(t) = \int_{-\infty}^{\infty} e^{it\lambda} F(d\lambda),
$$

we can assume without loss of generality that the mean is 0 and also that $r(t) = 1$, so that F is an ordinary distribution function. The F–measure is symmetric since the process, and hence, r is real.

To show that the condition stated above is necessary, is easy; we just consider the strictly stationary process $Y(t) = X^2(t)$. Its covariance function is

$$\rho(t) = E\{[X^2(s+t) - EX^2(s+t)][X^2(s) - EX^2(s)]\} = 2r^2(t),$$

using the fourth–order moments for Gaussian distributions. But the above equality implies that the spectral distribution F_Y of ρ is twice the convolution of F with itself. We know that

$$\lim_{L\to\infty} \frac{1}{L} \int_0^L Y(t)\, dt = 1 + \Delta Z_Y(0) \qquad \text{with variance} \quad \Delta F_Y(0).$$

If $X(t)$ is ergodic, the limit above should be a.e. constant and thus $\Delta F_Y(0) = 0$. But then Fourier inversion gives

$$0 = \Delta F_Y(0) = \lim_{L\to\infty} \frac{1}{2L} \int_{-L}^L \rho(t)\, dt = \lim_{L\to\infty} \frac{1}{L} \int_{-L}^L r^2(t)\, dt.$$

The last limit is well known from classical Fourier analysis (see, e.g. Bochner (1932)). It is twice the jump at $\lambda = 0$ of F_Y. Summing the convolutions F_κ of F with itself over all points of discontinuity of measure F, which is symmetric around 0, we get

$$\Delta F_Y(0) = \sum_\kappa [\Delta F(\lambda_\kappa)]^2.$$

Hence, the quantity above must be equal to 0, which says that F must be continuous everywhere.

To prove sufficiency we use the fact that ergodicity is equivalent to metric transitivity. The latter means that the only measurable sets that are invariant with respect to translations of the time axis are the trivial ones having P–measure 0 or 1. Let us assume for an indirect proof that there is an invariant set S with $P(S) = \rho$, $0 < \rho < 1$. Now we prove that F cannot be continuous.

For any given $\varepsilon > 0$ we can approximate S by a cylinder I with a finite dimensional base B such that

$$P(I) < \rho + \varepsilon \quad \text{and} \quad P(S \cap I^c) < \varepsilon.$$

Let the time point appearing in the definition of B be $\tau_1, \tau_2, ..., \tau_n$, and consider the translate $T_t I = I_t$. Then I_t has a base associated with the time points $\tau_1 + t, \tau_2 + t, ..., \tau_n + t$. Consider now $2n$ stochastic variables $x_i = X(\tau_i)$, $x_{n+i} = X(t + \tau_i)$, $i = 1, 2, ..., n$. They have a Gaussian probability distribution with the covariance matrix in the following block form

$$\Lambda(t) = \left\{ \begin{array}{cccc|cccc} 1 & r(\tau_1 - \tau_2) & \cdots & & r(\tau_1 - \tau_1 - t) & \cdots & \\ \cdots & & & & \cdots & & \\ r(\tau_n - \tau_1) & \cdots & 1 & & r(\tau_n - t - \tau_1) & \cdots & \\ \hline r(\tau_1 + t - \tau_1) & \cdots & & & 1 & r(\tau_1 - \tau_2) & \cdots \\ \cdots & & & & \cdots & & \\ r(\tau_n - t - \tau_1) & \cdots & & & r(\tau_n - \tau_1) & \cdots & 1 \end{array} \right\}$$

$$= \left\{ \begin{array}{cc} A & B(t) \\ B(t) & A \end{array} \right\}.$$

Recall that if a stationary process has a continuous spectrum, then all its finite dimensional covariance matrices are nonsingular. We now complete the proof of the theorem. The covariance matrices are nonsingular for large t–values when the sets $\{t_\kappa\}$ and $\{t + t_\kappa\}$ do not overlap. Hence we can write

$$P(I \cap I_t) = (2\pi)^{-n}[\det\Lambda(t)]^{-\frac{1}{2}} \int \ldots \int \exp\left[-\frac{1}{2}Q(x)\right] dx_1 dx_2 \ldots dx_{2n} \quad (9.2.1)$$

integrated over $(x_1, \ldots x_n) \in I$, $(x_{n+1}, \ldots, x_{2n}) \in I_t$, and where

$$Q(x) = x^T \Lambda^{-1}(t)x.$$

Let us arrange all the numbers $\tau_\kappa - \tau_\mu$ as $t_1, t_2, \ldots, t_N$. Direct evaluation gives

$$\frac{1}{2L} \int_{-L}^{L} \sum_{k=1}^{N} |r(t_k + t)|^2 \, dt$$

$$= \frac{1}{2L} \int_{-L}^{L} \sum_{k=1}^{N} \int_{-\infty}^{\infty} \int_{-\infty}^{\infty} e^{i(t_k+t)\lambda - i(t_k+t)\mu} F(d\lambda)F(d\mu) \, dt$$

$$= \sum_{k=1}^{N} \int_{-\infty}^{\infty} \int_{-\infty}^{\infty} \frac{e^{i(t_k+L)(\lambda-\mu)} - e^{i(t_k-L)(\lambda-\mu)}}{2Li(\lambda - \mu)} F(d\lambda) \, F(d\mu).$$

Let L tend to infinity in the right–hand side in the last expression above. For each k the integrand is bounded by 1 in absolute value and tends to 0 everywhere except on the diagonal $\lambda = \mu$ in the (λ, μ)–plane. Note that $F \times F$ is a bounded measure and apply the Lebesgue Bounded Convergence Theorem. The limit will be the mass of the product measure $F \times F$ on the diagonal. Since F is continuous, this mass is 0.

This implies that

$$\liminf_{t \to \infty} \sum_{1}^{N} |r(t_k + t)|^2 = 0,$$

so that there exists a sequence $s_\kappa \to +\infty$ such that the matrix introduced above $B(s_r) \to 0$. In other words,

$$\Lambda(s_r) \to \left\{ \begin{array}{cc} A & 0 \\ 0 & A \end{array} \right\}, \quad (9.2.2)$$

where the limiting matrix is also nonsingular. Using (9.2.2) in (9.2.1) and again the Lebesgue Bounded Convergence Theorem, we get

$$\lim_{\kappa \to \infty} P(I \cap I_{s_\kappa}) = P(I)^2 < (\rho + \varepsilon)^2.$$

For large values of κ this implies that with $\varepsilon > 0$,

$$
\begin{aligned}
(\rho+\varepsilon)^2 &> P(I \cap I_{s_\kappa}) \ge P(S \cap I \cap I_{s_\kappa}) \\
&\ge P(S) - P(S \cap I^c) - P(S \cap I^c_{s_\kappa}) > \rho - 2\varepsilon.
\end{aligned}
\tag{9.2.3}
$$

Here we have used the invariance of S and the stationarity of $X(t)$ to get

$$
P(S \cap I^c_{s_\kappa}) = P\left[(I^c_{s_\kappa}S) \cap I^c)\right] = P(S \cap I^c) < \varepsilon.
$$

But (9.2.3) is only possible for arbitrarily small ε if $\rho = 0$ or 1, contrary to our assumption.   □

For more details we refer the reader to Grenander (1981).

9.3   Basic Facts of General Ergodic Theory

Now, after recalling some basic facts and fixing notation, we present the Birkhoff Individual Ergodic Theorem.

Basic function spaces.   Let $(\Omega, \mathcal{F}, P)$ denote a probability space. By $L^0(\Omega, \mathcal{F}, P)$ we denote the space of all real random variables (measurable functions) on Ω, identifying two functions if they are equal with probability 1. The space $L^0(\Omega, \mathcal{F}, P)$ is a complete metric space with a topology of convergence in probability. We need also the function spaces $L^p(\Omega, \mathcal{F}, P)$ with $p \ge 1$, with two important special cases of the Banach space $L^1(\Omega, \mathcal{F}, P)$ and the Hilbert space $L^2(\Omega, \mathcal{F}, P)$.

Basic notions of ergodic theory.   Let $T : \Omega \to \Omega$ be a transformation that is *measurable* in the sense that $A \in \mathcal{F}$ implies $T^{-1}(A) \in \mathcal{F}$. If T is also one-to-one, $T(\Omega) = \Omega$ and $A \in \mathcal{F}$ implies $T(A) \in \mathcal{F}$ then we say that T is *invertible*. If $P(T^{-1}(A)) = P(A)$ for every $A \in \mathcal{F}$ then T is said to be *measure preserving*. If T is invertible then an equivalent requirement is that $P(T(A)) = P(A)$. Even if a measure–preserving transformation is not invertible, its range is essentially all of Ω, since $T(\Omega) \subset A \in \mathcal{F}$ implies $T^{-1}(A) = \Omega$ and hence $P(A) = 1$. In particular, if $T(\Omega)$ belongs to $\mathcal{F}$ then $P(T(\Omega)) = 1$.

Any transformation T of Ω induces a transformation U_T of the set of functions on Ω. Namely, for $f : \Omega \to \Omega$ we can define

$$
(U_T f)(\omega) = f(T(\omega)).
$$

Of course U_T is linear. If T is measure–preserving then U_T is an isometry on $L^1(\Omega, \mathcal{F}, P)$. The fact that U_T is an isometry on $L^1(\Omega, \mathcal{F}, P)$ implies immediately that U_T is an isometry on $L^2(\Omega, \mathcal{F}, P)$; all that is nedeed is the observation that the $L^2(\Omega, \mathcal{F}, P)$–norm of f is the square root of the $L^1(\Omega, \mathcal{F}, P)$–norm of f^2. If T is an invertible measure–preserving transformation then U_T is an invertible isometry.   An invertible isometry on a Hilbert space is a unitary operator. Thus the functional operator induced on $L^2(\Omega, \mathcal{F}, P)$ by an invertible measure–preserving transformation is unitary.

A basic asymptotic problem of ergodic theory reduces thus to studying the limiting behavior of the averages $\frac{1}{n}\sum_{j=0}^{n-1} U^j$, where U is an isometry on a Hilbert space. In Hilbert space terms, however, the natural question is not that of the pointwise convergence of $\frac{1}{n}\sum_{j=0}^{n-1} f(T^j(\omega))$, but rather its convergence in the mean of order two. The assertion that mean convergence always does take place is the first result of modern ergodic theory and was first proved by von Neumann. In this book, however, we are mostly interested in pointwise convergence of such sums. The solution to this problem is given by the Birkhoff Individual Ergodic Theorem. The proof presented here is due to A. M. Garsia (1965).

Theorem 9.3.1 *If T is a measure–preserving (not necessarily invertible) transformation on a probability space $(\Omega, \mathcal{F}, P)$ then for every $f \in L^1(\Omega, \mathcal{F}, P)$ the following limit exists almost everywhere*

$$\lim_{n\to\infty} \frac{1}{n} \sum_{j=0}^{n-1} f(T^j(\omega)) = f_*(\omega); \tag{9.3.1}$$

moreover,

$$f_* \in L^1(\Omega, \mathcal{F}, P), \quad \int f_* \, dP = \int f \, dP, \quad f_* \circ T = f_*. \tag{9.3.2}$$

We first prove the following lemma.

Lemma 9.3.1 *Let*

$$S_n(\omega) = \sum_{j=0}^{n-1} f(T^j(\omega)), \quad n \geq 1, \quad S_0(\omega) = 0,$$

$$B = \{\omega : \sup_{n\geq 1} S_n(\omega) > 0\}.$$

Then $\int_{A\cap B} f \, dP \geq 0$ for every set $A \in \mathcal{F}$ which is T–invariant.

PROOF. Write $S_n^+(\omega) = \max_{0\leq k\leq n} S_k(\omega)$. Then $S_n^+(\omega) \geq 0$ (since $S_0(\omega) = 0$) and

$$f(\omega) + S_n^+(T(\omega)) \geq f(\omega) + S_k(T(\omega)) = S_{k+1}(\omega), \quad \text{for} \quad k = 0,...,n.$$

Thus for any $\omega \in B_n \stackrel{df}{=} \{\omega : S_n^+(\omega) > 0\}$ we have

$$f(\omega) + S_n^+(T(\omega)) \geq \max_{1\leq k\leq n} S_k(\omega) = S_n^+(\omega).$$

Hence

$$\int_{A\cap B_n} f \, dP \geq \int_{A\cap B_n} \left(S_n^+(\omega) - S_n^+(T(\omega))\right) \, dP(\omega)$$

$$= \int_{A\cap B_n} S_n^+(\omega) \, dP(\omega) - \int_{A\cap B_n} S_n^+(T(\omega)) \, dP(\omega)$$

$$\geq \int_A S_n^+(\omega) \, dP(\omega) - \int_A S_n^+(T(\omega)) \, dP(\omega) = 0,$$

since T is measure–preserving, $B_n \subset B_{n+1}$ and A is T–invariant. As $\bigcup B_n = B$, we have

$$\int_{A \cap B} f \, dP = \lim_{n \to \infty} \int_{A \cap B_n} f \, dP \geq 0,$$

which completes the proof. $\square$

PROOF of Theorem 9.3.1. Write

$$B_u = \left\{ \omega : \sup_{n \geq 1} \frac{1}{n} \sum_{k=0}^{n-1} f(T^k(\omega)) > u \right\}.$$

Applying Lemma 9.3.1 to the function $f(\omega) - u$, we obtain

$$\int_{A \cap B_u} f \, dP \geq u \, P(A \cap B_u)$$

for every set $A \in \mathcal{F}$ which is T–invariant.

Further, set

$$\overline{f}(\omega) = \limsup_{n \to \infty} \frac{1}{n} \sum_{k=0}^{n-1} f(T^k(\omega)), \qquad \underline{f}(\omega) = \liminf_{n \to \infty} \frac{1}{n} \sum_{k=0}^{n-1} f(T^k(\omega)),$$

$$E_{u,v} = \left\{ \omega : \underline{f}(\omega) < v, \ \overline{f}(\omega) > u \right\}.$$

The set $E_{u,v}$ is measurable, T–invariant and contained in B_u. Hence, by Lemma 9.3.1

$$\int_{E_{u,v}} f \, dP \geq u \, P(E_{u,v}).$$

Repeating the argument with f, u, v replaced by $-f$, $-u$, $-v$, respectively, we obtain

$$\int_{E_{u,v}} f \, dP \leq v \, P(E_{u,v}).$$

Hence $P(E_{u,v}) = 0$ for $v < u$. Consequently,

$$P\{\omega : \underline{f}(\omega) < \overline{f}(\omega)\} = P \left(\bigcup_{\substack{v < u \\ u,v \ \text{rational}}} E_{u,v} \right) \leq \sum_{\substack{v < u \\ u,v \ \text{rational}}} P(E_{u,v}) = 0.$$

This proves the a.e. existence of the limit in (9.3.1).

We now pass to the proof of (9.3.2). Let $f = f^+ - f^-$ be the standard decomposition of f into the difference of non–negative functions. Limits analogous to (9.3.2) exist likewise for f^+ and f^-:

$$\frac{1}{n} \sum_{k=0}^{n-1} f^{\pm}(T^k(\omega)) \to f_*^{\pm}(\omega) \qquad a.e.$$

Introducing the truncated functions

$$f_m^{\pm}(\omega) = \begin{cases} f^{\pm}(\omega) & \text{if } f^{\pm}(\omega) \le m, \\ 0 & \text{if } f^{\pm}(\omega) > m \end{cases}$$

(with m natural), we have also

$$\frac{1}{n} \sum_{k=0}^{n-1} f_m^{\pm}(T^k(\omega)) \to f_{m,*}^{\pm}(\omega) \quad a.e.$$

and

$$\left| \frac{1}{n} \sum_{k=0}^{n-1} f(T^k(\omega)) \right| \le m \quad \text{for all } \omega.$$

By the Lebesgue Bounded Convergence Theorem ($P(\Omega) = 1$) we obtain

$$\int f_{m,*}^{\pm} dP = \lim_{n \to \infty} \int \frac{1}{n} \sum_{k=0}^{n-1} f_m^{\pm}(T^k) \, dP = \int f_m^{\pm} \, dP,$$

because T is measure preserving. The sequences $\{f_{m,*}^{\pm}\}$ and $\{f_m^{\pm}\}$ are monotonically convergent to $f_*^{\pm}$, $f^{\pm}$, respectively. Thus,

$$\int f_*^{\pm} \, dp = \int f^{\pm} \, dP,$$

showing that

$$f_* \in L^1(\Omega, \mathcal{F}, P) \quad \text{and} \quad \int f_* \, dP = \int f \, dP.$$

The equality

$$f_* \circ T = f_*.$$

is obvious. The proof is complete.   □

We must make some comments. The first is that this theorem is true not necessarily for finite measures. However, on the space of finite measure we have convergence in the mean (of order one), as well as almost everywhere convergence. If, in other words, T is a measure preserving transformation on Ω, with $P(\Omega) < \infty$, and if $f \in L^1(\Omega, \mathcal{F}, P)$then

$$\int \left| \frac{1}{n} \sum_{j=0}^{n-1} f(T^j(\omega)) - f_*(\omega) \right| dP(\omega) \to 0,$$

where, of course, $f_*(\omega) = \lim_{n \to \infty} \frac{1}{n} \sum_{j=0}^{n-1} f(T^j(\omega))$. If $P(\Omega) = \infty$ then we have convergence in the $L^1(\Omega, \mathcal{F}, P)$–norm. If f is bounded then the averages all have the same bound and the assertion follows from the Lebesque Bounded Convergence Theorem. If f is not bounded, the assertion follows from an approximation argument.

The second remark is about the generalization to a continuous–parameter group. The obvious thing is to consider a one–parameter group of measure preserving transformations $(T_t)_{t \in \mathbb{R}}$ such that $T_{s+t} = T_s T_t$. The sums over powers of a transformation that occur in the discrete ergodic theorems become integrals in the continuous case; the ergodic theorem asserts the convergence of $\frac{1}{T} \int_0^T f(T_t(\omega))\, dP$, where f is an arbitrary element of $L^1(\Omega, \mathcal{F}, P)$. In order that the above integral to make sense, some assumption has to be made on the way T_t depends on t. The natural assumption turns out to be that $T_t(\omega)$ should be a measurable function of its two arguments, where the measurability on the real axis is interpreted in the sense of Borel. Under this assumption the continuous ergodic theorem is meaningful and true. The proof is a straightforward imitation of the proof in the discrete case. The essential trick is to apply the discrete ergodic theorem to the transformation T_1 and to the function F defined by $F(\omega) = \int_0^1 f(T_t(\omega))\, dt$. We take advantage of this idea in the proof of the version of the Birkhoff Theorem concerning stationary processes.

The quadruple $(\Omega, \mathcal{F}, P, (T_t)_{t \in \mathbb{R}})$ is called a *dynamical system*.

9.4   Birkhoff Theorem for Stationary Processes

Before stating the main theorem, we describe some properties of stationary stochastic processes in terms of the theory of dynamical systems. Although most of definitions can be formulated in more general situation, we deal here only with real–valued stochastic processes.

Characterization of such processes is given by the Spectral Representation Theorem discussed in Chapter 5. Examples of computer simulation and visualization of stationary processes are presented in Chapter 10.

Stationary stochastic processes.   Let $\mathcal{B}_\infty$ denote the σ–field in $\mathbb{R}^{\mathbb{R}}$ generated by measurable cylinders. A real stochastic process $\{X(t) : t \in \mathbb{R}\}$ (or in full notation $\{X(t, \omega) : t \in \mathbb{R}, \, \omega \in \Omega\}$) can be treated as a measurable mapping from the probability space $(\Omega, \mathcal{F}, P)$ to $(\mathbb{R}^{\mathbb{R}}, \mathcal{B}_\infty)$ and as such a mapping is denoted by $\mathbf{X}$. The measure $P_{\mathbf{X}} = P \circ \mathbf{X}^{-1}$ on $\mathcal{B}_\infty$ is called the distribution of the process $\mathbf{X}$. On $\mathbb{R}^{\mathbb{R}}$ we can consider a group of left–shift transformations $(S_t)_{t \in \mathbb{R}}$ which are defined for $x \in \mathbb{R}^{\mathbb{R}}$ and for each $t \in \mathbb{R}$ by the equality

$$(S_t x)(s) = x(s + t).$$

Let us recall that a stochastic process is *stationary* if its finite dimensional distributions do not depend on time shift transformations, i.e., for each $n \in \mathbb{N}$, $s, t_1, \ldots, t_n \in \mathbb{R}$ and $A_1, \ldots, A_n \in \mathcal{B}_{\mathbb{R}}$,

$$P(X(t_1) \in A_1, \ldots, X(t_n) \in A_n) = P(X(t_1 + s) \in A_1, \ldots, X(t_n + s) \in A_n),$$

where $\mathcal{B}_{\mathbb{R}}$ denotes the Borel σ–field in $\mathbb{R}$.

If the process $\mathbf{X}$ is stationary then the group (S_t) preserves its distribution $P_{\mathbf{X}}$. Namely, for $A \in \mathcal{B}_\infty$ and $t \in \mathbb{R}$,

$$P_{\mathbf{X}}(A) = P_{\mathbf{X}}(S_t^{-1} A).$$

The triplet $(\mathbb{R}^{\mathbb{R}}, \mathcal{B}_\infty, P_{\mathbf{X}})$ together with the group $(S_t)_{t \in \mathbb{R}}$ is a typical object of study in the theory of dynamical systems. Thus, the ergodic theory of stochastic processes can be treated as a part of the ergodic theory of dynamical systems. But most ergodic properties of such systems possess their own, sometimes more intuitive meaning when they are expressed in the language of the theory of stochastic processes. In the continuous–time case there is another technical argument not to study ergodic properties of a stochastic process in terms of the group of shift transformations. To formulate such ergodic properties as weak– mixing or ergodicity of a given dynamical system with continuous time we need an assumption of measurability of an appropriate group of transformation.

Spaces of random variables corresponding to X.
For the stochastic process $\mathbf{X}$ by $L^o(\mathbf{X})$ we denote the closed subspace of $L^o(\Omega, \mathcal{F}, P)$ containing all $\mathcal{F}_{\mathbf{X}}$-measurable functions, where $\mathcal{F}_{\mathbf{X}} = \sigma\{X(t, \cdot) : t \in \mathbb{R}\}$. It is not difficult to prove that $\mathcal{F}_{\mathbf{X}} = \mathbf{X}^{-1}(\mathcal{B}_\infty)$.

We also introduce the space $L_0(\mathbf{X})$ which is the closure of the linear span $lin\{X(t) : t \in \mathbb{R}\}$ with respect to the topology of convergence in probability. Of course, $L_0(\mathbf{X}) \subseteq L^o(\mathbf{X}) \subseteq L^o(\Omega, \mathcal{F}, P)$. We also need the space $L_p(\mathbf{X})$, defined in an obvious way starting from the set $\{Y \in L_0(\mathbf{X}) : E|Y|^p < \infty\}$ for $p \geq 1$.

We can identify an element of $L^o(\mathbf{X})$ with a random variable $f(\mathbf{X})$, where f is some measurable function from $\mathbb{R}^{\mathbb{R}}$ to $\mathbb{R}$. This is a consequence of the following general lemma.

Lemma 9.4.1 *Let $\mathbf{X}$ be a random variable with values in a measurable space $(E, \mathcal{E})$ and Y – a real random variable measurable with respect to $\mathbf{X}^{-1}(\mathcal{E})$. Then there exists a measurable function $f : E \to \mathbb{R}$ such that $Y = f(\mathbf{X})$ a.e.*

PROOF. If Y is a nonnegative random variable, measurable with respect to $\mathbf{X}^{-1}(\mathcal{E})$ then the set function ν defined on $\mathcal{E}$ by the formula

$$\nu(A) = \int_{\mathbf{X}^{-1}(A)} Y \, dP$$

is a σ–finite measure which is absolutely continuous with respect to the distribution $P_{\mathbf{X}}$. By the Radon–Nikodym Theorem there exists a measurable function $f : E \to \mathbb{R}^+$ such that $\nu(A) = \int_A f \, dP_{\mathbf{X}}$. Thus,

$$\nu(A) = \int_{\mathbf{X}^{-1}(A)} f(\mathbf{X}) \, dP,$$

i.e., $Y = f(\mathbf{X})$ a.s. .

For Y not necessary non–negative we have $Y = Y^+ - Y^-$, where Y^+ and Y^- are non–negative random variables and it is enough to repeat the above argument. This ends the proof. □

Using this fact we can define on $L^o(\mathbf{X})$ the group $(T_t)_{t \in \mathbb{R}}$ of linear operators by the formula

$$T_t(f(\mathbf{X})) = f(S_t \circ \mathbf{X})$$

One can formulate some properties of stationary processes in terms of the theory of dynamical systems. For example, as we have already mentioned, stationarity of $\mathbf{X}$ means that the distribution $P_\mathbf{X}$ is invariant with respect to the group $(S_t)_{t \in \mathbb{R}}$ of shift transformations. Conversely, if the measure $P_\mathbf{X}$ on $\mathcal{B}_\infty$ is invariant with respect to $(S_t)_{t \in \mathbb{R}}$ then the process $\{X(t) : t \in \mathbb{R}\}$ is stationary. For an arbitrary dynamical system $(\Omega, \mathcal{F}, P, (W_t)_{t \in \mathbb{R}})$, where $(W_t)_{t \in \mathbb{R}}$ is a group of measure preserving transformations, and any real measurable function f on Ω the process $\{X(t) : t \in \mathbb{R}\}$ defined by $X(t, \omega) = f(W_t(\omega))$ is a measurable stationary stochastic process. Let us remark that, although to every stationary stochastic process there corresponds a group of measure preserving transformations $(S_t)_{t \in \mathbb{R}}$, this group, however, is not a dynamical system because the mapping $\mathbb{R} \times \mathbb{R}^\mathbb{R} \ni (t, x) \to S_t x \in \mathbb{R}^\mathbb{R}$ does not have to be measurable. Therefore, it is more convenient to formulate some general problems of ergodic theory of stochastic processes in terms of dynamical systems rather than in terms of the theory of stochastic processes. Of course, relations between suitable properties and basic facts are completely analogous in both settings. One of such facts is the Birkhoff Ergodic Theorem. Before stating this theorem we have to mention that if $\mathbf{X}$ is a measurable process and $Y \in L^0(\mathbf{X})$ then there exists a measurable stochastic process $\{Z(t) : t \in \mathbb{R}\}$ such that $Z(t) = T_t Y$ a.s. for every $t \in \mathbb{R}$. Namely, for $Y = f(\mathbf{X})$ (where $f : \mathbb{R}^\mathbb{R} \to \mathbb{R}$ is some measurable function) the process defined by $Z(t) = f(S_t \circ \mathbf{X})$ satisfies the above condition, which follows from the measurability of f.

According to what was said above $L^1(\mathbf{X})$ denotes the subspace of $L^0(\mathbf{X})$ consisting of random variables with finite expectation. It is a Banach space with the norm $\|Y\| = \|f(\mathbf{X})\| = E|f(\mathbf{X})|$. We shall assume that $\mathbf{X}$ is measurable, i.e., that the mapping

$$\Omega \times \mathbb{R} \ni (\omega, t) \to X(t, \omega) \in \mathbb{R}$$

is measurable with respect to $\mathcal{F} \times \mathcal{B}_\mathbb{R}$ and $\mathcal{B}_\mathbb{R}$. Let $Y \in L^1(\mathbf{X})$ and let $\{Z(t)\}_{t \in \mathbb{R}}$ be as above. By the Fubini Theorem, for every $T \in \mathbb{R}$ we have

$$\int_{[0,T] \times \Omega} |Z(t, \omega)| \, d(\lambda \times P)(t, \omega) = \int_0^T \left(\int_\Omega |Z(t, \omega)| \, dP(\omega) \right) \, d\lambda(t)$$

$$= \int_0^T E|Y| \, d\lambda(t) = T \, E|Y| < \infty,$$

where λ is the Lebesgue measure. It follows that for almost every $\omega \in \Omega$ there exists $\int_0^T Z(t, \omega) \, dt$. So, we can define $\int_0^T T_t Y \, dt \in L^1(\mathbf{X})$ as the equivalence class of all elements which are almost everywhere equal to $\int_0^T Z(t, \omega) \, dt$. It is not difficult to prove that this definition does not depend on the choice of the family $\{Z(t)\}_{t \in \mathbb{R}}$.

The following theorem is a version of the Birkhoff Ergodic Theorem for processes with continuous time (see also Cornfeld, Fomin and Sinai (1982)).

Theorem 9.4.1 *Let* **X** *be a stationary measurable process. Then for any* $Y \in L^1(\mathbf{X})$ *there exists* $Y_* \in L^1(\mathbf{X})$ *such that we have*

$$\lim_{T \to \infty} \frac{1}{T} \int_0^T T_t Y \, dt = Y_* \quad a.e. \quad \text{and} \quad T_t Y_* = Y_* \quad \text{forevery } t \in \mathbb{R}.$$

PROOF. The proof of this theorem in the case of a discrete parameter group was presented in Section 9.3. Now we extend it to the case of the continuous parameter t. Let $Y = f(\mathbf{X})$, where $f : \mathbb{R}^{\mathbb{R}} \to \mathbb{R}$ is a measurable function and define for $n \in \mathbb{N}$

$$Y_n = \int_0^1 f(S_1^{n-1} \circ S_t \circ \mathbf{X}) \, dt.$$

Then $(Y_n)_{n \in \mathbb{N}}$ is a stationary sequence of elements of $L^1(\mathbf{X})$. It follows from the Birkhoff Theorem (Theorem 9.4.1) that there exists $Y_* \in L^1(\mathbf{X})$ such that

$$\lim_{n \to \infty} \frac{1}{n} \sum_{k=1}^n Y_k = Y_* \quad a.e.$$

Similarly, for $V_n = \int_0^1 |f(S_1^{n-1} \circ S_t \circ \mathbf{X})| \, dt$ there exists $V_* \in L^1(\mathbf{X})$ such that

$$\lim_{n \to \infty} \frac{1}{n} \sum_{k=1}^n V_k = V_* \quad a.e.$$

Hence,

$$\lim_{n \to \infty} \frac{V_n}{n} = \lim_{n \to \infty} \frac{1}{n} \sum_{k=1}^n V_k - \lim_{n \to \infty} \frac{n-1}{n} \lim_{n \to \infty} \frac{1}{n-1} \sum_{k=1}^{n-1} V_k = 0 \quad a.e.$$

Now, let $T \in \mathbb{R}$ and n_T be a natural number such that $n_T < T \leq n_T + 1$. Then we have

$$\frac{1}{T} \int_0^T f(S_t \circ \mathbf{X}) \, dt = \frac{1}{T} \int_0^{n_T} f(S_t \circ \mathbf{X}) \, dt + \frac{1}{T} \int_{n_T}^T f(S_t \circ \mathbf{X}) \, dt$$

$$= \frac{1}{T} \sum_{k=1}^{n_T} \int_{k-1}^k f(S_t \circ \mathbf{X}) \, dt + \frac{1}{T} \int_{n_T}^T f(S_t \circ \mathbf{X}) dt = \frac{1}{T} \sum_{k=1}^{n_T} Y_k + \frac{1}{T} \int_{n_T}^T f(S_t \circ \mathbf{X}) \, dt.$$

But

$$\left| \frac{1}{T} \int_{n_T}^T f(S_t \circ \mathbf{X}) \, dt \right| \leq \frac{1}{T} \int_{n_T}^{n_T+1} |f(S_t \circ \mathbf{X})| \, dt = \frac{n_T + 1}{T} \frac{1}{n_T + 1} V_{n_T+1}$$

and so, $\lim_{T \to \infty} \frac{1}{T} \int_{n_T}^T f(S_t \circ \mathbf{X}) \, dt = 0$ *a.e.* Thus, we have

$$\lim_{T \to \infty} \frac{1}{T} \int_0^T f(S_t \circ \mathbf{X}) \, dt = \lim_{T \to \infty} \frac{n_T}{T} \frac{1}{n_T} \sum_{k=1}^{n_T} Y_k = Y* \quad a.e.$$

Moreover, for $s \in \mathbb{R}$ we have

$$
\begin{aligned}
T_s Y_* &= T_s \lim_{T \to \infty} \frac{1}{T} \int_0^T T_t Y \, dt = \lim_{T \to \infty} \frac{1}{T} \int_0^T T_{t+s} Y \, dt \\
&= \lim_{T \to \infty} \left[\frac{T+s}{T} \left(\frac{1}{T+s} \int_0^{T+s} T_t Y \, dt \right) - \frac{1}{T} \int_0^s T_t Y \, dt \right] \\
&= Y_* - \lim_{T \to \infty} \frac{1}{T} \int_0^s T_t Y \, dt = Y_* \quad a.e.
\end{aligned}
$$

This completes the proof. □

The following theorem (known as the von Neumann Theorem) is a consequence of the Birkhoff Theorem. Its proof can be found in Krengel (1985).

Theorem 9.4.2 *Let $\mathbf{X}$ be a stationary measurable process. Fix $p \in [1, \infty)$. If $Y \in L^p(\mathbf{X})$ then there exists $Y_* \in L^p(\mathbf{X})$ such that for every $t \in \mathbb{R}$ we have $T_t Y_* = Y_*$ a.e., and*

$$
\lim_{T \to \infty} E \left| \frac{1}{T} \int_0^T T_t Y \, dt - Y_* \right|^p = 0.
$$

One can see that for a stationary process $\mathbf{X}$ every operator S_t preserves distributions, i.e., for $Y \in L^0(\mathbf{X})$ we have $T_t(Y) \stackrel{d}{=} Y$. Thus, for a stationary process $\mathbf{X}$ we can define a σ-field $\mathcal{P}_{\mathbf{X}}$ of invariant sets

$$
\mathcal{P}_{\mathbf{X}} = \{ A \in \mathcal{F}_{\mathbf{X}} : I_A = T_t I_A \quad P - a.e. \}.
$$

Since $\mathcal{F}_{\mathbf{X}} = \mathbf{X}^{-1}(\mathcal{B}_\infty)$, one can equivalently write

$$
\mathcal{P}_{\mathbf{X}} = \left\{ A \in \mathcal{F}_{\mathbf{X}}; \ A = \mathbf{X}^{-1} B \ \text{and} \ \forall_{t \in \mathbb{R}} P \left(\{ \mathbf{X} \in B \} \triangle \{ S_t \circ \mathbf{X} \in B \} \right) = 0 \right\}.
$$

Remark 9.4.1 *Using the notion of conditional expectation one can identify the limit Y_* which appears in the above theorems. Namely, for $A \in \mathcal{P}_{\mathbf{X}}$ we have*

$$
\begin{aligned}
\int_A Y_* \, dP &= \int_A \left(\lim_{T \to \infty} \frac{1}{T} \int_0^T T_t Y \, dt \right) dP \\
&= \lim_{T \to \infty} \frac{1}{T} \int_0^T \left(\int_A T_t Y \, dP \right) dt = \lim_{T \to \infty} \frac{1}{T} \int_0^T E[T_t(I_A Y)] \, dt \\
&= \lim_{T \to \infty} \frac{1}{T} \int_0^T \left(\int_A Y \, dP \right) dt = \int_A Y \, dP.
\end{aligned}
$$

Hence $Y_* = E(Y | \mathcal{P}_{\mathbf{X}})$.

9.5   Hierarchy of Chaotic Properties

Let $\mathbf{X}$ be a measurable stationary stochastic process and let $\mathcal{P}_{\mathbf{X}}$ denote its σ-field of invariant sets. Of course, $\mathcal{P}_{\mathbf{X}} \subset \mathcal{F}$ and for $A \in \mathcal{P}_{\mathbf{X}}$, if $A = \mathbf{X}^{-1}B$, where $B \in \mathcal{B}_{\infty}$, then $P(\{\mathbf{X} \in B\} \triangle \{S_t \circ \mathbf{X} \in B\}) = 0$. Moreover, if $Y \in L^0(\mathbf{X})$ is $\mathcal{P}_{\mathbf{X}}$-measurable then $T_t Y = Y \; P - a.e.$.

Definition 9.5.1 *A stationary process* $\mathbf{X}$ *is called **ergodic** if for every* $A \in \mathcal{P}_{\mathbf{X}}$ *we have* $P(A) = 0$ *or* $P(A) = 1$.

Equivalently, $\mathbf{X}$ *is ergodic whenever for* $Y \in L^0(\mathbf{X})$ *from the condition* $T_t Y = Y$ *a.e. true for all* $t \in \mathbb{R}$ *it follows that* $Y = const$ *a.e.* .

Recall that $L^2(\mathbf{X})$ is a Hilbert space with the inner product defined by

$$(Y, Z) = E(YZ) \quad \text{for } Y, Z \in L^2(\mathbf{X}).$$

The Birkhoff Ergodic Theorem yields the following characterization of ergodic processes.

Theorem 9.5.1 *Let* $\mathbf{X}$ *be a stationary measurable stochastic process. The following conditions are equivalent:*

(i) $\mathbf{X}$ *is ergodic;*

(ii) for every $Y \in L^1(\mathbf{X})$ *we have*

$$\lim_{T \to \infty} \frac{1}{T} \int_0^T T_t Y \, dt = EY \quad a.e.;$$

(iii) for every $Y \in L^1(\mathbf{X})$ *we have*

$$\lim_{T \to \infty} E|\frac{1}{T} \int_0^T T_t Y \, dt - EY| = 0;$$

(iv) for every $Y \in L^2(\mathbf{X})$ *we have*

$$\lim_{T \to \infty} E(\frac{1}{T} \int_0^T T_t Y \, dt - EY)^2 = 0;$$

(v) for every $Y, Z \in L^2(\mathbf{X})$ *we have*

$$\lim_{T \to \infty} \frac{1}{T} \int_0^T E((T_t Y)Z) \, dt = EY \, EZ;$$

(vi) for every $Y \in L^2(\mathbf{X})$ *we have*

$$\lim_{T \to \infty} \frac{1}{T} \int_0^T E((T_t Y)Y) \, dt = (EY)^2;$$

(vii) for Y from some linearly dense subset of $L^2(X)$ we have

$$\lim_{T\to\infty} \frac{1}{T} \int_0^T E((T_tY)Y)\, dt = (EY)^2.$$

PROOF. The equivalence of first four conditions follows immediately from Theorems 9.4.1 and 9.4.2. Condition (iv) means that $\frac{1}{T}\int_0^T T_tY\, dt$ converges in the norm of $L^2(X)$, as $T \to \infty$. Thus, for $Z \in L^2(X)$ we have

$$\lim_{T\to\infty} E\left(\left(\frac{1}{T}\int_0^T T_tY\, dt\right) Z\right) = E((EY)Z) = EY\, EZ.$$

Now, using the Fubini Theorem we obtain (v). The implications (v)$\Rightarrow$(vi) and (vi)$\Rightarrow$(vii) are obvious so it is enough to prove that condition (vii) implies ergodicity of X. Let $Y \in L^2(X)$ and

$$\lim_{T\to\infty} \frac{1}{T} \int_0^T E((T_tY)Y)\, dt = (EY)^2.$$

Using the Fubini Theorem and Theorem 9.4.2 we get

$$E\left(\left(\lim_{T\to\infty}\frac{1}{T}\int_0^T T_tY\, dt\right) Y\right) = E(E(Y|\mathcal{P}_X)Y) = (EY)^2.$$

Thus, for $Z = E(Y|\mathcal{P}_X)$ we have

$$\begin{aligned} E(Z^2) &= E\{ E(Y|\mathcal{P}_X)\, E(Y|\mathcal{P}_X) = E\{ E[E(Y|\mathcal{P}_X)Y|\mathcal{P}_X]\} \\ &= E\{ E(Y|\mathcal{P}_X)Y\} = (EY)^2 = (EZ)^2. \end{aligned}$$

This means that $Var\, Z = 0$, so $Z = const$ $a.e.$. Thus, for Y belonging to linearly dense subset of $L^2(X)$ we have $E(Y|\mathcal{P}_X) = const a.e.$ and since the operator of conditional expectation is continuous on $L^2(X)$ we $E(Y|\mathcal{P}_X) = const$ $a.e.$ for each $Y \in L^2(X)$ implying ergodicity. $\square$

Definition 9.5.2 *A stationary process* X *is called* **weak mixing** *if for all* $Y, Z \in L^2(X)$,

$$\lim_{T\to\infty} \frac{1}{T} \int_0^T | E((T_tY)Z) - EY\, EZ|\, dt = 0.$$

We have the following characterization of stationary weak mixing processes.

Theorem 9.5.2 *Let* X *a be stationary measurable stochastic process. The following conditions are equivalent:*

(i) X *is weak mixing;*

(ii) for every $Y \in L^2(X)$ we have

$$\lim_{T\to\infty} \frac{1}{T} \int_0^T | E((T_tY)Y) - (EY)^2|\, dt = 0;$$

(iii) for Y from some linearly dense subset of $L^2(\mathbf{X})$ we have

$$\lim_{T\to\infty} \frac{1}{T} \int_0^T | E((T_tY)Y) - (EY)^2| \, dt = 0.$$

PROOF. Implications $(i) \Rightarrow (ii)$ and $(ii) \Rightarrow (iii)$ are obvious. We shall show that $(iii) \Rightarrow (i)$. Let $\mathcal{E}_0$ be a linearly dense subset of $L^2(\mathbf{X})$ such that for $Y \in \mathcal{E}_0$ we have

$$\lim_{T\to\infty} \frac{1}{T} \int_0^T | E((T_tY)Y) - (EY)^2| \, dt = 0.$$

Fix $Y \in \mathcal{E}_0$ and set

$$\mathcal{E}_Y = \{Z \in L^2(\mathbf{X}) : \lim_{T\to\infty} \frac{1}{T} \int_0^T | E((T_tY)Z) - EY \, EZ| \, dt = 0\}.$$

Direct verifications show that $\mathcal{E}_Y$ is a linear subspace of $L^2(\mathbf{X})$. Moreover, if the sequence $\{Z_n\}_{n \in \mathbb{N}}$ of elements of $\mathcal{E}_Y$ is norm convergent to $Z \in L^2(\mathbf{X})$ then

$$\frac{1}{T} \int_0^T | E((T_tY)Z) - EY \, EZ| \, dt = \frac{1}{T} \int_0^T |(T_tY, Z) - (Y,1)(1,Z)| \, dt$$

$$\leq \frac{1}{T} \int_0^T |(T_tY, Z_n - Z)| \, dt + |(Y,1)(1, Z - Z_n)| + \frac{1}{T} \int_0^T |(T_tY, Z_n) - (Y,1)(1,Z_n)| \, dt$$

$$\leq 2\|Z - Z_n\|_2\|Y\|_2 + \frac{1}{T} \int_0^T |(T_tY, Z_n) - (Y,1)(1,Z_n)| \, dt.$$

It means that $\mathcal{E}_Y$ is a closed linear subspace of the Hilbert space $L^2(\mathbf{X})$. Suppose that Z is orthogonal to $\mathcal{E}_Y$. It follows from the properties of the group $(T_t)_{t \in \mathbb{R}}$ that for $s \in \mathbb{R}$ and $Y \in L^2(\mathbf{X})$ we have

$$\int_0^T | E(T_tYT_sY) - EY \, ET_sY| \, dt = \int_0^T |(T_{t-s}Y, Y) - (Y,1)(1,Y)| \, dt$$

$$= \int_{-s}^{T-s} |(T_uY, Y) - (Y,1)(1,Y)| \, du \leq \int_0^T |(T_uY, Y) - (Y,1)(1,Y)| \, du$$

$$+ \int_{-s}^0 | E(T_uY, Y) - (Y,1)(1,Y)| \, du + \int_{T-s}^T |(T_uY, Y) - (Y,1)(1,Y)| \, du$$

$$\leq \int_0^T |(T_uY, Y) - (Y,1)(1,Y)| \, du + 4s\|Y\|_2^2.$$

Hence, for $Y \in \mathcal{E}_Y$ we have $T_sY \in \mathcal{E}_Y$, i.e. for each $s \in \mathbb{R}$, Z is orthogonal to T_sY. Let us also notice that

$$\int_0^T | E((T_tY)1) - EY \, E1| \, dt = \int_0^T | EY - EY| \, dt = 0$$

and thus $1 \in \mathcal{E}_Y$. So Z is orthogonal to 1. It follows that

$$\int_0^T | E((T_tY)Z) - EY \, EZ| \, dt = \int_0^T |(T_tY, Z) - (Y,1)(Z,1)| \, dt = 0$$

and consequently $Z \in \mathcal{E}_Y$. Since we have assumed that Z is orthogonal to $\mathcal{E}_Y$, it must be $Z = 0$. Thus $\mathcal{E}_Y = L^2(\mathbf{X})$. It follows that for $Y \in \mathcal{E}_0$ and $Z \in L^2(\mathbf{X})$ we have

$$\lim_{T\to\infty} \frac{1}{T} \int_0^T | E((T_tY)Z) - EY \, EZ| \, dt = 0.$$

Let

$$\mathcal{E} = \left\{ Y \in L^2(\mathbf{X}) : \forall_{Z \in L^2(\mathbf{X})} \lim_{T \to \infty} \frac{1}{T} \int_0^T | E((T_t Y)Z) - E Y E Z| \, dt = 0 \right\}.$$

Similarly to the case of $\mathcal{E}_Y$ one can prove that $\mathcal{E}$ is a closed subspace of $L^2(\mathbf{X})$. Since $\mathcal{E}_0 \subset \mathcal{E}$ and $\mathcal{E}_0$ is a linearly dense subset of $L^2(\mathbf{X})$, thus $\mathcal{E} = L^2(\mathbf{X})$. This completes the proof. □

Definition 9.5.3 *A stationary process* $\mathbf{X}$ *is called **mixing of order** p, with $p \in \mathbb{N}$, if for all $Y_0, Y_1, ..., Y_p \in L^{p+1}(\mathbf{X})$ and $0 = t_0 \leq t_1 \leq ... \leq t_p$,*

$$E(T_{t_0} Y_0 T_{t_1} Y_1 \cdots T_{t_p} Y_p) \to E Y_0 \, E Y_1 \cdots E Y_p,$$

when $\min_{1 \leq j \leq p}(t_j - t_{j-1}) \to \infty$. *Mixing of order 1 is simply called mixing.*

Of course, every mixing process is weak mixing and each one of these properties implies ergodicity. In general, the contrary is not true (see Walters (1982)). Similarly to the proof of Theorem 9.5.2 we can prove the following characterization of mixing processes.

Theorem 9.5.3 *Let* $\mathbf{X}$ *be a stationary measurable stochastic process. The following conditions are equivalent:*

 (i) $\mathbf{X}$ *is mixing;*

 (ii) for every $Y \in L^2(\mathbf{X})$ *we have*

$$\lim_{t \to \infty} E((T_t Y)Y) = (E Y)^2;$$

 (iii) for every Y *from some linearly dense subset of* $L^2(\mathbf{X})$ *we have*

$$\lim_{t \to \infty} E((T_t Y)Y) = (E Y)^2.$$

Notice that the mixing and weak mixing properties describe to some extent the asymptotic independence of $\mathbf{X}$.

Definition 9.5.4 *We say that a stochastic process* $\mathbf{X}$ *has the **Kolmogorov property** or **K-property** if there exists a σ-field $\mathcal{F}_0 \subseteq \mathcal{F}_\mathbf{X}$ such that $\mathcal{F}_0 \subseteq T_t \mathcal{F}_0$ for all $t \in \mathbb{R}$, the σ-field generated by $\bigcup_{t \in \mathbb{R}} T_t \mathcal{F}_0$ is equal to $\mathcal{F}_\mathbf{X}$ and $\bigcap_{t \in \mathbb{R}} T_t \mathcal{F}_0$ is a trivial σ-field, where the action of T_t on a measurable set is defined by the action of T_t on its indicator function.*

Definition 9.5.5 *A positive time process* $\mathbf{X}$ *has **exactness property** or shortly **is exact** if the σ-field $\bigcap_{t \in (0,\infty)} T_t \sigma\{X(t) : t \geq 0\}$ is trivial.*

See Rochlin (1964).

It is clear that, if a positive time stationary process $\{X(t)\}_{t \geq 0}$ has a stationary extension to $\{X(t)\}_{t \in \mathbb{R}}$, then its exactness implies the K–property of $\{X(t)\}_{t \in \mathbb{R}}$ (by taking $\mathcal{F}_0 = \sigma\{X(t) : t \geq 0\}$).

9.6   Dynamical Functional

Suppose that $\mathbf{X}$ is a measurable stationary stochastic process.

Definition 9.6.1 *The map* $\Phi : L^o(\mathbf{X}) \times \mathbb{R} \to \mathbb{C}$ *defined by*

$$\Phi(Y,t) = E \exp\{i(T_t Y - Y)\}$$

*is called the **dynamical functional** of the stochastic process* $\mathbf{X}$.

The dynamical functional was introduced in Podgórski and Weron (1991).

For each $Y \in L^o(\mathbf{X})$ the function $\Phi(Y,\cdot)$ is positive definite. If the process $\mathbf{X}$ is in addition stochastically continuous, then the group $(T_t)_{t \in \mathbb{R}}$ is continuous on $L^o(\mathbf{X})$ with respect to the topology of convergence in probability. Consequently, Φ is continuous in the product topology on $L^o(\mathbf{X}) \times \mathbb{R}$. By stationarity we have for $Y \in L^o(\mathbf{X})$,

$$\Phi(Y,-t) = E \exp\{i(T_{-t} Y - Y)\} = E \exp\{i(Y - T_t Y)\} = \overline{\Phi(Y,t)}$$

and thus, if $\mathbf{X}$ is symmetric then Φ is real and $\Phi(Y,-t) = \Phi(Y,t)$.
Now we shall characterize ergodicity, weak mixing and mixing in terms of the dynamical functional. To this aim we need the following lemma.

Lemma 9.6.1 *For any random variable* $Y \in L^1(\mathbf{X})$ *there exists a sequence* $\{Z_n\}_{n \in \mathbb{N}}$ *of elements of* $\lin_{\mathbb{C}}\{e^{iY} : Y \in \lin\{X(t) : t \in \mathbb{R}\}\}$ *such that*

$$\lim_{n \to \infty} E|Y - Z_n| = 0.$$

Similarly, for every random variable $Y \in L^2(\mathbf{X})$ *there exists a sequence* $\{Z_n\}_{n \in \mathbb{N}}$ *of elements of* $\lin_{\mathbb{C}}\{e^{iY} : Y \in \lin\{X(t) : t \in \mathbb{R}\}\}$ *such that*

$$\lim_{n \to \infty} E|Y - Z_n|^2 = 0.$$

PROOF. Since every continuous function defined on a closed interval is a uniform limit of functions from $\lin_{\mathbb{C}}\{e^{iax} : a \in \mathbb{R}\}$, it follows that every characteristic function of a measurable set is a pointwise limit of such a sequence. So for each cylindrical set $A = \{x \in \mathbb{R}^{\mathbb{R}} : x(t_i) \in I_i, i = 1,...,n\}$, where $n \in \mathbb{N}$ and for $i = 1,...,n$ I_i denotes an interval, the random variable $Y = I_A(\mathbf{X})$ is a P-a.s. limit of an uniformly bounded sequence from $\lin_{\mathbb{C}}\{e^{iY} : Y \in \lin\{X(t) : t \in \mathbb{R}\}\}$. Since $\lin\{I_A(\mathbf{X}) : A - \text{cylindrical set}\}$ is dense in $L^1(\mathbf{X})$ as well as in $L^2(\mathbf{X})$, the rest follows from the Lebesgue Bounded Convergence Theorem. $\square$

Now we are in a position to prove the mentioned results.

Proposition 9.6.1 *Let* $\mathbf{X}$ *be a stationary stochastic process. The following conditions are equivalent*

(i) $\mathbf{X}$ *is ergodic;*

(ii) for each $Y \in L^o(\mathbf{X})$ we have

$$\lim_{T \to \infty} \frac{1}{T} \int_0^T \Phi(Y, t) \, dt = |\, E e^{iY}|^2;$$

(iii) for each $Y \in \mathrm{lin}\{X(t) : t \in \mathbb{R}\}$ we have

$$\lim_{T \to \infty} \frac{1}{T} \int_0^T \Phi(Y, t) \, dt = |\, E e^{iY}|^2.$$

PROOF. Suppose that $\mathbf{X}$ is ergodic. Let $X, Y \in L^2(\mathbf{X})$ and $Z = X + iY$. Then

$\lim_{T \to \infty} \frac{1}{T} \int_0^T E((T_t Z)\overline{Z} \, dt = \lim_{T \to \infty} \frac{1}{T} \int_0^T E((T_t X + iT_t Y)(X - iY)) \, dt$

$= \lim_{T \to \infty} \frac{1}{T} \int_0^T E((T_t X)X) \, dt + \lim_{T \to \infty} \frac{1}{T} \int_0^T E((T_t Y)Y) \, dt$

$+ i \left(\lim_{T \to \infty} \frac{1}{T} \int_0^T E((T_t Y)X) \, dt - \lim_{T \to \infty} \frac{1}{T} \int_0^T E((T_t X)Y) \, dt \right)$

$= (\, EX)^2 + (\, EY)^2 + i(\, EY \, EX - EX \, EY) = |\, EZ|^2.$

Since for $Y \in L^o(\mathbf{X})$ we have

$$\mathrm{Im}\, e^{iY}, \mathrm{Re}\, e^{iY} \in L^2(\mathbf{X}) \quad \text{and} \quad \Phi(Y, t) = E((T_t e^{iY})e^{-iY}),$$

the implication $(i) \Rightarrow (ii)$ is proved. The implication $(ii) \Rightarrow (iii)$ is obvious. Now, it follows from Lemma 9.6.1 that the set

$$\mathcal{E} = \mathrm{Re}\, (\mathrm{lin}_{\mathbb{C}}\{e^{iY} : Y \in \mathrm{lin}\{X(t) : t \in \mathbb{R}\}\})$$

is a linearly dense subset of $L^2(\mathbf{X})$. On the other hand the condition (iii) means that for $Y \in \mathcal{E}$ we have

$$\lim_{T \to \infty} \frac{1}{T} \int_0^T E((T_t Y)Y) \, dt = (\, EY)^2,$$

which implies (by Theorem 9.5.1) ergodicity of $\mathbf{X}$. $\square$

Proposition 9.6.2 *Let $\mathbf{X}$ be a stationary stochastic process. The following conditions are equivalent*

(i) $\mathbf{X}$ is weak mixing;

(ii) for each $Y \in L^o(\mathbf{X})$ we have

$$\lim_{T \to \infty} \frac{1}{T} \int_0^T \left| \Phi(Y, t) - |\, E e^{iY}|^2 \right| \, dt = 0;$$

(iii) for each $Y \in \lin\{X(t) : t \in \mathbb{R}\}$ we have

$$\lim_{T \to \infty} \frac{1}{T} \int_0^T \left| \Phi(Y, t) - | E e^{iY}|^2 \right| \, dt = 0.$$

PROOF. Notice that for $X, Y \in L^2(\mathbf{X})$ and $Z = X + iY$ we have

$$\int_0^T | E((T_t Z)\overline{Z}) - | E Z|^2 | \, dt$$

$$\leq \int_0^T | E((T_t X)X) - (E X)^2 | \, dt + \int_0^T E((T_t Y)Y) - (E Y)^2 | \, dt$$

$$+ \int_0^T | E((T_t Y)X) - E Y \, E X | \, dt + \int_0^T | E((T_t X)Y) + E X \, E Y | \, dt.$$

This and the weak mixing of $\mathbf{X}$ give the equality

$$\lim_{T \to \infty} \frac{1}{T} \int_0^T | E((T_t e^{iY})e^{-iY}) - | E e^{iY}|^2 | \, dt = 0,$$

for all $Y \in L^o(\mathbf{X})$, which proves the implication $(i) \Rightarrow (ii)$, because we have $\Phi(Y, t) = E((T_t e^{iY})e^{-iY})$.

The implication $(ii) \Rightarrow (iii)$ is obvious.

Now we prove the implication $(iii) \Rightarrow (i)$. Since any element of $L^2(\mathbf{X})$ can be approximated by linear combinations of random variables of the form $\exp(iY)$, where $Y \in \lin\{X(t) : t \in \mathbb{R}\}$, it suffices to prove that if for all Y from a linearly dense subset $\mathcal{E}$ of $L^2(\mathbf{X})$ we have

$$\lim_{T \to \infty} \frac{1}{T} \int_0^T | E((T_t Y)Y) - (E Y)^2 | \, dt = 0,$$

then $\mathbf{X}$ is weak mixing. For $Y \in \mathcal{E}$ we define

$$\mathcal{E}_Y = \left\{ Z \in L^2(\mathbf{X}) : \lim_{T \to \infty} \frac{1}{T} \int_0^T | E((T_t Y)Z) - E Y \, E Z | \, dt = 0 \right\}.$$

Then $\mathcal{E}_Y$ is a closed subspace of $L^2(\mathbf{X})$ and by assumption (i), $T_t Y \in \mathcal{E}_Y$, for each $t \in \mathbb{R}$. Thus, if Z is orthogonal to $\mathcal{E}_Y$ then $E((T_t Y)Z) = 0$ and $E Z = 0$ and $Z \in \mathcal{E}_Y$. It implies that $\mathcal{E}_Y = L^2(\mathbf{X})$ for each $Y \in \mathcal{E}$. Consequently, if $Y \in \mathcal{E}$ then

$$\lim_{T \to \infty} \frac{1}{T} \int_0^T | E((T_t Y)Z) - E Y \, E Z | \, dt = 0, \quad \text{for all } Z \in L^2(\mathbf{X}).$$

Now, from the fact that $\mathcal{E}$ is a linear dense subset of $L^2(X)$ and the set on the right hand side of the above inclusion is a closed subspace of $L^2(X)$ it follows that it is equal to the whole space $L^2(X)$. Thus, for all $Y, Z \in L^2(X)$ we have

$$\lim_{T \to \infty} \frac{1}{T} \int_0^T | E((T_t Y)Z) - EY\, EZ|\ dt = 0.$$

This completes the proof.   □

Proposition 9.6.3 *Let* X *be a stationary stochastic process. The following conditions are equivalent*

(i) X *is mixing;*

(ii) *for each* $Y \in L^o(X)$ *we have*

$$\lim_{T \to \infty} \Phi(Y, T) = | E e^{iY}|^2;$$

(iii) *for each* $Y \in \mathrm{lin}\{X(t) : t \in \mathbb{R}\}$ *we have*

$$\lim_{T \to \infty} \Phi(Y, T) = | E e^{iY}|^2.$$

The proof of this proposition is very similar to the proofs of two previous propositions, so we omit it here.

Proposition 9.6.4 *Let* X *be a stationary stochastic process. The following conditions are equivalent*

(i) X *is mixing of order* p;

(ii) *for all* $Y_0, ..., Y_p$ *from* $L^o(X)$ *we have*

$$\Phi_p(Y, t) \to E \exp(iY_0)... E \exp(iY_p), \quad \text{when} \quad \min_{1 \leq j \leq p}(t_j - t_{j-1}) \to \infty,$$

where $\Phi_p(Y, t) = E \exp\{i(T_{t_0}Y_0 + ... + T_{t_p}Y_p)\};$

(iii) *for all* $Y_0, ..., Y_t$ *from* $\mathrm{lin}\{X(t) : t \in \mathbb{R}\}$ *and* $0 = t_0 \leq t_1 \leq ... \leq t_p$ *we have*

$$\Phi_p(Y, t) \to E \exp(iY_0)... E \exp(iY_p), \quad \text{when} \quad \min_{1 \leq j \leq p}(t_j - t_{j-1}) \to \infty.$$

The proof of this proposition follows simply from Definition 9.5.3 by an approximation argument, so we omit it here.

Example 9.6.1 *Dynamical functional for the SαS Ornstein–Uhlenbeck process.*

As we known from the spectral representation theorem for SαS stationary processes (Section 5.3) the Ornstein–Uhlenbeck process $\{X(t) : t \in [0, \infty)\}$ as a moving average process can be represented on the corresponding space L^α by the function $f_o(x) = e^{-x}I_{[0,\infty)}(x)$ and the group of shift operators $U_t g(x) = g(x - t)$. Let Y and T_t correspond via the spectral representation theorem to h and U_t, respectively. Then we have

$$\Phi(Y, t) = E \exp\{(T_t Y - Y)\} = \exp\{-\|U_t h - h\|_\alpha^\alpha\}. \qquad (9.6.1)$$

Define four functions from the linear span $lin\{U_t f_o(x) : t \geq 0\}$

$$
\begin{aligned}
h_1(x) &= f_o(x), \\
h_2(x) &= f_o(x) - \tfrac{1}{2}U_2 f_o(x), \\
h_3(x) &= f_o(x) - \tfrac{1}{2}U_2 f_o(x) + \tfrac{1}{2}U_4 f_o(x), \\
h_4(x) &= f_o(x) - \tfrac{1}{2}U_2 f_o(x) + \tfrac{1}{2}U_4 f_o(x) - U_6 f_o(x).
\end{aligned}
$$

The graphs of these functions versus $x \in [0, 10]$ are presented in Figure 9.6.1. Next, in Figure 9.6.2 we plotted $U_t h_i(x) - h_i(x)$ for $i = 1, ..., 4$ and $t = \frac{1}{2}$. According to (9.6.1) the dynamical functional $\Phi(Y, t)$ for the SαS Ornstein–Uhlenbeck process $\{X(t)\}$ can be evaluated as

$$\Phi^*(h, t) = \exp\{-\|U_t h - h\|_\alpha^\alpha\},$$

where functions h correspond to random variables Y. Figure 9.6.3 contains numerical evaluations of $\Phi^*(h_i, t)$ for $\alpha = 1.7$ and the above defined functions $h_i(x)$, where $i = 1, ..., 4$. Plotted is $\Phi^*(h_i, t)$ versus $t \in [0, 10]$. Note the different asymptotic values of the dynamical functional represented by dotted lines for the functions $h_i(x)$.

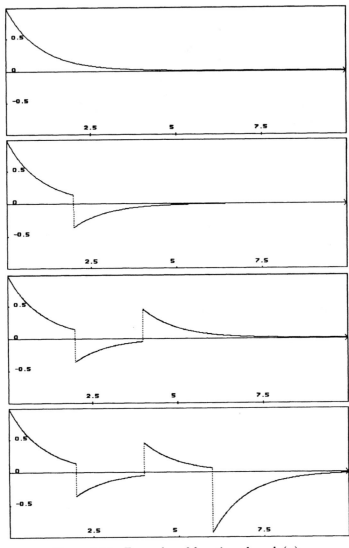

Figure 9.6.1. Examples of functions $h_i = h_i(x)$.

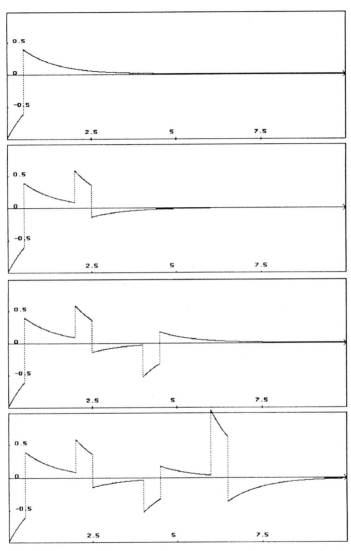

Figure 9.6.2. Examples of $U_t h(x) - h(x)$, for $h_i = h_i(x)$ defined above and $t = 0.5$.

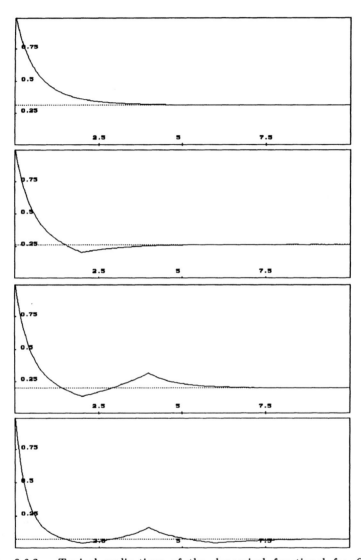

Figure 9.6.3.   Typical realizations of the dynamical functional for $S1.7S$ Ornstein–Uhlenbeck process. Plotted is $\Phi^*(h_i, t)$ versus $t \in [0, 10]$ for the same functions h_i as in two previous figures.

Chapter 10

Hierarchy of Chaos for Stable and ID Stationary Processes

10.1 Introduction

A large number of papers on chaotic properties of stochastic processes have been devoted to Gaussian processes, starting from Maruyama (1949), Grenander (1950) and Fomin (1950).

As a source of information on chaotic behavior of stable processes should be regarded the paper by Cambanis, Hardin and Weron (1987). See also Weron (1984), (1985), Podgórski and Weron (1991) and Gross (1992b).

For infinitely divisible processes we refer the reader to the pioneering work of Maruyama (1970), where he introduced an analytical approach to the study of ID processes, based on the Lévy–Khintchine representation. For harmonizable ID processes he proved that they are never ergodic, gave necessary and sufficient conditions for mixing, and pointed out that mixing and mixing of all orders are equivalent. For stationary Gaussian processes this was already established by Leonov (1960). See also Cambanis et al. (1991) and Gross (1992a) for a recent development.

Here we are concerned with two basic questions: how are the ergodic and mixing properties of stationary symmetric stable (and, more generally, infinitely divisible) processes related to the spectral representation? And how analogous is the general symmetric stable or infinitely divisible situation to the Gaussian?

All infinitely divisible processes share with the Gaussians the property that pairwise independence implies mutual independence (Maruyama (1970)); this follows from the fact that integrals of deterministic functions with respect to a Poisson measure are independent if and only if the functions have disjoint supports. It is reasonable, therefore, to guess that mixing (asymptotic independence) would be equivalent to asymptotic pairwise independence of random variables in an infinitely divisible process.

While Maruyama did not state the following result explicitly, it is contained in the proof of his theorem characterizing mixing of infinitely divisible processes.

Proposition 10.1.1 *A stationary infinitely divisible process* $\{X(t)\}$ *is mixing if and only if for all* $\theta_1, \theta_2 \in \mathbb{R}$ *we have*

$$\lim_{t \to \infty} \left\langle e^{i\theta_1 X(t)}, e^{i\theta_2 X(0)} \right\rangle = \left\langle e^{i\theta_1 X(0)}, 1 \right\rangle \left\langle 1, e^{i\theta_2 X(0)} \right\rangle.$$

This means that all infinitely divisible processes are like the Gaussians in the sense that mixing is determined by the bivariate marginal distributions. With Gaussian processes, however, it suffices to take $\theta_1 = \theta_2 = 1$, as Gaussian processes are mixing if and only if the covariances converge to zero (see Cornfeld, Fomin and Sinai (1982)). One open question is whether the non–Gaussian ID processes share this property with the Gaussian processes. The results in this direction are so far incomplete. See Gross (1992) for $S\alpha S$ processes and Kokoszka and Taqqu (1993) for the class of processes of type G.

In contrast to Maruyama (1970), we employ here, as a simple tool, the concept of the dynamical functional and combine it with the spectral representation of ID processes developed by Rajput and Rosinski (1989). As a result we are able to present, in a rather simple way, a systematic study of the chaotic behavior of non–Gaussian ID stationary processes. In Section 10.5 we give a characterization of ergodic ID processes (Theorem 10.5.1) and prove that ergodicity and weak mixing are equivalent (Theorem 10.5.2). Section 10.6 contains a new characterization of mixing for ID processes (Theorem 10.6.1) and Maruyama's result that mixing and mixing of all orders are equivalent. We also discuss some examples. It turns out that each ID moving average process is mixing (Example 10.7.1) and there exists a non–Gaussian moving average process which has the Kolmogorov property, or is exact (Example 10.7.3). In Section 10.8 we study a simple class of random sequences and their chaotic properties as they relate to the spectral representation. We also present some examples, including an example of a weakly mixing ID sequence which is not mixing (Example 10.8.2).

A schematic representation of the hierarchy of chaotic properties: ergodicity, weak mixing, p–mixing, Kolmogorov property and exactness is presented in Figure 10.1.1 on the next page.

The discussed hierarchy exhibits gradually stronger chaotic properties. Kolmogorov flows, which are invertible and therefore can not be exact, are stronger than mixing. To some extent they are parallel to exact flows. (The dotted lines show the case where the question of proper inclusion is still open.)

The significance of these properties for studying and modeling the chaotic behavior of physical systems is discussed in Lasota and Mackey (1985) and Devaney (1989).

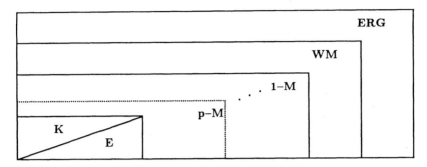

(i) General dynamical systems.

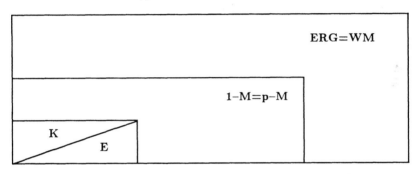

(ii) Stationary Gaussian process.

Figure 10.1.1. Schematic illustration of the hierarchy of gradually stronger chaotic properties: ergodicity (ERG), weak mixing (WM), p–mixing (p–M), exactness (E) and the Kolmogorov property (K).

10.2   Ergodicity of Stable Processes

Let us recall once more that a real random variable Y has a stable distribution if for every $a, b > 0$ and independent copies Y_1, Y_2 of Y there exists $c > 0$ such that

$$aY_1 + bY_2 = cY.$$

For every stable random variable Y there exists a unique $\alpha \in (0, 2]$ (the *index of stability*) such that the number c which appears in the above definition is uniquely determined by the equality $c = (a^\alpha + b^\alpha)^{\frac{1}{\alpha}}$. If the random variable Y has a *symmetric stable distribution* with index α then its characteristic function is of the form

$$\phi(\theta) = exp(-c_Y |\theta|^\alpha),$$

where c_Y is some positive constant.

Definition 10.2.1 *A stochastic process* **X** *is called* **symmetric** α**-stable** *or* **Lévy** **SαS** *or, shortly,* **SαS** **process** *for* $\alpha \in (0,2]$, *if for every* $n \in \mathbb{N}$ *and any* $a_1, ..., a_n \in \mathbb{R}$, $t_1, ..., t_n \in \mathbb{R}$, *the random variable* $Y = \sum_{i=1}^{n} a_i X(t_i)$ *has a symmetric stable distribution with index* α.

Let **X** be an $S\alpha S$ process, $\alpha \in (0,2]$. For an $S\alpha S$ random variable Y, set $\|Y\|_\alpha = c_Y^{1/\alpha}$. Then $\| \cdot \|_\alpha^{1 \wedge \alpha}$ defines a norm in the case $1 \leq \alpha \leq 2$ and a quasi–norm in the case $0 < \alpha < 1$ on the space $\lin\{X(t) : t \in \mathbb{R}\}$, metrizing the convergence in probability. Then, for $Y \in \lin\{X(t) : t \in \mathbb{R}\}$ we have

$$\mathsf{E} e^{i\theta Y} = \exp(-|\theta|^\alpha \|Y\|_\alpha^\alpha).$$

Taking the closure of the linear span $lin\{X(t) : t \in \mathbb{R}\}$ with respect to the norm (quasi–norm) $\| \cdot \|_\alpha$ in the space $L_0(\mathbf{X}) \subseteq L^o(\Omega, \mathcal{F}, P)$ we obtain the space $L_\alpha(\mathbf{X})$.

If **X** is a stationary process then for $Y \in L^\alpha$ and $t \in \mathbb{R}$ we have $\|T_t Y\|_\alpha = \|Y\|_\alpha$. Hence, $(T_t)_{t \in \mathbb{R}}$ is a group of isometries on L^α.

Definition 10.2.2 *Let* $(E, \mathcal{E}, \lambda)$ *be a measure space. Let us introduce the family of sets* $\mathcal{E}_0 \stackrel{df}{=} \{A \in \mathcal{E} : \lambda(A) < \infty\}$. *The map* $\mathbf{Z} : \mathcal{E}_0 \to L^o$ *is called a* **stochastic SαS measure** *with a* **control measure** λ *if:*

(i) $Z(\emptyset) = 0$ *with probability 1;*

(ii) *for* $A \in \mathcal{E}_0$ *the law of* $Z(A)$ *is described by* $\mathsf{E} e^{i\theta Z(A)} = e^{-|\theta|^\alpha \lambda(A)}$;

(iii) *for every sequence* $\{A_n\}_{n \in \mathbb{N}}$ *of pairwise disjoint sets from* $\mathcal{E}_0$ *the sequence of random variables* $\{Z(A_n)\}_{n \in \mathbb{N}}$ *is independent and such that* $Z\left(\bigcup_{n=1}^{\infty}\right) = \sum_{n=1}^{\infty} Z(A_n)$ *with probability 1.*

According to Definition 4.3.1, for every function $f \in L^\alpha(E, \mathcal{E}, \lambda)$ one can define a stochastic integral $\int_E f \, d\mathbf{Z}$ as an α–stable random variable with the law given by

$$\mathsf{E} \exp\left(i\theta \int_E f \, d\mathbf{Z}\right) = \exp(-|\theta|^\alpha \|f\|_\alpha^\alpha).$$

The Spectral Representation Theorem for a stationary stochastic $S\alpha S$ process $\mathbf{X} = \{X(t)\}_{t \in \mathbb{R}}$ (cf. Theorems 5.3.1, 5.4.1 and 5.4.2) says that there exist a measure space $(E, \mathcal{E}, \lambda)$ and a group (U_t) of isometries of $L^\alpha(E, \mathcal{E}, \lambda)$ described by a function $f_0 \in L^\alpha(E, \mathcal{E}, \lambda)$, such that

$$X(t) = \int_E U_t f_0 \, d\mathbf{Z} \quad \text{for all} \quad t \in \mathbb{R}.$$

If **X** is measurable then the above group of isometries is strongly continuous (see Cambanis, Hardin and Weron (1987), Theorem 6).

The characterization of ergodic processes in terms of their spectral representation plays an important role in the ergodic theory of $S\alpha S$ processes. The characterization given below was established in Cambanis, Hardin and Weron (1987, Theorem 1).

Theorem 10.2.1 *Let* $\mathbf{X}$ *be a stationary stochastic* $S\alpha S$ *process with spectral representation of the form* $\{\int_E U_t f_0 \, dZ\}$. *Then* $\mathbf{X}$ *is ergodic if and only if for every function* $h \in \overline{\mathrm{lin}}\{U_t f_0 : t \in \mathbb{R}\}$ *we have*

$$\lim_{T\to\infty} \frac{1}{T} \int_0^T \|U_t h - h\|_\alpha^{2\alpha} dt = 4\|h\|_\alpha^{2\alpha} \tag{10.2.1}$$

and

$$\lim_{T\to\infty} \frac{1}{T} \int_0^T \|U_t h - h\|_\alpha^\alpha dt = 2\|h\|_\alpha^\alpha. \tag{10.2.2}$$

PROOF. According to Theorem 9.5.1 and Propositon 9.6.1, the ergodicity of $\mathbf{X}$ is equivalent to the condition

$$\lim_{T\to\infty} \frac{1}{T} \int_0^T T_t Y \, dt = \boldsymbol{E} Y \quad a.s. \quad \text{for each } Y \in L^1(\mathbf{X}). \tag{10.2.3}$$

As in the Gaussian case (see e.g. Dym and McKean (1976)), it is enough to check this condition for random variables Y of the form $Y = \exp[i \sum_{n=1}^N a_n X(t_n)]$. Then, putting $h = (\sum_{n=1}^N a_n U_{t_n})\phi$, we have $T_\tau Y = \exp[i \int_E U_\tau h \, dZ]$ and

$$\Upsilon_T \overset{\mathrm{df}}{=} \frac{1}{T}\int_0^T T_\tau Y \, d\tau = \frac{1}{T}\int_0^T \exp\left[i\int_E U_\tau h \, dZ\right] d\tau.$$

By the Birkhoff Theorem (see Section 9.4), $\Upsilon_T \to \boldsymbol{E}(Y|\mathcal{P}_\mathbf{X}) \overset{\mathrm{df}}{=} \Upsilon_\infty$ *a.s.* Thus (10.2.3) is satisfied, i.e. $\Upsilon_\infty = \boldsymbol{E} Y$ if and only if $\boldsymbol{E}|\Upsilon_\infty|^2 = |\boldsymbol{E}\Upsilon_\infty|^2$ if and only if $\lim_{T\to\infty} \boldsymbol{E}|\Upsilon_T|^2 = \lim_{T\to\infty} |\boldsymbol{E}\Upsilon_T|^2$. But

$$\boldsymbol{E}\Upsilon_T = \frac{1}{T} \int_0^T \exp(-\|U_\tau h\|_\alpha^\alpha) \, d\tau = \exp(-\|h\|_\alpha^\alpha),$$

and

$$\boldsymbol{E}|\Upsilon_T|^2 = \frac{1}{T^2} \int_0^T \int_0^T \exp(-\|(U_\tau - U_\sigma)h\|_\alpha^\alpha) \, d\tau \, d\sigma.$$

Notice that $|x|^\alpha + |y|^\alpha - |x - y|^\alpha$ is a positive definite function of x and y. To see this, observe that we can easily evaluate a covariance function for a H–self–similar $(H - ss)$ process $\{X(t)\}$ with stationary increments (si) and finite second moment $\boldsymbol{E}|X(t)|^2 < \infty$, cf. Section 5.5. Namely, for $s = x$ and $t = y$,

$$
\begin{aligned}
\boldsymbol{E}X(t)X(s) &= \frac{1}{2}\left[\boldsymbol{E}X(t)^2 + \boldsymbol{E}X(s)^2 - \boldsymbol{E}(X(t) - X(s))^2\right] \\
&= \frac{1}{2}\left[\boldsymbol{E}X(t)^2 + \boldsymbol{E}X(s)^2 - \boldsymbol{E}X(t - s)^2\right] \\
&= \frac{1}{2}\left[|t|^{2H}\boldsymbol{E}X(1)^2 + |s|^{2H}\boldsymbol{E}X(1)^2 - |t - s|^{2H}\boldsymbol{E}X(1)^2\right] \\
&= \frac{1}{2}\boldsymbol{E}X(1)^2\left[|t|^{2H} + |s|^{2H} - |t - s|^{2H}\right],
\end{aligned}
$$

where we have used si and $H - ss$ properties of the process $\{X(t)\}$. Since the left hand side of the above formula is positive definite, the function

$$[|x|^\alpha + |y|^\alpha - |x - y|^\alpha]$$

is also positive definite, where we put $H = \alpha/2$.

Thus $|(U_\tau h)(\theta)|^\alpha + |(U_\sigma h)(\theta)|^\alpha - |[(U_\tau - U_\sigma)h](\theta)|^\alpha$ is a positive definite function of τ and σ for each θ, and thus so is its λ-integral over E: $2\|h\|_\alpha^\alpha - \|(U_\tau - U_\sigma)h\|_\alpha^\alpha$. Since the latter depends only on the difference $\tau - \sigma$, and is continuous, we have by the Bochner Theorem

$$2\|h\|_\alpha^\alpha - \|(U_\tau - U_\sigma)h\|_\alpha^\alpha = \int_{-\infty}^\infty e^{i(\tau-\sigma)u}\, d\nu(u),$$

where ν is a finite symmetric measure. Then we obtain

$$\frac{E|\Upsilon_T|^2}{|E\Upsilon_T|^2} = \frac{1}{T^2} \int_0^T \int_0^T \exp\left[\int_{-\infty}^\infty e^{i(\tau-\sigma)u}\, d\nu(u)\right]\, d\tau\, d\sigma$$

$$= 1 + \sum_{k=1}^\infty \frac{1}{k!}\frac{1}{T^2} \int_0^T \int_0^T \left[\int_{-\infty}^\infty e^{i(\tau-\sigma)u}\, d\nu^{*k}(u)\right]\, d\tau\, d\sigma$$

and so,

$$\lim_{T\to\infty} \frac{E|\Upsilon_T|^2}{|E\Upsilon_T|^2} = 1 + \sum_{k=1}^\infty \frac{1}{k!}\nu^{*k}(\{0\}),$$

where ν^{*k} means k-fold convolution of ν. It follows that $\mathbf{X}$ is ergodic if and only if $\nu^{*k}(\{0\}) = 0$ for all $k \geq 0$ (and all $h \in \overline{\lin}\{U_t\phi : t \in \mathbb{R}\} \subset L^\alpha(\mu)$).

Since the function $\int_{-\infty}^\infty e^{i\tau u}\, d\nu(u) = 2\|h\|_\alpha^\alpha - \|U_\tau h - h\|_\alpha^\alpha$ is even, we have by the inversion formula

$$\nu(\{0\}) = 2\|h\|_\alpha^\alpha - \lim_{T\to\infty} \frac{1}{T} \int_0^T \|U_\tau h - h\|_\alpha^\alpha\, d\tau$$

and thus $\nu(\{0\}) = 0$ if and only if (10.2.2) is satisfied. Also, by the Wiener Theorem,

$$\nu^{*2}(\{0\}) = \int_{-\infty}^\infty \nu(\{-x\})\, d\nu(x) = \sum_x \nu(\{-x\})\nu(\{x\}) = \sum_x \nu^2(\{x\})$$

$$= \lim_{T\to\infty} \frac{1}{T} \int_0^T (2\|h\|_\alpha^\alpha - \|U_\tau h - h\|_\alpha^\alpha)^2\, d\tau$$

$$= 4\|h\|_\alpha^{2\alpha} - 4\|h\|_\alpha^\alpha \lim_{T\to\infty} \frac{1}{T} \int_0^T \|(U_\tau h - h\|_\alpha^\alpha\, d\tau$$

$$+ \lim_{T\to\infty} \frac{1}{T} \int_0^T \|U_\tau h - h\|_\alpha^{2\alpha}\, d\tau,$$

from which it follows that $\nu^{*k}(\{0\}) = 0$ for $k = 1, 2$ if and only if (10.2.2) and (10.2.3) are satisfied. The proof is completed by noting that (from the above calculations) $\nu^{*2}(\{0\}) = 0$ implies that ν has no atoms and thus $\nu^{*k}(\{0\}) = 0$ for all $k > 2$. $\qquad\square$

Remark 10.2.1 *When a stationary SαS process* **X** *is ergodic, we can use the Birkhoff Theorem to estimate its covariation function (for the definition consult Section 2.4; see also Theorem 4.3.3), which plays a role analogous to that of the covariance when $\alpha = 2$. Indeed, when $1 < p < \alpha < 2$, we have*

$$\lim_{T \to \infty} \frac{1}{T} \int_0^T X(t) X^{<p-1>}(t + \tau)\, dt = E\{X(0) X^{<p-1>}(\tau)\}$$

$$= C^p(p, \alpha) \frac{[X(0), X(\tau)]_\alpha}{\|X(0)\|_\alpha^{\alpha-p}} \quad a.e.,$$

where $x^{<\beta>} \stackrel{df}{=} |x|^\beta sign(x)$ and $[X_1, X_2]_\alpha$ denotes the covariation of the jointly SαS vector (X_1, X_2). The equality follows from Cambanis, Hardin and Weron (1987). For $\tau = 0$ this gives the scaling constant of the process:

$$\lim_{T \to \infty} \frac{1}{T} \int_0^T |X(t)|^p\, dt = E|X(0)|^p = C^p(p, \alpha)\|X(0)\|_\alpha^p \quad a.e.\, .$$

Now we present another characterization of ergodicity, also in terms of the spectral representation. We also give some examples demonstrating the usefulness of these results. But first, let us state the following lemma.

Lemma 10.2.1 *For every $h \in \overline{\lin}\{U_t f_0 : t \in \mathbb{R}\}$, the function $\psi_h : \mathbb{R} \to \mathbb{R}$ defined by*

$$\psi_h(t) = exp(-\|U_t h - h\|_\alpha^\alpha)$$

is continuous, positive definite and $\psi_h(0) = 1$.

PROOF. The continuity of ψ_h follows from the strong continuity of the group $(U_t)_{t \in \mathbb{R}}$. Moreover $\psi_h(0) = \exp(-\|h - h\|_\alpha^\alpha)$. So it remains to verify that ψ_h is positive definite. Let $n \in \mathbb{N}$, $t_1, ..., t_n \in \mathbb{R}, a_1, ..., a_n \in \mathbb{R}$ and $Y \in L^\alpha$ be the random variable represented by h. Then

$$\sum_{i=1}^n \sum_{j=1}^n a_i a_j \psi_h(t_i - t_j) = \sum_{i=1}^n \sum_{j=1}^n a_i a_j \exp\left(-\|U_{t_i - t_j} h - h\|_\alpha^\alpha\right)$$

$$= \sum_{i=1}^n \sum_{j=1}^n a_i a_j \exp\left(-\|U_{t_i} h - U_{t_j} h\|_\alpha^\alpha\right)$$

$$= \sum_{i=1}^n \sum_{j=1}^n a_i a_j E\left\{\exp\left(iT_{t_i} Y\right) \cdot \exp\left(-iT_{t_j} Y\right)\right\}$$

$$= E\left\{\sum_{i=1}^n a_i \exp\left(iT_{t_i} Y\right) \cdot \sum_{j=1}^n a_j \exp\left(-iT_{t_j} Y\right)\right\}$$

$$= E\left|\sum_{i=1}^n a_i \exp\left(iT_{t_i} Y\right)\right|^2 \geq 0,$$

which ends the proof.   □

Now we formulate and prove another theorem characterizing ergodicity of $S\alpha S$ processes, due to Podgórski (1992).

Theorem 10.2.2 *Let* **X** *be a stationary stochastic* $S\alpha S$ *process with the spectral representation of the following form*

$$\{X(t)\}_{t \in \mathbb{R}} \overset{d}{=} \left\{ \int_E U_t f_0 d\mathbf{Z} \right\}_{t \in \mathbb{R}}.$$

Then **X** *is ergodic if and only if for every function* $h \in \overline{\lin}\{U_t f_0 : t \in \mathbb{R}\}$

$$\lim_{T \to \infty} \frac{1}{T} \int_0^T \exp\left(2\|h\|_\alpha^\alpha - \|U_t h - h\|_\alpha^\alpha\right) dt = 1.$$

PROOF. Let $Y \in \overline{\lin}\{X(t) : t \in \mathbb{R}\}$ and let h correspond to Y in the spectral representation. It follows from Lemma 10.2.1 and the Bochner Theorem that there exists a probability measure ν on $\mathcal{B}_{\mathbb{R}}$ such that for $t \in \mathbb{R}$ we have

$$\psi_h(t) = \exp(-\|U_t h - h\|_\alpha^\alpha) = \int_{\mathbb{R}} e^{itx} d\nu(x).$$

Thus,

$$
\begin{aligned}
E\left| \frac{1}{T} \int_0^T \exp(iT_t Y) dt \right|^2 &= E\left\{ \frac{1}{T^2} \int_0^T \int_0^T \exp(iT_t Y) \exp(-iT_u Y) dt du \right\} \\
&= \frac{1}{T^2} \int_0^T \int_0^T E\left(\exp[i(T_{t-u} Y - Y)]\right) dt\, du \\
&= \frac{1}{T^2} \int_0^T \int_0^T \psi_h(t - u)\, dt\, du \\
&= \frac{1}{T^2} \int_0^T \int_0^T \int_{\mathbb{R}} e^{i(t-u)x}\, d\nu(x)\, dt\, du \\
&= \int_{\mathbb{R}} \left| \frac{1}{T} \int_0^T e^{itx} dt \right|^2 d\nu(x) \\
&= \int_{\mathbb{R}} \frac{(\sin \frac{Tx}{2})^2}{(\frac{Tx}{2})^2}\, d\nu(x)
\end{aligned}
$$

and

$$
\begin{aligned}
\frac{1}{T} \int_0^T \exp(-\|U_t h - h\|_\alpha^\alpha)\, dt &= \frac{1}{T} \int_0^T \int_{\mathbb{R}} e^{itx}\, d\nu(x)\, dt \\
&= \int_{\mathbb{R}} \frac{1}{T} \int_0^T e^{itx}\, dt\, d\nu(x) = \int_{\mathbb{R}} \frac{e^{iTx} - 1}{iTx}\, d\nu(x).
\end{aligned}
$$

Since $\left|\frac{e^{iTx}-1}{iTx}\right|^2 = \frac{(\sin\frac{Tx}{2})^2}{(\frac{Tx}{2})^2} < 1$ and

$$\lim_{T\to\infty}\frac{e^{iTx}-1}{iTx} = \lim_{T\to\infty}\frac{(\sin\frac{Tx}{2})^2}{(\frac{Tx}{2})^2} = \delta_{\{0\}}(x),$$

we have by the Lebesgue Bounded Convergence Theorem

$$\lim_{T\to\infty}E\left|\frac{1}{T}\int_0^T \exp(iT_tY)\,dt\right|^2 = \lim_{T\to\infty}\frac{1}{T}\int_0^T \exp(-\|U_th-h\|_\alpha^\alpha)\,dt = \nu(\{0\}).$$

From the Birkhoff Theorem it follows that

$$\lim_{T\to\infty}\frac{1}{T}\int_0^T \exp(iT_tY)\,dt = E(e^{iY}|\mathcal{P}_\mathbf{X})$$

with probability 1. Applying once more the Lebesgue Bounded Convergence Theorem we obtain

$$\lim_{T\to\infty}E\left|\frac{1}{T}\int_0^T \exp(iT_tY)\,dt\right|^2 = E\left(E(e^{iY}|\mathcal{P}_\mathbf{X})\,E(e^{-iY}|\mathcal{P}_\mathbf{X})\right)$$

$$= E\left(E(e^{iY}|\mathcal{P}_\mathbf{X})\,e^{-iY}\right).$$

From this and previous equalities it follows that

$$\lim_{T\to\infty}\frac{1}{T}\int_0^T \exp(-\|U_th-h\|_\alpha^\alpha)\,dt = E\left(E(e^{iY}|\mathcal{P}_\mathbf{X})\,e^{-iY}\right).$$

Let us notice that the stochastic process $\mathbf{X}$ is ergodic if and only if for every $Y \in \mathrm{lin}\{X(t) : t \in \mathbb{R}\}$ we have

$$E(E(e^{iY}|\mathcal{P}_\mathbf{X})e^{-iY}) = |Ee^{iY}|^2.$$

Indeed, if $\mathbf{X}$ is ergodic then of course $E(e^{iY}|\mathcal{P}_\mathbf{X}) = Ee^{iY}$. On the other hand, if the above equality is true for $Y \in \mathrm{lin}\{X(t) : t \in \mathbb{R}\}$ then, for $Z = E(e^{iY}|\mathcal{P}_\mathbf{X})$, we have

$$E|Z|^2 = E(E\{e^{-iY}E(e^{iY}|\mathcal{P}_\mathbf{X})|\mathcal{P}_\mathbf{X}\}) = E(E(e^{iY}|\mathcal{P}_\mathbf{X})e^{-iY})$$

$$= |Ee^{iY}|^2 = |EZ|^2.$$

This means that $E|Z-EZ|^2 = 0$ or, equivalently, that $E(e^{iY}|\mathcal{P}_\mathbf{X}) = Ee^{iY}$ with probability 1. Since conditional expectation is a linear continuous operator, it follows from Lemma 9.6.1 that for each $Y \in L^1(\mathbf{X})$ we have $E(Y|\mathcal{P}_\mathbf{X}) = EY$ with probability 1, which implies ergodicity. □

Theorems 10.2.1 and 10.2.2 are generalized in Section 10.5 to cover stationary symmetric infinitely divisible processes.

Now we present a few examples of application of Theorems 10.2.1 and 10.2.2. First, we show that any real–valued $S\alpha S$ process with harmonic spectral representation is not ergodic.

Let us recall that an $S\alpha S$ process $\{X(t)\}_{t \in \mathbb{R}}$ has a harmonic spectral representation if there exists a complex stochastic measure W defined on $(\mathbb{R}, \mathcal{B}_{\mathbb{R}}, \mu)$ with finite control measure μ such that

$$X(t) = \text{Re} \int_{\mathbb{R}} e^{it\theta} \, dW(\theta), \qquad t \in \mathbb{R}.$$

For a stationary $S\alpha S$ process $\{X(t)\}_{t \in \mathbb{R}}$ with such a representation there exists a positive constant c_α such that for $n \in \mathbb{N}$, $a_1, ..., a_n \in \mathbb{R}$, $t_1, ..., t_n \in \mathbb{R}$, we have

$$\left\| \sum_{l=1}^{n} a_l U_{t_l} f_0 \right\|_\alpha^\alpha = c_\alpha \int_{\mathbb{R}} \left| \sum_{l=1}^{n} a_l \exp(it_l\theta) \right|^\alpha \, d\mu(\theta)$$

(see Weron (1984)).

We have the following statement.

Proposition 10.2.1 *A real–valued stationary $S\alpha S$ process with harmonic spectral representation is never ergodic for $\alpha \in (0, 2)$.*

PROOF. With some minor adjustment in the proof of Theorem 10.2.1 allowing complex–valued functions in the spectral representation, we have that $\mathbf{X}$ is ergodic if and only if

$$\frac{1}{T} \int_0^T \|(e^{i\tau\theta} - 1)h(\theta)\|_\alpha^\alpha d\tau \to 2\|h\|_\alpha^\alpha$$

and

$$\frac{1}{T} \int_0^T \|(e^{i\tau\theta} - 1)h(\theta)\|_\alpha^{2\alpha} d\tau \to 4\|h\|_\alpha^{2\alpha},$$

for all complex $h \in \mathbf{L}^\alpha(E, \mathcal{E}, m)$. But

$$\lim_{T \to \infty} \frac{1}{T} \int_0^T \|(e^{i\tau\theta} - 1)h(\theta)\|_\alpha^\alpha \, d\tau$$

$$= \lim_{T \to \infty} \int_{-\infty}^{\infty} |h(\theta)|^\alpha \left\{ \frac{1}{T} \int_0^T \left| 2\sin\frac{\tau\theta}{2} \right|^\alpha d\tau \right\} d\mu(\theta)$$

$$= \lim_{T \to \infty} \int_{\mathbb{R}\backslash\{0\}} |h(\theta)|^\alpha \left\{ \frac{2}{T\theta} \int_0^{T\theta/2} |2\sin u|^\alpha du \right\} d\mu(\theta)$$

$$= \frac{1}{\pi} \int_0^\pi |2\sin u|^\alpha du \int_{\mathbb{R}\backslash\{0\}} |h(\theta)|^\alpha \, d\mu$$

$$= D_\alpha \int_{\mathbb{R}\backslash\{0\}} |h(\theta)|^\alpha \, d\mu.$$

Note that, when $\alpha = 2, D_2 = 2$ and thus the first of the pair of the necessary and sufficient conditions for ergodicity is satisfied provided $\mu(\{0\}) = 0$. We now show that when $0 < \alpha < 2$, then $D_\alpha < 2$, and thus this condition is not satisfied and **X** is not ergodic. Indeed, by the Jensen Inequality we have

$$
\begin{aligned}
D_\alpha &= \frac{1}{\pi} \int_0^\pi |2 \sin u|^\alpha du \\
&= \frac{1}{\pi} \int_0^\pi (|2 \sin u|^2)^{\alpha/2} du \leq (\frac{1}{\pi} \int_0^\pi |2 \sin u|^2 du)^{\alpha/2} \\
&= 2^{\alpha/2}.
\end{aligned}
$$

This ends the proof.   □

It turns out that stationary α–sub–Gaussian processes are not ergodic for either $\alpha \in (0, 2)$.

Let us recall that a nontrivial stochastic process **X** is α–sub–Gaussian if there exists a positive definite function $R : \mathbb{R} \to \mathbb{R}$ such that

$$
E \exp\left(i \sum_{l=1}^n a_l X(t_l) \right) = \exp\left(- \left(\frac{1}{2} \sum_{k=1}^n \sum_{j=1}^n a_k a_j R(t_k - t_j) \right)^{\alpha/2} \right)
$$

(see Example 5.3.3).

From this definition it follows that any α–sub–Gaussian process is $S\alpha S$ and

$$
\| \sum_{l=1}^n a_l X(t_l) \|_\alpha^\alpha = \left(\frac{1}{2} \sum_{l=1}^n \sum_{j=1}^n a_l a_j R(t_l - t_j) \right)^{\alpha/2}.
$$

Proposition 10.2.2 *A nontrivial stationary α–sub–Gaussian process is not ergodic.*

PROOF. It is enough to show that

$$
\liminf_{T \to \infty} \frac{1}{T} \int_0^T \exp(2\|X(0)\|_\alpha^\alpha - \|X(t) - X(0)\|_\alpha^\alpha) \, dt > 1.
$$

The function R is positive definite, $R(0) \geq 0$ and $R(t) \leq R(0)$ for all $t \in \mathbb{R}$. Thus, since **X** is non-trivial, we have $R(0) > 0$. Consequently, for each $T \in \mathbb{R}$ we have

$$
\frac{1}{T} \int_0^T \exp(2\|X(0)\|_\alpha^\alpha - \|X(t) - X(0)\|_\alpha^\alpha) \, dt
$$

$$
= \frac{1}{T} \int_0^T \exp(2(\tfrac{1}{2}R(0))^{\alpha/2} - (R(0) - R(t))^{\alpha/2}) \, dt
$$

$$
exp(2^{1-\alpha/2} R(0)^{\alpha/2}) > 1.
$$

This ends the proof.   □

We can make use of Theorem 10.2.1 or Theorem 10.2.2 to obtain another proof of the classical Maruyama–Grenander–Fomin Theorem, characterizing ergodic Gaussian processes (cf. Section 9.2).

Any real–valued stationary Gaussian process $\mathbf{X}$ with mean 0, variance 1 and with correlation function R is $S\alpha S$ with $\alpha = 2$ and

$$\left\| \sum_{l=1}^{n} a_l X(t_l) \right\|_2^2 = \frac{1}{2} \sum_{l=1}^{n} \sum_{j=1}^{n} a_l a_j R(t_l - t_j).$$

The correlation function can be obtained as the Fourier transform of the symmetric finite measure $\mu_{\mathbf{X}}$ on $\mathbb{R}$, called the *spectral measure of the process* $\mathbf{X}$. There exists an one–to–one correspondence between the space $L^2(\mathbf{X})$ and the set of all measures on $\mathbb{R}$ that are absolutely continuous with respect to the probability measure $\mu_{\mathbf{X}}$. Namely, for any element $Y \in L^2(\mathbf{X})$ we have a measure μ_Y such that

$$\operatorname{Cov}(T_t Y, Y) = \int_{\mathbb{R}} e^{it\theta} \, d\mu_Y(\theta). \tag{10.2.4}$$

Proposition 10.2.3 *A real–valued stationary Gaussian process* $\mathbf{X}$ *is ergodic if and only if its spectral measure* $\mu_{\mathbf{X}}$ *has no atoms.*

PROOF. Let μ_Y be the spectral measure related to Y by (10.2.4). We can assume that $\mathbf{E}\, X(t) \equiv 0$ because $\mathbf{X}$ is stationary. Let $h \in \overline{\operatorname{lin}}\{U_t f_0 : t \in \mathbb{R}\}$ and let $Y \in L^2$ correspond to h through the spectral representation of stable processes. Thus

$$
\begin{aligned}
\exp(2\|h\|_2^2 - \|U_t h - h\|_2^2) &= \exp(2\|Y\|_2^2 - \|T_t Y - Y\|_2^2) \\
&= \exp(2\operatorname{Var}(Y) - \operatorname{Var}(T_t Y - Y)) \\
&= \exp(2\operatorname{Var}(Y) - \operatorname{Var}(T_t Y) + 2\operatorname{Cov}(T_t Y, Y) - \operatorname{Var}(Y)) \\
&= \exp(2\operatorname{Cov}(T_t Y, Y)) \\
&= \exp\left(\int_{\mathbb{R}} e^{it\theta}\, 2 d\mu_Y(\theta)\right) \\
&= \exp((2\mu_Y)\hat{}(t)) = (\exp(2\mu_Y))\hat{}(t).
\end{aligned}
$$

Since

$$\lim_{T \to \infty} \frac{1}{T} \int_0^T (\exp(2\mu_Y))\hat{}(t)\, dt = \exp(2\mu_Y)(\{0\}),$$

thus from Theorem 10.2.2 it follows that the process $\mathbf{X}$ is ergodic if and only if the measure $\exp(2\mu_Y)$ has no atoms at 0 equal to 1 or, by Lemma 10.4.2, if μ_Y has no atoms. A symmetric mesure has no atoms if and only if any symmetric measure absolutly continuous with respect to it has no atoms. This completes the proof. □

Let us notice that if a process $\mathbf{X}$ is stationary and $S\alpha S$ then for $t \in \mathbb{R}$ and $Y \in L^\alpha$ the dynamical functional Φ takes the form

$$\Phi(Y, t) = \exp\left(-\|T_t Y - Y\|_\alpha^\alpha\right) = \exp\left(-\|U_t h - h\|_\alpha^\alpha\right), \tag{10.2.5}$$

where $h \in \overline{\operatorname{lin}}\{U_t f_0 : t \in \mathbb{R}\}$ corresponds to Y in the spectral representation.

Now we give a characterization of ergodicity of stationary $S\alpha S$ processes in terms of the dynamical functional, further developing the results contained in Theorems 10.2.1 and 10.2.2 and providing an insight into relations between different conditions characterizing ergodicity, see Podgórski and Weron (1991).

Theorem 10.2.3 *Let* **X** *be a stationary $S\alpha S$ process. Then the following conditions are equivalent:*

(i) process **X** *is ergodic;*

(ii) for any function $h \in \text{lin}\{U_t f_0 : t \in \mathbb{R}\}$ we have

$$\lim_{T \to \infty} \frac{1}{T} \int_0^T \exp\left(2\|h\|_\alpha^\alpha - \|U_t h - h\|_\alpha^\alpha\right)\, dt = 1;$$

(iii) there exist positive integers n, k, $n \neq k$ such that for $h \in \text{lin}\{U_t f_0 : t \in \mathbb{R}\}$ we have

$$\lim_{T \to \infty} \frac{1}{T} \int_0^T \|U_t h - h\|_\alpha^{n\alpha}\, dt = 2^n \|h\|_\alpha^{n\alpha}$$

and

$$\lim_{T \to \infty} \frac{1}{T} \int_0^T \|U_t h - h\|_\alpha^{k\alpha}\, dt = 2^k \|h\|_\alpha^{k\alpha};$$

(iv) for any natural number n and any function $h \in \text{lin}\{U_t f_0 : t \in \mathbb{R}\}$ we have

$$\lim_{T \to \infty} \frac{1}{T} \int_0^T \|U_t h - h\|_\alpha^{n\alpha}\, dt = 2^n \|h\|_\alpha^{n\alpha};$$

(v) for any function $h \in \text{lin}\{U_t f_0 : t \in \mathbb{R}\}$ we have

$$\lim_{T \to \infty} \frac{1}{T} \int_0^T \left|\|U_t h - h\|_\alpha^\alpha - 2\|h\|_\alpha^\alpha\right| = 0.$$

PROOF. According to (10.2.5), the equivalence $(i) \Leftrightarrow (ii)$ follows immediately from Proposition 9.6.1. Let us notice that since for $c > 0$ we have $\|c^{1/\alpha} h\|_\alpha^\alpha = c\|h\|_\alpha^\alpha$, hence in condition (ii) we can write $\exp\left(c\left(2\|h\|_\alpha^\alpha - \|U_t h - h\|_\alpha^\alpha\right)\right)$. Thus, the equivalence of the next conditions follows from Lemma 10.4.3. $\square$

Example 10.2.1 *Numerical illustration of the ergodic property for the $S\alpha S$ Ornstein-Ulenbeck process.*

This is a continuation of Example 9.6.1. The numerical evaluation of the both time averages which appear in Theorem 10.2.1 is presented in Figure 10.2.1 for the $S\alpha S$ Ornstein-Uhlenbeck process with $\alpha = 1.7$. Here the same functions h_l as in Figure 9.6.1 are used.

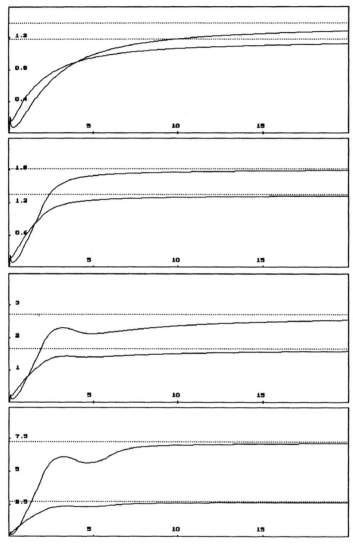

Figure 10.2.1. Illustration of the ergodic property of the symmetric 1.7–stable Ornstein–Uhlenbeck process via Theorem 10.2.1.

Plotted are the both time averages, see Equations (10.2.1) and (10.2.2), versus $T \in [0, 20]$, as indicated. The lower dotted line represents the theoretical value of the limit $2\|h_l\|_\alpha^\alpha$ and the upper dotted line $4\|h_l\|_\alpha^{2\alpha}$, respectively. Their values are presented in the following table.

THEORETICAL VALUES		
function	1st value	2nd value
h_1	1.177	1.384
h_2	1.349	1.820
h_3	1.645	2.707
h_4	2.692	7.248

In Figure 10.2.1 the upper curves indicate the numerical results of the 1st time average versus T (equation (10.2.1)) and the lower curves the numerical results of the 2nd time average (equation (10.2.2)), respectively. Note that these curves approach the theoretical values of the both limits for all four given functions $h_1, ..., h_4$. By Theorem 10.2.1 this is a typical behavior of any ergodic $S\alpha S$ stationary process. We will see later that the $S\alpha S$ Ornstin–Uhlenbeck process has even more stronger chaotic properties than ergodicity.

Example 10.2.2 *Numerical illustration of the lack of ergodic property for the $S\alpha S$ harmonizable process.*

In this example we would like to examine the behavior of the $S\alpha S$ harmonizable process. Let us recall that its spectral representation is given by $f_o(x) = I_{[0,\infty)}(x)$ and $U_t g(x) = \cos(tx)g(x)$. Take $W(dx) = e^{-x}L_\alpha(dx)$, where $L_\alpha(\cdot)$ stands for the $S\alpha S$ Lévy motion with $\alpha = 1.7$.

Similarly as in Example 9.6.1 we define four functions from the linear span $lin\{U_t f_o(x) : t \geq 0\}$

$$h_1(x) = f_o(x),$$

$$h_2(x) = f_o(x) - \tfrac{1}{2}U_2 f_o(x),$$

$$h_3(x) = f_o(x) - \tfrac{1}{2}U_2 f_o(x) + \tfrac{1}{2}U_4 f_o(x),$$

$$h_4(x) = f_o(x) - \tfrac{1}{2}U_2 f_o(x) + \tfrac{1}{2}U_4 f_o(x) - U_6 f_o(x).$$

In Figure 10.2.2 we illustrate the numerical results for the $S\alpha S$ harmonizable process and the above choice of functions $h_l(x)$. Plotted are the both time averages, corresponding to equations (10.2.1) and (10.2.2), versus $T \in [0, 20]$, as indicated. The lower dotted line represents the theoretical value of the limit $2\|h_l\|_\alpha^\alpha$ and the upper dotted line the second limit $4\|h_l\|_\alpha^{2\alpha}$, respectively. Their values are presented in the following table.

THEORETICAL VALUES		
function	1st value	2nd value
h_1	2.000	4.000
h_2	1.821	3.310
h_3	1.998	3.990
h_4	2.447	5.960

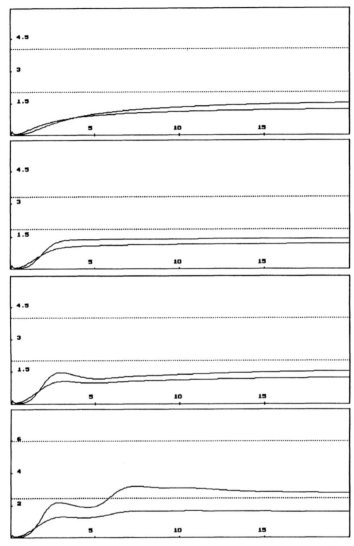

Figure 10.2.2. Illustration of the fact that the symmetric 1.7–stable harmoniz-able process is not ergodic via Theorem 10.2.1.

In contrast to the previous example, in all four cases the curves which repre-sent the time averages of the $S1.7S$ harmonizable process do not approach the corresponding theoretical limit values indicated by the dotted lines. In Propo-sition 10.2.1 we prove that $S\alpha S$ harmonizable processes are never ergodic for $\alpha < 2$. Thus, this example illustrates a typical non–ergodic behavior of the $S\alpha S$ harmonizable processes. Note that in a sharp contrast to the Gaussian case, for $\alpha < 2$ the Ornstein–Uhlenbeck process is never harmonizable since the first one is ergodic and the second not.

10.3 Mixing and Other Chaotic Properties of Stable Processes

It is clear that mixing is a stronger property than ergodicity. A stationary Gaussian process with the harmonic spectral representation

$$X(t) = \operatorname{Re} \int_{-\infty}^{\infty} e^{it\vartheta} \, dW(\vartheta)$$

is mixing if and only if its covariance $R(u) = \int_{-\infty}^{\infty} e^{iu\vartheta} \, d\lambda(\vartheta)$ tends to 0 as $u \to \infty$. For non–Gaussian stationary stable processes Cambanis, Hardin and Weron (1987), Theorem 2, obtained the following characterization.

Theorem 10.3.1 *A stationary $S\alpha S$ process* $\mathbf{X}$ *with $0 < \alpha \le 2$ and spectral representation*

$$\{X(t) : t \in \mathbb{R}\} \overset{d}{=} \left\{ \int_E (U_t f)(\vartheta) \, dZ(\vartheta) : t \in \mathbb{R} \right\} \tag{10.3.1}$$

is mixing if and only if

$$\lim_{t \to \infty} \|g + U_t h\|_\alpha^\alpha = \|g\|_\alpha^\alpha + \|h\|_\alpha^\alpha \tag{10.3.2}$$

for every $g \in \overline{\operatorname{lin}}\{U_t f : t \le 0\} \subset \boldsymbol{L}^\alpha(E, \mathcal{E}, \lambda)$ and $h \in \overline{\operatorname{lin}}\{U_t f : t \ge 0\} \subset \boldsymbol{L}^\alpha(E, \mathcal{E}, \lambda)$

PROOF. The process $\{X(t)\}$ is mixing if and only if

$$\lim_{t \to \infty} \boldsymbol{E}((T_t Y)X) = \boldsymbol{E} Y \, \boldsymbol{E} X,$$

where X is $\sigma\{X(t) : t \le 0\}$–measurable and Y is $\sigma\{X(t) : t \ge 0\}$–measurable, $\boldsymbol{E} X^2 < \infty$ and $\boldsymbol{E} Y^2 < \infty$. It suffices to have this equality for random variables $Y = \exp\left[i \sum_{n=1}^N a_n X_{t_n}\right]$, $t_n \ge 0$ and $X = \exp\left[i \sum_{m=1}^M b_m X_{s_m}\right]$, $s_m \le 0$. Putting $h = \sum_{n=1}^N a_n U_{t_n} f$, $g = \sum_{m=1}^M b_m U_{s_m} f$, we have

$$\boldsymbol{E} X = \boldsymbol{E} \exp\left[i \int_E g \, dZ(\vartheta)\right] = \exp[-\|g\|_\alpha^\alpha],$$

$$\boldsymbol{E}(T_t Y) = \boldsymbol{E} \exp\left[i \int_E U_t h \, dZ(\vartheta)\right] = \exp[-\|U_t h\|_\alpha^\alpha],$$

$$\boldsymbol{E}((T_t Y)X) = \boldsymbol{E} \exp\left[i \int_E (g + U_t h) \, dZ(\vartheta)\right] = \exp[-\|g + U_t h\|_\alpha^\alpha],$$

which ends the proof.   □

Applying once more the dynamical functional of an $S\alpha S$ stationary process (see (10.2.5)) one can obtain another characterization of the mixing property.

Theorem 10.3.2 *Let* **X** *be a stationary $S\alpha S$ process. The process* **X** *is mixing if and only if for any function $h \in \lin\{U_t f_0 : t \in \mathbb{R}\}$*

$$\lim_{t\to\infty} \|U_t h - h\|_\alpha^\alpha = 2\|h\|_\alpha^\alpha.$$

PROOF. Thanks to Theorem 9.5.3, the process **X** is mixing if and only if for any $Y \in \lin\{X_t : t \in \mathbb{R}\}$ we have

$$\lim_{t\to\infty} \Phi(Y, t) = E\left|e^{iY}\right|^2.$$

From the spectral representation of **X** and the form (10.2.5) of the dynamical functional it follows that the above statement is equivalent to the fact that for any $h \in \lin\{U_t f_0 : t \in \mathbb{R}\}$ we have

$$\lim_{t\to\infty} \exp\left(-\|U_t h - h\|_\alpha^\alpha\right) = \exp\left(-2\|h\|_\alpha^\alpha\right),$$

which ends the proof.   □

Example 10.3.1 *Numerical illustration of the mixing property for the $S\alpha S$ Ornstein–Uhlenbeck process.*

This is a continuation of Examples 9.6.1 and 10.2.1. Consider the same four functions $h_1, ..., h_4$. In order to check the mixing property by Theorem 10.3.1 it is enough to evaluate $\lim_{t\to\infty} \Phi^*(h, t)$.

The numerical evaluation of the dynamical functional for the $S1.7S$ Ornstein–Uhlenbeck process for the given functions h_i is presented in Figure 10.3.1. The theoretical limits $\exp(-2\|h_i\|_{1.7}^{1.7})$ are denoted by the dotted lines and their values can be calculated from the table presented in Example 10.2.1. They take the following values from the top to the bottom

$$0.308, \quad 0.259, \quad 0.193, \quad 0.067.$$

It is clear that in all four cases the curves representing the dynamical functional approaches well the theoretical limits even on the interval $[0, 10]$. Figure 10.3.1 illustrates a typical behavior of any mixing $S\alpha S$ stationary process.

Doubly stationary processes. In what follows we consider the so called *doubly stationary $S\alpha S$ process* $\mathbf{X} = \{X(t) : t \in \mathbb{R}\}$, i.e. the process with spectral representation of the form

$$\{X(t)\} \overset{d}{=} \left\{\int U_t f \, d\mathbf{Z}\right\},$$

where $\{U_t f\}$ is stationary, cf. Example 5.4.1.

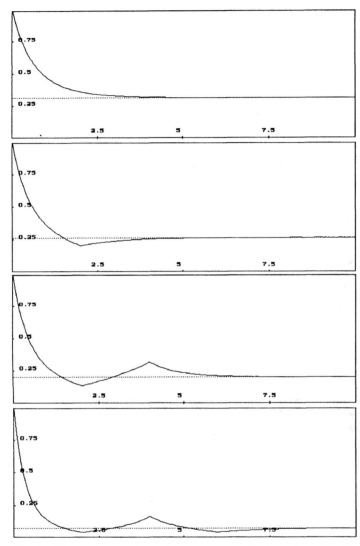

Figure 10.3.1. Illustration of the mixing property of the 1.7–stable Ornstein–Uhlenbeck process via the dynamical functional.

Suppose that $\{U_t f\}$ is stationary. Since the distribution of $\{\int U_t f \, d\mathbf{Z}\}$ does not depend on what happens outside the support of the $U_t f$'s, it is natural to restrict our attention to the σ-ring $\mathcal{E}'$ generated by cylinder sets of the form $\cap_{t \in F} U_t f^{-1}(B_t)$, where $F \subset \mathbf{R}$ is finite and $B_t \in \mathcal{B}_{\mathbf{R}}$, with at least one B_t not containing 0.

In all interesting cases where $(U_t f)$ is stationary, (U_t) restricted to the $\mathcal{E}'$-measurable functions will be induced by a flow on $(E, \mathcal{E}', \lambda)$; i.e., $U_t g = g \circ \tau_t$. (A *flow* is a group $(\tau_t)_{t \in \mathbf{R}}$ of measure-preserving point transformations such that for any $\mathcal{E}'$-measurable f, the function $(x, t) \mapsto f(\tau_t(x))$ is $\mathcal{E}' \times \mathcal{B}_{\mathbf{R}}$-measurable.) Conversely, if U_t is given by $U_t g = g \circ \tau_t$ for some flow $(\tau_t)_{t \in \mathbf{R}}$ then $(U_t f)$ will be stationary—just apply τ_t to cylinder sets to verify this.

For convenience, we will assume that (U_t) is induced by a flow. For the pathological cases in which this assumption does not hold, one can easily modify the proof of Theorem 10.3.3 and the statement and proof of Theorem 10.3.4 to deal with set transformations acting on the measure algebra of equivalence classes of sets modulo null sets.

Let us recall that the *density* of a subset S of the positive reals is the limit (if it exists)

$$\lim_{T \to \infty} \frac{|S \cap [0, T]|}{T},$$

where $|A|$ denotes the Lebesgue measure of a set $A \subset \mathbf{R}$.

Some of the results in this section follow directly from Lemma 10.3.1.

Lemma 10.3.1 *Let $(U_t f)$ be functions in some $L^\alpha(E, \mathcal{E}, \lambda)$. If, either $\alpha \in (0, 2)$ and*

$$\|U_t f - f\|_\alpha^\alpha + \|U_t f + f\|_\alpha^\alpha - 2\|U_t f\|_\alpha^\alpha - 2\|f\|_\alpha^\alpha \to 0,$$

or $\alpha \in (0, 1)$ and

$$\|U_t f - f\|_\alpha^\alpha - \|U_t f\|^\alpha - \|f\|_\alpha^\alpha \to 0,$$

then for all $c, C > 0$,

$$\lambda(c \le |f| \le C, c \le |U_t f| \le C) \to 0.$$

The similar statement holds for convergence as $t \to \infty$ outside a set of density 0.

PROOF. We will prove the convergence as t goes to infinity; the convergence as t goes to infinity outside a set of density 0 is proven similarly. Consider the case $\alpha \in (0, 2)$. Fix some positive $c < C$. Let $h(x, y) = |x - y|^\alpha + |x + y|^\alpha - 2|x|^\alpha - 2|y|^\alpha$, and let $E_t = \{c \le |f| \le C, c \le |U_t f| \le C\}$. It is well known that $h \le 0$ and that $h(x, y) < 0$ for all nonzero x and y when $\alpha \in (0, 2)$ (modify slightly, for instance, a part of the proof of Theorem 10.2.1). Therefore, h is bounded away from 0 on the compact set $\{(x, y) \in \mathbf{R}^2 : c \le |x| \le C, c \le |y| \le C\}$; say $h < -\delta$ on this set. Hence,

$$\left| \int_E h(f, U_t f) \, d\lambda \right| \ge \left| \int_{E_t} h(f, U_t f) \, d\lambda \right| \ge \delta \lambda(E_t),$$

and the integral on the left goes to 0 by hypothesis. Therefore $\lambda(E_t)$ goes to 0.

For the case $\alpha \in (0, 1)$, replace h by $h(x, y) = |x - y|^\alpha - |x|^\alpha - |y|^\alpha$ and repeat the argument above.   □

Lemma 10.3.1 does not hold for $\alpha = 2$; for instance, take $f = I_{\{1\}} + I_{\{2\}}$ and $U_t f = I_{\{1\}} - I_{\{2\}}$ for all $t > 0$ on the two–point space with uniform measure. The following theorem (Gross (1992)) does, however, hold for Gaussian processes.

Theorem 10.3.3 *Assume that* $\{X(t)\} \overset{d}{=} \{\int U_t f \, d\mathbf{Z}\}$ *is a doubly stationary* $S\alpha S$ *process for some* $\alpha \in (0, 2]$.

Then $\{X(t)\}$ *is mixing if and only if*

$$\|U_t f - f\|_\alpha^\alpha \to 2 \|f\|_\alpha^\alpha \qquad (10.3.3)$$

and

$$\|U_t f + f\|_\alpha^\alpha \to 2 \|f\|_\alpha^\alpha . \qquad (10.3.4)$$

If $\alpha \in (0, 1)$ *or* $\alpha = 2$ *then* $\{X(t)\}$ *is mixing if and only if (10.3.3) holds.*

Similarly, $\{X(t)$ *is weakly mixing in each case if and only if the corresponding convergence holds outside a set of density 0.*

PROOF. We will prove the mixing property of $\{X(t)\}$; the weak mixing case is proven with the obvious modifications.

If $\{X(t)\}$ is mixing then conditions (10.3.3) and (10.3.4) are obtained by taking g and h in Theorem 10.3.1 to be $-f$ and f, and then f and f, respectively.

Now, assume that $\alpha \in (0, 2)$ and conditions (10.3.3) and (10.3.4) hold. We claim that in order to show mixing it suffices to show that

$$\|g + U_t h\|_\alpha^\alpha \to \|g\|_\alpha^\alpha + \|h\|_\alpha^\alpha \qquad (10.3.5)$$

for all g and h of the form

$$\sum_{s \in S} \beta_s U_s f I_{\{c \leq |U_s f| \leq C\}}, \qquad (10.3.6)$$

where $S \subset \mathbb{R}$ is a finite index set, β_s are real numbers, and c, C are positive numbers. This follows from the fact that functions of the form (10.3.6) approximate elements of the closed linear span of the $U_s f$'s in the L^α metric. Therefore, by Theorem 10.3.1 it will follow that $\{X(t)\}$ is mixing. Actually, by Proposition 10.1.1, it would suffice to take S to be a singleton; however, this would not make the proof of the theorem any simpler.

Consider functions g and h of the form in Equation (10.3.6). Double stationarity implies that

$$U_t(\beta_s U_s f I_{\{c \leq |U_s f| \leq C\}}) = (\beta_s U_s f I_{\{c \leq |U_s f| \leq C\}}) \circ \tau_t$$
$$= \beta_s U_{s+t} f I_{\{c \leq |U_{s+t} f| \leq C\}},$$

so that $g + U_t h$ is a sum of terms of the form

$$\beta_s U_{s+t} f I_{\{c \leq |U_{s+t} f| \leq C\}} + \beta_{s'} U_{s'} f I_{\{c \leq |U_{s'} f| \leq C\}}. \qquad (10.3.7)$$

Lemma 10.3.1 says that

$$\lambda(\{c \le |U_s f| \le C, c \le |U_{s+t} f| \le C\}) \to 0$$

for all positive c and C; note that this holds for all s, not just $s = 0$, by the assumption of stationarity of $(U_t f)$. Thus, for any g and h of the form in equation (10.3.6), the supports of g and $U_t h$ are asymptotically disjoint. As these functions are bounded and the supports have finite measure, it follows that condition (10.3.2) holds for all such g and h, and $\{X(t)\}$ is mixing.

For $\alpha = 2$, the statement follows from the well-known fact that a Gaussian process is mixing if and only if its covariance function converges to 0 (e.g. Cornfeld, Fomin and Sinai (1982)).

For $\alpha \in (0, 1)$, repeat the above argument with the fact that

$$\|U_t f - f\|_\alpha^\alpha - \|U_t f\|_\alpha^\alpha - \|f\|_\alpha^\alpha \to 0$$

and apply Lemma 10.3.1 as above.   □

Remark 10.3.1 *Conditions (10.3.3) and (10.3.4) are equivalent to the conditions*

$$\langle e^{iX(t)}, e^{iX(0)} \rangle \to \langle e^{iX(0)}, 1 \rangle \langle 1, e^{iX(0)} \rangle$$

and

$$\langle e^{iX(t)}, e^{-iX(0)} \rangle \to \langle e^{iX(0)}, 1 \rangle \langle 1, e^{-iX(0)} \rangle.$$

Theorem 10.3.3 says that if the process is doubly stationary then in order to check mixing or weak mixing one needs only to look at the inner products of the $e^{iX(t)}$'s with $e^{iX(0)}$ and $e^{-iX(0)}$.

Recently, Gross (1993) extended this result to general stationary $S\alpha S$ processes and Kokoszka and Taqqu (1993) characterized mixing processes of type G, i.e. infinitely divisible processes which are mixtures of Gaussian processes.

The next theorem gives another characterization of mixing property of α-stable processes. The "if" part of this theorem was proven in Cambanis, Hardin and Weron (1987, Theorem 7) for $\alpha \in (0, 2]$. The other was recently proven by Gross (1992) for $\alpha \in (0, 2)$.

Theorem 10.3.4 *Assume that $\{X(t)\} \stackrel{\mathrm{d}}{=} \{\int f \circ \tau_t \, d\mathbf{M}\}$ is an $S\alpha S$ process for some $\alpha \in (0, 2)$ and some flow (τ_t).*
Then $\{X(t)\}$ is mixing if and only if

$$\lambda(\tau_t A_1 \cap A_2) \to 0 \tag{10.3.8}$$

for all $A_1, A_2 \in \mathcal{E}'$ of finite measure, where $\mathcal{E}'$ is as defined above.
The similar result holds for weak mixing and convergence outside a set of density zero.

PROOF. We will prove the result for the mixing case; the proof can be modified in the obvious way for weak mixing.

Suppose (10.3.8) holds. Since the $U_t f$'s can be approximated in $\boldsymbol{L}^\alpha$ by $\mathcal{E}'$-measurable simple functions, it suffices to show that condition 10.3.2 holds for simple $\mathcal{E}'$-measurable g and h. Then by approximation, convergence will hold for all g and h in the closed linear span of the $U_t f$'s. Now,

$$(U_t h)^{-1}(B) = \tau_{-t} h^{-1}(B),$$

so for simple g and h the supports of g and h are asymptotically disjoint. Condition 10.3.2 follows. Hence $\{X(t)\}$ is mixing by Theorem 10.3.1.

Conversely, suppose $\{X(t)\}$ is mixing. We will need to approximate the elements of $\mathcal{E}'_f$ by the elements of a certain subclass. Let $\mathcal{K}$ denote the ring generated by sets of the form $\cap_{t \in F} U_t f^{-1}(B_t)$, where $F \subset \mathbb{R}$ is finite and each B_t is in $\mathcal{B}_{\mathbb{R}}$ with at least one B_t bounded and bounded away from zero. We claim $\mathcal{K}$ to be dense in $\mathcal{E}'_f$, where the distance between two sets is defined to be the measure of their symmetric difference. To see this, observe that any cylinder set of the form $\cap_{t \in F} U_t f^{-1}(B_t)$, where $F \subset \mathbb{R}$ is finite and each B_t is in $\mathcal{B}_{\mathbb{R}}$ with at least one B_t not containing zero, can be written as a countable union of sets in $\mathcal{K}$; this is because for any $B \in \mathcal{B}_{\mathbb{R}}$ not containing zero,

$$\{f \in B\} = \cup_{m \geq 1} \{f \in B, 1/m \leq |f| \leq m\}.$$

As the family of cylinder sets of the form $\cap_{t \in F} U_t f^{-1}(B_t)$ (some B_t not containing zero) generates $\mathcal{E}'$ by definition, the ring $\mathbb{R}$ generates $\mathcal{E}'$ as well. The measure λ is σ-finite on $\mathcal{K}$; indeed, λ is finite on $\mathbb{R}$ since $U_t f \in \boldsymbol{L}^\alpha$ implies that $\lambda(U_t f^{-1}(B_t)) < \infty$ whenever B_t is bounded away from zero. Therefore, by a well-known result from measure theory (e.g., Billingsley (1986)) $\mathcal{K}$ is dense in $\mathcal{E}'_f$. Now, by Theorem 10.3.1,

$$\|U_t f - U_0 f\|_\alpha^\alpha \to 2 \|U_0 f\|_\alpha^\alpha$$

and

$$\|U_t f + U_0 f\|_\alpha^\alpha \to 2 \|U_0 f\|_\alpha^\alpha .$$

Adding and using stationarity of $(U_t f)$,

$$\|U_t f - U_0 f\|_\alpha^\alpha + \|U_t f + U_0 f\|_\alpha^\alpha - 2 \|U_t f\|_\alpha^\alpha - 2 \|U_0 f\|_\alpha^\alpha \to 0.$$

Thus, by Lemma 10.3.1, for every $c, C > 0$ convergence in (10.3.8) holds for

$$A_i = \{c \leq |U_0 f| \leq C\}, \quad i = 1, 2,$$

and, by stationarity of $(U_t f)$, (10.3.8) holds for

$$A_i = \{c \leq |U_{s(i)} f| \leq C\}, \quad i = 1, 2,$$

where $s(1)$ and $s(2)$ are any real numbers.

Now, for any $B_1, B_2 \in \mathcal{B}_{\mathbb{R}}$ bounded and bounded away from zero, we can choose $c < C$ such that for $i = 1, 2$,

$$\{U_{s(i)}f \in B_i\} \subset \{c \le |f_{s(i)}| \le C\};$$

therefore convergence in (10.3.8) holds for

$$A_1 = (U_{s(1)}f)^{-1}(B_1), \quad A_2 = (U_{s(2)}f)^{-1}(B_2).$$

It follows that (10.3.8) holds for all $A_1, A_2 \in \mathbb{R}$. But $\mathbb{R}$ is dense in $\mathcal{E}'_f$, so convergence in (10.3.8) holds for all $A_1, A_2 \in \mathcal{E}'_f$.

Proposition 10.3.1 *Assume that $\alpha \in (0,2)$, $\{X(t)\} \stackrel{\mathrm{d}}{=} \{\int U_t f \, dM\}$, $(U_t f)$ is stationary, and $\{X(t)\}$ is not identically zero.*
If $\lambda(E) < \infty$ then $\{X(t)\}$ is not ergodic.

PROOF. As (U_t) is strongly continuous, there is a countable dense set $S \subset \mathbb{R}$ such that $E_0 \stackrel{\mathrm{df}}{=} \cup_{s \in S}(\mathrm{supp}(U_s f))$ is essentially all of $\cup_{t \in \mathbb{R}}(\mathrm{supp}(U_t f))$, in the sense that any $\mathrm{supp}(U_t f)$, $t \in \mathbb{R}$, is contained in E_0 a.e. Thus E_0 is invariant modulo null sets and it belongs to $\mathcal{E}'$. If $\lambda(E) < \infty$, take $A_1 = A_2 = E_0$ in Theorem 10.3.4 to conclude that $\{X(t)\}$ is not weakly mixing. But $\{X(t)\}$ is symmetric infinitely divisible, and by Cambanis *et al.* (1991), ergodicity and weak mixing are equivalent for such processes. $\square$

The following example shows that neither Theorem 10.3.4 nor Theorem 10.3.3 hold for sequences with $\alpha = 2$.

Example 10.3.2 *There is a mixing Gaussian sequence with a stationary spectral representation on a space of finite measure.*

Let $(E, \mathcal{E}, \lambda)$ be $[0,1)^2$ with Lebesgue measure, and let S_b be the baker transformation, which is defined and graphically presented in Section 9.1.
Transformation S_b is an automorphism (Cornfeld, Fomin and Sinai (1982)). Let $f(x,y) = \sin(2\pi x)$. The sequence $(f \circ S_b^n)_{n \in \mathbf{Z}}$ is stationary because S_b is measure–preserving, and it can be checked directly that the $f \circ S_b^n$'s are orthogonal. Therefore, if M is an independently scattered Gaussian measure on the unit square, the sequence $(\int f \circ S_b^n \, dM)_{n \in \mathbf{Z}}$ is i.i.d.

The following example shows that the condition of stationarity of $(U_t f)$ in Theorem 10.3.4 and Proposition 10.3.1 cannot be dispensed with.

Example 10.3.3 *There is a mixing $S\alpha S$ process, $\alpha \in (0,2)$, with a spectral representation on a space of finite measure.*

This example is just our Example 5.4.1 *(iv)* of a stationary $S1S$ process. Here $(E, \mathcal{E}, \lambda)$ is the unit interval with Lebesgue measure, and

$$U_t f(x) = 2^t x^{2^t - 1} f(x^{2^t}), \qquad x \in [0,1], t \in \mathbb{R}.$$

Take $U_t f = U_t 1$. Note that the distribution of $U_t f$ depends on t.

It is easy to verify that if g and h are any two finite real linear combinations of the $U_t f$'s, then

$$\int |g + U_t h| \, d\lambda \to \int |g| \, d\lambda + \int |h| \, d\lambda,$$

so by Theorem 10.3.1 the process $\{X(t)\}$ is mixing.

10.4   Introduction to Stationary ID Processes

Let us recall that a random vector V in $\mathbb{R}^d$ with characteristic function ϕ_V has *infinitely divisible distribution* or is *ID random vector*, if for each $n \in \mathbb{N}$ there exists a characteristic function ϕ_n, such that $\phi_V = (\phi_n)^N$. A stochastic process $\{X(t)\}_{t \in \mathbb{R}}$ is called *infinitely divisible (ID)*, if for each $n \in \mathbb{N}$ and $(t_1, .., t_n) \in \mathbb{R}^n$ the vector $(X(t_1), .., X(t_n))$ has infinitely divisible distribution. In this paper we will only deal with symmetric and stochastically continuous ID processes, i.e., for each $n \in \mathbb{N}$ and $(t_1, .., t_n) \in \mathbb{R}^n$ the vector $(X(t_1), .., X(t_n))$ is a symmetric ID random vector and the sequence $\{X(u_n)\}$ converges in probability to $X(u_0)$, whenever $\{u_n\}$ converges to u_0. Since stochastically continuous processes have measurable modifications, we assume (without mention it further) that all processes under consideration are measurable and stochastically continuous.

The Lévy–Khintchine representation of the characteristic function of an ID random vector $Z \in \mathbb{R}^m$ has the form

$$\phi_Z(t) = E e^{i(Z,t)} = \exp\left(-\frac{(Rt,t)}{2} + \int (1 - \cos(x,t))Q(dx)\right),$$

where R is a positive definite $m \times m$ matrix and Q is a symmetric σ-finite measure on $\mathbb{R}^m$ such that $\int (1 \wedge |x|^2) Q(dx) < \infty$ or, equivalently, $\int (|x|^2/(1+|x|^2)) Q(dx) < \infty$. Further we will refer to (Q, R) as characteristics of a vector Z. In Maruyama (1970); Proposition 5.1 is given the full description of weak convergence of ID multidimensional laws by their characteristics.

We will formulate it for symmetric multidimensional ID distributions. Let us define $\phi : \mathbb{R}^m \to \mathbb{R}^{m+1}$ by

$$\phi(x) = \begin{cases} (|x|, x/|x|) & x \neq 0, \\ 0 & x = 0. \end{cases}$$

Let $\mathcal{S}_{m-1} = \{x \in \mathbb{R}^m : |x| = 1\}$ and S be a finite measure on Borel subsets of $[0, \infty) \times \mathcal{S}_{m-1}$ defined by

$$S(A) = \int_{\phi^{-1}(A)} |x|^2/(1 + |x|^2) Q(dx).$$

Thus the pair (S, R) can be considered as equivalent characteristics of ID multidimensional law. In a symmetric case Maruyama result states that the laws of ID random variables Z_n with characteristics (S_n, R_n) are weakly convergent if and only if R_n is convergent as an element of a finite dimensional vector space and S_n is weakly convergent as a measure on $[0, \infty) \times S_m$ i.e., there exists a measure S_0 on $[0, \infty) \times S_m$ such that for any continuous and bounded function g we have $\lim_{n \to \infty} \int g \, dS_n = \int g \, dS_0$. The limit distribution will be also ID with characteristics (S_0, R_0). But R_0 is not just a limit of R_n. We have

$$R_0 = \lim_{n \to \infty} R_n + \tilde{R},$$

where $\tilde{R}_{i,j} = \int_{S_{m-1}} x_i x_j \, S_0(dx)$. For a measurable $A \subseteq \mathbb{R}^{m+1}$ not including 0 we have also $Q_0(A) = \int_{\phi(A)} (1 + |\phi^{-1}(x)|^2)/|\phi^{-1}(x)|^2 S_0(dx)$ (notice that the inverse function ϕ^{-1} is well defined on $\phi(\mathbb{R}^m \setminus \{0\}) = (0, \infty) \times S_m$). However in this paper we will only deal with the cases when $\tilde{R} = 0$. In such a case the convergence in distribution of an ID distribution to a distribution with characteristics (R_0, S_0) is simply characterized by

$$R_0 = \lim_{n \to \infty} R_n, \qquad S_0 = \lim_{n \to \infty} S_n.$$

The spectral representation of symmetric ID processes is the basic tool used here. We first introduce some basic notation and properties needed to formulate it. For details we refer the interested reader to Rajput and Rosiński (1989).

Let $(S, \mathcal{S})$ be a measurable space and let Λ be a symmetric ID independently scattered stochastic measure on a σ-ring, which generates $\mathcal{S}$. There is a one to one correspondence between Λ and a triple $(\lambda, \sigma^2, \rho)$, where λ is a σ-finite measure on $\mathcal{S}$, called the *control measure of Λ*, σ is a non-negative function from $L^2(S, \mathcal{S}, \lambda)$ and the function $\rho : S \times \mathcal{B}_\mathbb{R} \to [0, \infty]$, where $\mathcal{B}_\mathbb{R}$ is the σ-field of Borel sets of the real line. This correspondence is such that for any fixed $s \in S$ a measure $\rho(s, \cdot)$ is a symmetric Lévy measure and for fixed $B \in \mathcal{B}_\mathbb{R}$ is a function $\rho(\cdot, B)$ measurable and finite, whenever 0 does not belong to the closure of B. The correspondence is given by the form of the characteristic function

$$\phi_{\Lambda(A)}(t) = \exp \left\{ -\int_A \left[\frac{t^2}{2} \sigma^2(s) + \int (1 - \cos(tx)) \, \rho(s, dx) \right] \lambda(ds) \right\}.$$

For $t \in \mathbb{R}$ and $s \in S$ we define

$$K(t, s) = \frac{t^2}{2} \sigma^2(s) + \int (1 - \cos(tx)) \, \rho(s, dx)$$

and

$$\Psi(t, s) = t^2 \sigma^2(s) + \int \left(1 \wedge (tx)^2 \right) \rho(s, dx).$$

The function Ψ generates the Musielak–Orlicz space $L^\Psi(S, \lambda)$ consisting of all measurable functions $f : S \to \mathbb{R}$ such that $\int_S \Psi(|f(s)|, s) \, \lambda(ds) < \infty$ with a Frechet norm defined by

$$\|f\|_\Psi = \inf \{ c > 0; \int_S \Psi(|f(s)|/c, s) \, \lambda(ds) \le c \}.$$

The proof that Ψ satisfies the conditions which guarantees the appropriate structure of $\mathbf{L}_\Psi(S, \lambda)$ with $\|f\|_\Psi$ is given in Lemma 3.1 of Rajput and Rosiński (1989). More detailed information on Musielak–Orlicz spaces, called also the generalized Orlicz spaces, can be found in Musielak (1983). Let us only mention here that $\mathbf{L}^\Psi(S, \lambda)$ is a complete linear metric space such that $\lim_{n\to\infty} f_n = 0$ if and only if $\lim_{n\to\infty} \int_S \Psi(|f_n(s)|, s)\lambda(ds) = 0$. A measurable function f on S is integrable with respect to the symmetric ID random measure Λ, if and only if $f \in \mathbf{L}^\Psi(S, \lambda)$. Then

$$\phi_{\int_S f \, d\Lambda}(t) = \exp\left\{-\int_S K(tf(s), s) \, \lambda(ds)\right\}.$$

We now formulate the spectral representation of a symmetric ID process derived in Rajput and Rosiński (1989). Let $\mathbf{X}$ be a symmetric stochastically continuous ID process. Then there exist a measurable space $(S, \mathcal{S})$, a symmetric ID independently scattered random measure Λ on $(S, \mathcal{S})$ with corresponding triple (λ, σ, ρ), a closed subspace $\mathbf{L}^\Psi(\mathbf{X})$ of $\mathbf{L}_\Psi(S, \lambda)$ and a linear topological isomorphism of $\mathbf{L}^0(\mathbf{X})$ onto $\mathbf{L}^\Psi(\mathbf{X})$, such that the processes $\{X(t)\}_{t \in \mathbb{R}}$ and $\{\int_S f_t \, d\Lambda\}_{t \in \mathbb{R}}$ have the same finite dimensional distributions, where for each $t \in \mathbb{R}$ f_t corresponds to $X(t)$ by the above isomorphism.

It follows from this representation that if $Y \in \mathbf{L}_0(\mathbf{X})$ and $f \in \mathbf{L}^\Psi(S, \lambda)$ corresponds to Y then

$$\phi_Y(t) = \exp\left\{-\int_S K(tf(s), s) \, \lambda(ds)\right\}.$$

By $(\mathbf{T}_t)_{t \in \mathbb{R}}$ we denote the group of transformations of $\mathbf{L}^\Psi(\mathbf{X})$, which corresponds to $(T_t)_{t \in \mathbb{R}}$ by the spectral representation. If the process $\mathbf{X}$ is stationary then for each $Y \in \mathbf{L}_0(\mathbf{X})$ and $u \in \mathbb{R}$ we have $\phi_Y = \phi_{T_u Y}$ and consequently, for each $f \in \mathbf{L}^\Psi(\mathbf{X})$ and $u, t \in \mathbb{R}$,

$$\int_S K(tf(s), s) \, \lambda(ds) = \int_S K(t\mathbf{T}_u f(s), s) \, \lambda(ds).$$

From now on we will always assume that $\mathbf{X}$ is a symmetric stochastically continuous ID process with the spectral representation as defined above.

Let us notice that for an ID stochastic process $\mathbf{X}$ with the spectral representation of the form

$$\{X(t) : t \in \mathbb{R}\} \overset{\mathrm{d}}{=} \left\{\int_S f_t(\theta)d\Lambda(\theta) : t \in \mathbb{R}\right\},$$

the dynamical functional is given for $Y \in \mathbf{L}^0(\mathbf{X})$ and $t \in \mathbb{R}$ by the formula

$$\Phi(Y, t) = \exp\left(-\int_S K((\mathbf{T}_t f - f)(s), s) \, \lambda(ds)\right),$$

where $f \in \mathbf{L}^\psi(\mathbf{X})$ corresponds to Y.

When $\mathbf{X}$ is a symmetric α–stable process ($0 < \alpha \leq 2$), then the dynamical functional for $Y \in \mathbf{L}^0(\mathbf{X})$ takes the form

$$\Phi(Y, t) = \exp\left\{-\|U_t f - f\|_\alpha^\alpha\right\},$$

where f is a function in some L^α-space, which corresponds to Y by the spectral representation of $\mathbf{X}$, and $(U_t)_{t \in \mathbb{R}}$ is a group of isometries on this space. In the Gaussian case of $(\alpha = 2)$ we have again for $Y \in \mathbf{L}^0(\mathbf{X})$,

$$\Phi(Y, t) = \exp \{\text{Cov}(T_t Y, Y) - \text{Var } Y\}.$$

Let $\mathbf{X}$ be a symmetric ID process as above. It is convenient to introduce

$$N_\Psi(f) = \int_S K(f(s), s)\lambda(ds),$$

where $f \in \mathbf{L}_\Psi(\mathbf{X})$. Then $\phi_{\int_S f(\theta)d\Lambda(\theta)}(t) = \exp(-N_\Psi(tf))$ and for $Y \in \mathbf{L}^0(\mathbf{X})$,

$$\Phi(Y, t) = \exp(-N_\Psi(\mathbf{T}_t f - f)),$$

where f corresponds to Y by the spectral representation.

Technical Lemmas. In order to prove the main results of this chapter we need the following three technical lemmas.

Let $\mathbf{X}$ be a stationary ID process and let $\left(\int_S f_t \, d\Lambda\right)_{t \in \mathbb{R}}$ be its spectral representation. For any fixed f from the space $\mathbf{L}_\Psi(\mathbf{X})$ we define the functions $R_f, R_f^G, R_f^P : \mathbb{R} \to \mathbb{R}$:

$$
\begin{aligned}
R_f(t) &= \int_S [2K(f(s), s) - K((\mathbf{T}_t f - f)(s), s)] \ \lambda(ds), \\
R_f^G(t) &= \int_S T_t f \sigma d\lambda, \hspace{3cm} (10.4.1) \\
R_f^P(t) &= R_f(t) - R_f^G(t).
\end{aligned}
$$

Lemma 10.4.1 *For each $f \in \mathbf{L}_\Psi(\mathbf{X})$ the functions R_f, R_f^G, R_f^P are continuous symmetric and positive definite.*

PROOF. Let us notice that $\exp\{R_f(t)\} = \Phi(Y, t)|Ee^{iY}|^2$. Thus R_f is continuous and symmetric by properties of $\Phi(Y, \cdot)$.

Since $\mathbf{X}$ is stochastically continuous i.e., $\lim_{t \to t_0} X(t) = X(t_0)$ in probability thus $\lim_{t \to t_0} f_t = f_{t_0}$ in the Frechet norm $\| \cdot \|_\Psi$. Thus

$$\lim_{t \to t_0} \int_S \Psi(|f_t(s) - f_{t_0}(s)|, s) \ \lambda(ds) = 0.$$

Consequently $\lim_{t \to t_0} \int_S (f_t - f_{t_0})\sigma^2 d\lambda = 0$. This implies continuity of R_f^G and thus R_f^P.

Now, since R_f^G is clearly positive definite it is enough to show that R_f^P is positive definite. By stationarity and symmetry of $\rho(s, dx)$ for $u, t \in \mathbb{R}$ we have

$$R_f^P(t - u) = \int_S \int \left(1 - e^{ix\mathbf{T}_t f}\right) \left(1 - e^{-ix\mathbf{T}_u f}\right) (s) \ \rho(s, dx) \ \lambda(ds).$$

Thus for $n \in \mathbb{N}$ and $a_1, ..., a_n \in \mathbb{R}$,

$$\sum_{i,j=1}^{n} a_i a_j R_f^P(t_i - t_j) = \int_S \int \left| \sum_{i=1}^{n} a_i \left(1 - e^{-ixT_{t_i}f} \right) \right|^2 (s) \, \rho(s, dx) \, \lambda(ds) \geq 0.$$

This completes the proof.   □

We use the following standard notation. If ν is a finite measure then $\hat{\nu}$ denotes its Fourier transform and $e^{\nu} = \sum_{k=0}^{\infty} \nu^{*k}/k!$, where ν^{*k} denotes the k-fold convolution of the measure ν. With this notation we have $(e^{\nu})\hat{} = e^{\hat{\nu}}$.

Lemma 10.4.2 *If ν is a symmetric finite measure on $\mathcal{B}_{\mathbb{R}}$ then the following conditions are equivalent:*

(i) $e^{\nu}(\{0\}) = 1$;

(ii) $\nu^{*2}(\{0\}) = 0$;

(iii) ν has no atoms.

PROOF. Since $e^{\nu}(\{0\}) = \sum_{k=0}^{\infty} \frac{\nu^{*k}}{k!}(\{0\})$ and ν^{*k} is a positive measure, clearly (i) $\Rightarrow$ (ii).

Since ν is symmetric, if a is an atom of ν then so is $-a$ and $\nu(-a) = \nu(a)$. Let $\mathcal{A}$ be the set of atoms of ν. Since $\nu^{*2}(0) = \sum_{a \in \mathcal{A}} \nu(-a)\nu(a) = \sum_{a \in \mathcal{A}} [\nu(a)]^2$, condition (ii) implies that the set $\mathcal{A}$ is empty.

If ν has no atoms then for each $k \in \mathbb{N}$, the measure ν^{*k} has no atoms, so we obtain (iii) $\Rightarrow$ (i).   □

Lemma 10.4.3 *For each bounded measurable function $\phi : \mathbb{R} \to \mathbb{R}$ the following conditions are equivalent:*

(i) *for each positive number c*

$$\lim_{T \to \infty} \frac{1}{T} \int_0^T e^{c\phi(t)} \, dt = 1;$$

(ii) *for any $\epsilon > 0$,* $\lim_{T \to \infty} \frac{1}{T} |\{t \in [0, T] : |\phi(t)| > \epsilon\}| = 0$;

(iii) *for each positive number c,* $\lim_{T \to \infty} \frac{1}{T} \int_0^T \left| e^{c\phi(t)} - 1 \right| \, dt = 0$;

(iv) $\lim_{T \to \infty} \frac{1}{T} \int_0^T |\phi(t)| \, dt = 0$;

(v) *for every $b > \sup\{-\phi(t) : t \in \mathbb{R}\}$ and any $n \in \mathbb{N}$ we have*

$$\lim_{T \to \infty} \frac{1}{T} \int_0^T (\phi(t) + b)^n \, dt = b^n \qquad \text{for all } n \in \mathbb{N};$$

(vi) there exist natural numbers $k, j, k \neq j$ and $b \in \mathbb{R}$ such that

$$\lim_{T \to \infty} \frac{1}{T} \int_0^T (\phi(t) + b)^n \, dt = b^n \qquad for \ n = k, j.$$

PROOF. We proceed as follows: $(i) \Rightarrow (ii) \Rightarrow (iii) \Rightarrow (i)$, $(ii) \Leftrightarrow (iv)$, $(ii) \Rightarrow$ $(v) \Rightarrow (vi) \Rightarrow (ii)$. Before proving implication $(i) \Rightarrow (ii)$ we shall reformulate the condition (ii).

For fixed $T > 0$ define a probability measure λ_T on $\mathcal{B}_{\mathbb{R}}$ by

$$\lambda_T(A) = \frac{|A \cap [0, T]|}{T} \qquad for \ A \in \mathcal{B}_{\mathbb{R}},$$

where $|A|$ denotes the Lebesgue measure of A. Since ϕ is bounded, for each $T > 0$ we have $\lambda_T \circ \phi^{-1}([-M, M]^c) = 0$, where $M = \sup_{t \in \mathbb{R}} |\phi(t)|$. It follows that the family of probability measures $\{\lambda_T \circ \phi^{-1}\}_{T>0}$ is tight. Therefore (see, e.g., Billingsley (1986)), for every sequence $(T_n)_{n \in \mathbb{N}}$ there exists a subsequence $(T'_k)_{k \in \mathbb{N}}$ and a measure ω such that the sequence $(\lambda_{T_k} \circ \phi^{-1})_{k \in \mathbb{N}}$ is weakly convergent to ω, i.e., for each continuous bounded function f on $\mathbb{R}$ we have

$$\lim_{k \to \infty} \int f \, d\lambda_{T'_k} \circ \phi^{-1} = \lim_{k \to \infty} \frac{1}{T_k} \int_0^{T_k} f(\phi(t)) \, dt = \int f \, d\omega.$$

Notice also that, since for $T > 0$, we have $\lambda_T \circ \phi^{-1}([-M, M]^c) = 0$, hence the function f satisfying the above condition can be unbounded.

Suppose now that the condition (i) is satisfied. So, we have for $c, \ c' \in \mathbb{R}$ such that $c \neq c'$, the following relations:

$$\int e^{cu} \, d\omega(u) = 1, \qquad \int e^{c'u} \, d\omega(u) = 1,$$

and thus,

$$\left(\int e^{cu} \, d\omega(u) \right)^{\frac{c'}{c}} = \int (e^{cu})^{\frac{c'}{c}} \, d\omega(u) = 1.$$

But for convex functions the Jensen Inequality becomes equality only for constant functions, i.e., $e^{cu} = const$ ω-a.e. Thus $\omega = \delta_{\{0\}}$ and so ω does not depend on the choice of sequence $(T'_k)_{k \in \mathbb{N}}$. Consequently, $\lambda_T \circ \phi^{-1}$ converges weakly to $\delta_{\{0\}}$, i.e.,

$$\lim_{T \to \infty} \lambda_T \circ \phi^{-1}\{t \in \mathbb{R} : |t| > \epsilon\}) = 0, \qquad \qquad .$$

which is equivalent to condition (ii). Before proving the implication $(ii) \Rightarrow (iii)$, let us make the following remark. Let f be a nondecreasing and η a measurable function such that $|\eta| \leq M$. Then the following inequalities hold

$$f(-\epsilon)\frac{1}{T} | \{t \in [0, T] : \eta(t) > -\epsilon\}| + f(-M)\frac{1}{T}|\{t \in [0, T] : \eta(t) \leq -\epsilon\}|$$

$$\leq \frac{1}{T} \int_0^T f(\eta(t)) \, dt$$

$$\leq f(\epsilon)\frac{1}{T}|\{t \in [0, T] : \eta(t) < \epsilon\}| + f(M)\frac{1}{T}|\{t \in [0, T] : \eta(t) \geq \epsilon\}|.$$

If $\lim_{T\to\infty} \frac{1}{T}|\{t \in [0,T] : |\eta(t)| \geq \epsilon\}| = 0$ then

$$f(-\epsilon) \leq \liminf_{T\to\infty} \frac{1}{T} \int_0^T f(\eta(t))\, dt \leq \limsup_{T\to\infty} \frac{1}{T} \int_0^T f(\eta(t))\, dt \leq f(\epsilon).$$

If f is nonincreasing on $(-\infty, 0)$ and nondecreasing on $(0, \infty)$ then

$$f(0) \leq \frac{1}{T} \int_0^T f(\eta(t))\, dt$$

$$\leq f(M)\frac{1}{T}|\{t \in [0,T] : \eta(t) > \epsilon\}| + f(\epsilon)\frac{1}{T}|\{t \in [0,T] : 0 \leq \eta(t) \leq \epsilon\}|$$

$$+ f(-\epsilon)\frac{1}{T}|\{t \in [0,T] : 0 \geq \eta(t) \geq -\epsilon\}|$$

$$+ f(-M)\frac{1}{T}|\{t \in [0,T] : \eta(t) < -\epsilon\}|$$

$$\leq f(M)\frac{1}{T}|\{t \in [0,T] : \eta(t) > \epsilon\}| + f(-M)\frac{1}{T}|\{t \in [0,T] : \eta(t) < -\epsilon\}|$$

$$+ \max\{f(\epsilon), f(-\epsilon)\}\frac{1}{T}|\{t \in [0,T] : |\eta(t)| \leq \epsilon\}|.$$

So, if we assume that $\lim_{T\to\infty} \frac{1}{T}|\{t \in [0,T] : \eta(t) \geq \epsilon\}| = 0$ then, by the above inequalities, we have

$$f(0) \leq \liminf_{T\to\infty} \frac{1}{T} \int_0^T f(\eta(t))\, dt$$

$$\leq \limsup_{T\to\infty} \frac{1}{T} \int_0^T f(\eta(t))\, dt \leq \max(f(\epsilon), f(-\epsilon)).$$

Now suppose that f is continuous at 0. Then we have

$$\lim_{T\to\infty} \frac{1}{T} \int_0^T f(\eta(t))\, dt = f(0).$$

Now, let us suppose that condition (ii) is satisfied. Defining

$$f(u) = |e^{cu} - 1| \quad \text{for } u, t \in \mathbb{R} \quad \text{and} \quad \eta(t) = \phi(t),$$

we have f decreasing on $(-\infty, 0)$ and increasing on $(0, \infty)$ so that the last equality gives us condition (iii). Implication $(iii) \Rightarrow (i)$ is obvious.

Now, taking $f(u) = u$ and $\eta(t) = |\phi(t)|$ for $u,\ t \in \mathbb{R}$ we obtain condition (iv). It is easy to see that (iv) implies (ii).

Now, let us take

$$f(u) = (u + b)^n \quad \text{for} \quad u \geq -b \quad \text{and} \quad f(u) = 0 \quad \text{for} \quad u < -b,$$

where $b > \sup\{-\phi(t) : t \in \mathbb{R}\}$,

$$\eta(t) = \phi(t), \quad t, u \in \mathbb{R}.$$

Then the above considerations give us implication $(ii) \Rightarrow (v)$. In turn, $(v) \Rightarrow (vi)$ is evident.

The proof of the implication $(vi) \Rightarrow (ii)$ is similar to to that of $(i) \Rightarrow (ii)$. It is enough to replace functions e^{cu}, $e^{c'u}$ by $(u + b)^k$ and $(u + b)^j$, respectively. Then, from equalities

$$\int (u + b)^k \, d\omega(u) = b^k, \quad \int (u + b)^j \, d\omega(u) = b^j$$

we get

$$\left(\int (u + b)^k \, d\omega(u) \right)^{j/k} = \int ((u + b)^k)^{j/k} \, d\omega(u).$$

Thus, for convex function $(u + b)^k$ the Jensen Inequality becomes equality, i.e., this function must be constant ω-a.e. As before, we conclude that $\lambda_T \circ \phi^{-1}$ converges weakly to $\delta_{\{0\}}$ and we get

$$\lim_{T \to \infty} \lambda_T \circ \phi^{-1} \{ t \in \mathbb{R} : |t| > \epsilon \}) = 0,$$

which is equivalent to (ii).   □

Let us note here a slightly different, probabilistic way of looking at Lemma 10.4.3. Since $|\phi(t)| \le M$ for all $t \in \mathbb{R}$, its normalized occupation measure over each interval $[0, T]$

$$\mu_T(B) = \frac{1}{T} |\{ t \in [0, T] : \phi(t) \in B \}|, \quad B \in \mathcal{B}_{[-M,M]},$$

is a probability measure on Borel subsets of $[-M, M]$, corresponding to a random variable X_T with $|X_T| \le M$. By the transformation theorem

$$\lim_{T \to \infty} \frac{1}{T} \int_0^T F(\phi(t)) \, dt = \int_{-M}^M F(x) \, d\mu_T(x) = \mathbf{E} F(X_T)$$

for any measurable function F for which either integral exists.

Remark 10.4.1 *In the probabilistic framework Lemma 10.4.3 can be re-stated as follows:*

(i) for each positive number c, $\lim_{T \to \infty} \mathbf{E} \exp(cX_T) = 1$;

(ii) for each positive number c, $\lim_{T \to \infty} \exp(cX_T) = 1$ in L^1-norm;

(iii) $\lim_{T \to \infty} X_T = 0$ in L^1-norm;

(iv) $\lim_{T \to \infty} X_T = 0$ in probability;

(v) there exist natural numbers $k, j, k \ne j$ and $b \in \mathbb{R}$ such that

$$\lim_{T \to \infty} \mathbf{E} (X_T + b)^n = b^n \quad \text{for} \quad n = k, j;$$

(vi) there exists $b \in \mathbb{R}$ such that

$$\lim_{T \to \infty} E(X_T + b)^n = b^n \quad \text{for all} \quad n \in \mathbb{N}.$$

We end this section by a remark on the relation between characteristics of finite dimensional distributions of $\mathbf{X}$ and their description in terms of $\mathbf{L_\Psi(X)}$.

Remark 10.4.2 *Let $X = (X_{s_1}, \dots, X_{s_m})$ and let (Q, R) be its characteristics then*

$$R_{ij} = \int f_{s_i} f_{s_j} \sigma^2 d\lambda = \int f_{s_i-s_j} f_0 \sigma^2 d\lambda$$

and for a Borel set $A \in \mathbb{R}^m$ not including zero

$$Q(A) = \int_S \left[\int I_{B_A}(s, x) \, \rho(s, dx) \right] \lambda(ds), \tag{10.4.2}$$

where $B_A = \{(s, x) : (xf_{s_1}(s), \dots, xf_{s_n}(s)) \in A\} \subseteq S \times \mathbb{R}$.

10.5 Ergodic Properties of ID Processes

Now we present systematically a study of the ergodic behavior of infinitely divisible (ID) stationary processes by means of the so–called dynamical functional which was introduced in Section 9.6.

Theorem 10.5.1 *Let $\mathbf{X}$ be a stationary symmetric ID stochastic process. Then the following conditions are equivalent:*

(i) $\mathbf{X}$ is ergodic,

(ii) for each $f \in \mathbf{L_\Psi(X)}$, or equivalently, for each $f \in \lin\{f_t : t \in \mathbb{R}\}$

$$\lim_{T \to \infty} \frac{1}{T} \int_0^T \exp\left\{ 2N_\Psi(f) - N_\Psi(\mathbf{T}_t f - f) \right\} \, dt = 1,$$

(iii) for each natural number n, or equivalently, for $n = 1, 2$, and for each $f \in \mathbf{L_\Psi(X)}$, or equivalently, for $f \in \lin\{f_t : t \in \mathbb{R}\}$

$$\lim_{T \to \infty} \frac{1}{T} \int_0^T N_\Psi^n(\mathbf{T}_t f - f) \, dt = 2^n N_\Psi^n(f),$$

(iv) for each $f \in \mathbf{L_\Psi(X)}$, or equivalently, for $f \in \lin\{f_t : t \in \mathbb{R}\}$

$$\lim_{T \to \infty} \frac{1}{T} \int_0^T |N_\Psi(\mathbf{T}_t f - f) - 2N_\Psi(f)| \, dt = 0.$$

PROOF. By Proposition 9.6.1 and the form of the dynamical functional for ID processes, $\mathbf{X}$ is ergodic if and only if for each $f \in \mathbf{L}_{\Psi}(\mathbf{X})$,

$$\lim_{T \to \infty} \frac{1}{T} \int_0^T \exp(-N_{\Psi}(\mathbf{T}_t f - f))\, dt = \exp(-2N_{\Psi}(f)),$$

which proves equivalence $(i) \Leftrightarrow (ii)$.

By Lemma 10.4.1 and the Bochner Theorem, for each $f \in \mathbf{L}_{\Psi}(\mathbf{X})$ there exists a finite symmetric measure ν_f such that $\hat{\nu}_f(t) = R_f(t) = 2N_{\Psi}(f) - N_{\Psi}(\mathbf{T}_t f - f)$. Condition (ii) and the fact that for any finite measure ν defined on $\mathcal{B}_{\mathbb{R}}$,

$$\lim_{T \to \infty} \frac{1}{T} \int_0^T \hat{\nu}(t)\, dt = \nu(\{0\}), \tag{10.5.1}$$

imply that $e^{\nu_f}(\{0\}) = 1$. By Lemma 10.4.2, this is equivalent to $\nu_f^{*2}(\{0\}) = 0$, which implies $\nu_f(\{0\}) = 0$. Applying equality (10.5.1) to ν_f and ν_f^{*2}, we obtain

$$\lim_{T \to \infty} \frac{1}{T} \int_0^T \{N_{\Psi}(\mathbf{T}_t f - f) - 2N_{\Psi}(f)\}\, dt = 0$$

and

$$\lim_{T \to \infty} \frac{1}{T} \int_0^T \{N_{\Psi}(\mathbf{T}_t f - f) - 2N_{\Psi}(f)\}^2\, dt = 0,$$

respectively. Thus,

$$
\begin{aligned}
0 &= \lim_{T \to \infty} \frac{1}{T} \int_0^T \left\{ N_{\Psi}^2(\mathbf{T}_t f - f) - 4N_{\Psi}(\mathbf{T}_t f - f)N_{\Psi}(f) + 4N_{\Psi}^2(f) \right\}\, dt \\
&= \lim_{T \to \infty} \frac{1}{T} \int_0^T N_{\Psi}^2(\mathbf{T}_t f - f)\, dt - 4\, N_{\Psi}(f) \lim_{T \to \infty} \frac{1}{T} \int_0^T N_{\Psi}(\mathbf{T}_t f - f)\, dt + 4N_{\Psi}^2(f) \\
&= \lim_{T \to \infty} \frac{1}{T} \int_0^T N_{\Psi}^2(\mathbf{T}_t f - f)\, dt - 4N_{\Psi}^2(f)
\end{aligned}
$$

and we have obtained condition (iii) for $n = 1$ and $n = 2$. Consequently, by Lemma 10.4.3, condition (iii) holds for each natural number n.

The other implications $((iii) \Rightarrow (iv)$ and $(iv) \Rightarrow (ii))$ follow immediately from Lemma 10.4.3.

It is easy to notice from the way we used Proposition 9.6.1 that in the above argument we can replace the space $\mathbf{L}_{\Psi}(\mathbf{X})$ by $lin\{f_t : t \in \mathbb{R}\}$.   □

While, in general, weak mixing is a stronger condition than ergodicity, we shall prove that for ID stochastic processes they coincide.

Theorem 10.5.2 *For stationary symmetric ID processes ergodicity implies weak mixing.*

PROOF. By Proposition 9.6.2, a symmetric ID process is weak mixing if and only if for each $f \in \mathbf{L}_{\Psi}(\mathbf{X})$,

$$\lim_{T \to \infty} \frac{1}{T} \int_0^T \left| \exp\left(2N_{\Psi}(f) - N_{\Psi}(\mathbf{T}_t f - f) \right) - 1 \right|\, dt.$$

If $\mathbf{X}$ is ergodic, condition (iii) of Theorem 10.5.1 shows that condition (vi) of Lemma 10.4.3 is satisfied, hence so is condition (ii) of Lemma 10.4.3 and $\mathbf{X}$ is weak mixing.   □

10.6   Mixing Properties of ID Processes

In this section, following Cambanis *et al.* (1991), we study mixing properties of symmetric *ID* processes. We present a new proof of equivalence of mixing and *p*–mixing which was first proved by Maruyama (1970). An application of the integral representation enables us to avoid all problems caused by presence of the limits in distribution in the original proof of Maruyama which blur the idea of the proof. We also give two characterizations of mixing in terms of the integral representation. One of them, involving the dynamical functional and discussed in Proposition 9.6.3, can be considered as an alternative to that given in Maruyama (1970).

Before proving the main theorem we need one technical lemma. Let us define the following functions:

$$S_1(u_1, u_2) = \sin u_1 + \sin u_2 - \sin(u_1 + u_2)$$

and

$$C_1(u_1, u_2) = 1 - \cos u_1 - \cos u_2 + \cos(u_1 + u_2).$$

Lemma 10.6.1 *Let $k, l \in \mathbb{N}$ and $Y_n = \{Y_{n1}, \ldots, Y_{n,k+l}\}$ be a sequence of ID vectors in $\mathbb{R}^{k+l}$ with characteristics (Q_n, R_n). If distributions of Y_n converge weakly to a ID distribution with characteristics (Q_0, R_0) then we have*

$$\int \left| S_1 \left(\sum_{i=1}^{k} x_i, \sum_{i=k+1}^{k+l} x_i \right) \right| Q_n(dx) \to \int \left| S_1 \left(\sum_{i=1}^{k} x_i, \sum_{i=k+1}^{k+l} x_i \right) \right| Q_0(dx),$$

$$\int \left| C_1 \left(\sum_{i=1}^{k} x_i, \sum_{i=k+1}^{k+l} x_i \right) \right| Q_n(dx) \to \int \left| C_1 \left(\sum_{i=1}^{k} x_i, \sum_{i=k+1}^{k+l} x_i \right) \right| Q_0(dx),$$

when $n \to \infty$.

PROOF. Let $h : [0, \infty) \times \mathcal{S}_{k+l} \to \mathbb{R}^{k+l}$ by defined by

$$h(r, \omega) = r\omega.$$

Define a measure $\tilde{S}_n$ on Borel sets of $\mathbb{R}^{k+l}$ by

$$\tilde{S}_n(A) = S_n(h^{-1}A).$$

Notice that since h is a continuous function thus for any continuous and bounded function g on $\mathbb{R}^{k+l}$ a function $g \circ h$ is also continuous and bounded. So, if S_n converges weakly to S_0 then $\int g d\tilde{S}_n = \int g \circ h dS_n$ converge to $\int g d\tilde{S}_0 = \int g \circ h dS_0$ and thus $\tilde{S}_n$ is weakly convergent to $\tilde{S}_0$. Notice also that

$$\tilde{S}_n(A) = \int_A \frac{|x|^2}{1 + |x|^2} Q_n(dx).$$

Thus in view of remarks on weak convergence of multidimensional ID distributions in Section 10.4 it is enough to prove that

$$\left| S_1 \left(\sum_{i=1}^{k} x_i, \sum_{i=k+1}^{k+l} x_i \right) \right| \frac{1 + |x|^2}{|x|^2}$$

and

$$\left| C_1 \left(\sum_{i=1}^{k} x_i, \sum_{i=k+1}^{k+l} x_i \right) \right| \frac{1 + |x|^2}{|x|^2}$$

are bounded and continuous functions on $\mathbb{R}^{k+l}$. Since we have

$$\left| S_1 \left(\sum_{i=1}^{k} x_i, \sum_{i=k+1}^{k+l} x_i \right) \right| \frac{1 + |x|^2}{|x|^2}$$

$$= \left(o_1(|x_1|^2) + o_2(|x_1|^2) + o_3(|x_1|^2) \right) \frac{1 + |x|^2}{|x|^2} I_{\{u:|u|\leq 1\}}(x)$$

$$+ \left| S_1 \left(\sum_{i=1}^{k} x_i, \sum_{i=k+1}^{k+l} x_i \right) \right| \frac{1 + |x|^2}{|x|^2} I_{\{u:|u|>1\}}(x)$$

$$\leq \frac{o(|x|^2)}{|x|^2} I_{\{u:|u|\leq 1\}}(x) + 4\, I_{\{u:|u|>1\}}(x),$$

hence, both continuity and boundedness are obvious. The similar arguments apply to the second part of the lemma.   □

Theorem 10.6.1 *Let* **X** *be a stationary symmetric ID process with the integral representation* $\left(\int_S T_t f_0 \, d\Lambda \right)_{t \in \mathbb{R}}$. *Then the following conditions are equivalent:*

(i) **X** *is mixing,*

(ii) *for each* $f \in L_\Psi(\mathbf{X})$, *or equivalently for each* $f \in lin\{f_t : t \in \mathbb{R}\}$,

$$\lim_{t\to\infty} N_\Psi(T_t f - f) = 2N_\Psi(f),$$

(iii) **X** *is p–mixing.*

PROOF. From the Proposition 9.6.3 and the form of the dynamical functional it follows immediately that **X** is mixing if and only if for each f in $L_\Psi(\mathbf{X})$ or in $lin\{f_t : t \in \mathbb{R}\}$

$$\lim_{t\to\infty} \exp\{-N_\Psi(T_t f - f)\} = \exp\{-2N_\Psi(f)\},$$

which means that conditions (i) and (ii) are equivalent.

Since (iii) implies (i) in the general case, it remains to prove that (ii) $\Rightarrow$ (iii). Thus, it remains to prove (iii) $\Rightarrow$ (iv).

Let $Y = (Y_0, \ldots, Y_p) \in \text{lin}\{X_t : t \in \mathbb{R}\}^{p+1}$ and let $g = (g_0, \ldots, g_p) \in \text{lin}\{f_t : t \in \mathbb{R}\}^{p+1}$ corresponds to Y through the spectral representation. Then for $t = (t_0, \ldots, t_p) \in \mathbb{R}^{p+1}$ we have

$$
\begin{aligned}
\Phi_p(Y, t) &= \exp(N_\Psi(\mathbf{T}_{t_0} g_0 + \ldots + \mathbf{T}_{t_p} g_p)) \\
&= \exp\left\{ \int_S (\textstyle\sum_{i=0}^p \mathbf{T}_{t_i} g_i)^2 \frac{\sigma^2}{2} d\lambda \right. \\
&\quad \left. + \int_S \left[\int (1 - \cos(\textstyle\sum_{i=0}^p \mathbf{T}_{t_i} g_i(s)x))\, \rho(s, dx) \right] \lambda(ds) \right\}
\end{aligned}
$$

and

$$
\begin{aligned}
E e^{iY_0} \ldots E e^{iY_p} &= \exp(N_\Psi(g_0) + \ldots + N_\Psi(g_p)) \\
&= \exp\left\{ \int_S \left(\textstyle\sum_{i=0}^p g_i^2 \frac{\sigma^2}{2} \right) d\lambda \right. \\
&\quad \left. + \int_S \left[\int \left(\textstyle\sum_{i=0}^p (1 - \cos(g_i(s)x)) \right) \rho(s, dx) \right] \lambda(ds) \right\}.
\end{aligned}
$$

Thus the condition for p-mixing described in Proposition 9.6.4 can be written as

$$
\lim_{\delta(t) \to \infty} G(t) + P(t) = 0,
$$

where

$$
G(t) = \int_S \left(\textstyle\sum_{i=0}^p g_i^2 - (\textstyle\sum_{i=0}^p \mathbf{T}_{t_i} g_i)^2 \right) \frac{\sigma^2}{2} d\lambda,
$$

and

$$
P(t) = \int_S \left[\int \left(p - \textstyle\sum_{i=0}^p \cos(g_i(s)x) + \cos(\textstyle\sum_{i=0}^p \mathbf{T}_{t_i} g_i(s)x) \right) \rho(s, dx) \right] \lambda(ds).
$$

Now we show that (i) and (ii) imply that

$$
\lim_{\delta(t) \to \infty} G(t) = 0 \tag{10.6.1}
$$

and

$$
\lim_{\delta(t) \to \infty} P(t) = 0. \tag{10.6.2}
$$

The mixing condition implies that the distribution of an ID vector (X_t, X_0) is convergent to a distribution of (X_0', X_0''), where X_0' and X_0'' are independent copies of X_0. Hence, for functions R_{f_0} and $R_{f_0}^G$ defined by (10.4.1) we have

$$
\lim_{t \to \infty} R_{f_0}(t) = 0 \quad \text{and} \quad \lim_{t \to \infty} R_{f_0}^G(t) = 0. \tag{10.6.3}
$$

Let $g_i = \sum_{k=1}^n a_{ik} f_{s_{ik}}$ for $i = 0, \ldots, p$. By stationarity of $\mathbf{X}$ we have

$$
G(t) = \int_S \left(2 \textstyle\sum_{i>j}^p \mathbf{T}_{t_i} g_i \mathbf{T}_{t_j} g_j \sigma^2 \right) d\lambda
$$

$$= 2\sum_{i>j}^{p} \int_S \left(\sum_{k,l}^{n} a_{ik}a_{jl}\mathbf{T}_{t_i-t_j+s_{ik}-s_{jl}}f_0 \cdot f_0\, \sigma^2\right) d\lambda$$

$$= 2\sum_{i>j}^{p}\sum_{k,l}^{n} R^G_{f_0}(t_i - t_j + s_{ik} - s_{jl}).$$

Since $t_i - t_j + s_{ik} - s_{jl}$ for $i > j$ tends to infinity when $\delta(t) \to \infty$ thus (10.6.1) follows by (10.6.3).

To prove (10.6.2) we make us of the mathematical induction with respect to the order of mixing. Assume that the process $\mathbf{X}$ is mixing of order $p - 1$.

Let $Z_t = (\mathbf{T}_{t_0}Y_0, \dots, \mathbf{T}_{t_{p-1}}Y_{p-1})$ and $Z_0 = (\tilde{Y}_0, \dots, \tilde{Y}_{p-1})$ be random vectors composed of independent random variables such that $\tilde{Y}_i \overset{d}{=} Y_i$ for $i = 0, \dots, p$. By the assumption we know that for any $\xi \in \mathbb{R}^p$ we have

$$\lim_{\delta(t)\to\infty} Ee^{i(\xi,Z_t)} = Ee^{i(\xi,Z_0)}$$

(cf. Proposition 9.6.4)). In other words the distributions of Z_t are weakly convergent to the distribution of Z_0. Further let (Q_t, R_t) denote characteristics of Z_t.

Two auxiliary functions:

$$C_r(u) = r - \sum_{i=0}^{r} \cos(u_i) + \cos\left(\sum_{i=0}^{p} u_i\right)$$

$$S_r(u) = \sum_{i=0}^{r} \sin(u_i) - \sin\left(\sum_{i=0}^{p} u_i\right),$$

defined on $\mathbb{R}^{r+1}$, for $r \in \mathbb{N}$, allow to shorten the final part of the proof.

Notice that if $t \in \mathbb{R}^{p+1}$ and $x \in \mathbb{R}$ then we have

$$C_p(xt) = C_{p-1}\left((xt_i)_{i\neq p-1}^p\right) + C_1\left(xt_{p-1}, x\sum_{i\neq p-1}^p t_i\right).$$

By symmetry of the measure $\rho(s, dx)$ we have

$$\int C_1\left(xt_{p-1}, x\sum_{i\neq p-1}^p t_i\right)\, \rho(s, dx)$$

$$= \int (1 - e^{ixt_{p-1}})(1 - e^{ix\sum_{i\neq p-1}^p t_i})\, \rho(s, dx)$$

$$= \int (1 - e^{ixt_{p-1}})(1 - e^{ixt_p})\, \rho(s, dx)$$

$$+ \int e^{ixt_p}(1 - e^{ixt_{p-1}})(1 - e^{ix\sum_{i=1}^{p-2} t_i})\, \rho(s, dx)$$

$$= \int C_1(xt_{p-1}, xt_p)\, \rho(s, dx) + \int \cos(t_p x)\, C_1\left(xt_{p-1}, x\sum_{i=1}^{p-2} t_i\right)\, \rho(s, dx)$$

$$+ \int \sin(t_p x)\, S_1\left(xt_{p-1}, x\sum_{i=1}^{p-2} t_i\right)\, \rho(s, dx).$$

Thus

$$P(t) =$$

$$= \int_S \int C_p \left(\mathbf{T}_{t_0} g_0(s)x, \ldots, \mathbf{T}_{t_p} g_p(s)x \right) \rho(s, dx)\lambda(ds)$$

$$= \int_S \int C_{p-1} \left((\mathbf{T}_{t_i} g_i(s)x)_{i \neq p-1}^p \right) \rho(s, dx)\lambda(ds)$$

$$+ \int_S \int C_1(\mathbf{T}_{t_{p-1}} g_{p-1}(s)x, \mathbf{T}_{t_p} g_p(s)x) \, \rho(s, dx)\lambda(ds)$$

$$+ \int_S \int \cos(\mathbf{T}_{t_p} g_p(s)x) \, C_1 \left(\mathbf{T}_{t_{p-1}} g_{p-1}(s)x, \sum_{i=0}^{p-2} \mathbf{T}_{t_i} g_i(s)x \right) \, \rho(s, dx)\lambda(ds)$$

$$+ \int_S \int \sin(\mathbf{T}_{t_p} g_p(s)x) \, S_1 \left(\mathbf{T}_{t_{p-1}} g_{p-1}(s)x, \sum_{i=0}^{p-2} \mathbf{T}_{t_i} g_i(s)x \right) \, \rho(s, dx)\lambda(ds).$$

The first two summands converge to zero by the assumption. For the other two by relation (10.4.2) we have

$$\left| \int_S \int \cos(\mathbf{T}_{t_p} g_p(s)x) \, C_1 \left(\mathbf{T}_{t_{p-1}} g_{p-1}(s)x, \sum_{i=0}^{p-2} \mathbf{T}_{t_i} g_i(s)x \right) \, \rho(s, dx)\lambda(ds) \right|$$

$$\leq \int_S \int \left| C_1 \left(\mathbf{T}_{t_{p-1}} g_{p-1}(s)x, \sum_{i=0}^{p-2} \mathbf{T}_{t_i} g_i(s)x \right) \right| \, \rho(s, dx)\lambda(ds)$$

$$= \int_{\mathbb{R}^{p+1}} \left| C_1 \left(x_{p-1}, \sum_{i=1}^{p-2} x_i \right) \right| Q_t(dx)$$

and similarly

$$\left| \int_S \int \sin(\mathbf{T}_{t_p} g_p(s)x) \, S_1 \left(\mathbf{T}_{t_{p-1}} g_{p-1}(s)x, \sum_{i=0}^{p-2} \mathbf{T}_{t_i} g_i(s)x \right) \, \rho(s, dx)\lambda(ds) \right|$$

$$\leq \int_S \int \left| S_1 \left(\mathbf{T}_{t_{p-1}} g_{p-1}(s)x, \sum_{i=0}^{p-2} \mathbf{T}_{t_i} g_i(s)x \right) \right| \, \rho(s, dx)\lambda(ds)$$

$$= \int_{\mathbb{R}^{p+1}} \left| S_1 \left(x_{p-1}, \sum_{i=1}^{p-2} x_i \right) \right| Q_t(dx)$$

Now Lemma 10.6.1 implies that the last expressions in these two estimations converge to

$$\int \left| 1 - \cos(x_{p-1}) - \cos \left(\sum_{i=1}^{p-2} x_i \right) + \cos \left(\sum_{i=1}^{p-1} x_i \right) \right| \, Q_0(dx)$$

and

$$\int \left| \sin(x_{p-1}) + \sin \left(\sum_{i=1}^{p-2} x_i \right) - \sin \left(\sum_{i=1}^{p-1} x_i \right) \right| \, Q_0(dx),$$

when $\delta(t) \to \infty$, respectively. But since Q_0 is a characteristic of a vector of independent ID random variables thus $Q_0(\{x \in \mathbb{R}^p : x_j x_i \neq 0\}) = 0$ for any i, j.

This yields $Q_0(\{x \in \mathbb{R}^p : x_{p-1}\sum_{i=1}^{p-2} x_i \neq 0\}) = 0$. Consequently, using Taylor expansions for $\sin x$, $\cos x$, we obtain

$$1 - \cos(x_{p-1}) - \cos\left(\sum_{i=1}^{p-2} x_i\right) = -\cos\left(\sum_{i=1}^{p-1} x_i\right)$$

and

$$\sin(x_{p-1}) + \sin\left(\sum_{i=1}^{p-2} x_i\right) = \sin\left(\sum_{i=1}^{p-1} x_i\right)$$

$Q_0 - a.e.$. This gives (10.6.2) and completes the proof . $\square$

Remark 10.6.1 *Slightly modifying the proof in Maruyama (1970) it is easy to derive another characterization of the mixing property of stationary ID processes. Namely, for any $a_1, a_2 \in \mathbb{R}$,*

$$\lim_{t \to \infty} N_\Psi(a_1 f_0 + a_2 \mathbf{T}_t f_0) = N_\Psi(a_1 f_0) + N_\Psi(a_2 f_0).$$

10.7 Examples of Chaotic Behavior of ID Processes

Now we give the explicit form of the conditions which appear in Theorems 10.5.1 and 10.6.1 for some stationary ID processes. In all these examples the symmetric ID random measure Λ on $(S, \mathcal{S})$ with control measure λ will have ρ and σ of the form described below.

Let r be a symmetric Lévy measure on $\mathbb{R}$ and let $\rho(s, A) = r(A)$ for all $A \in \mathcal{B}_{\mathbb{R}}$ and each $s \in \mathbb{R}$. Let $\sigma^2(s) \equiv \sigma_0^2 \geq 0$. Then $f \in \mathbf{L}_\Psi(\mathbf{X})$ if and only if

$$\sigma_0^2 \int_S |f(s)|^2 \lambda(ds) < \infty \text{ and } \int_S \int_{\mathcal{R}} \{1 \wedge |xf(s)|^2\} r(dx)\lambda(ds) < \infty$$

Notice that if $\sigma_0^2 > 0$ then $\mathbf{L}^\psi \subseteq \mathbf{L}^2(S, \lambda)$. For $A \in \mathcal{S}$ we have

$$Ee^{it\Lambda(A)} = \exp\left\{\lambda(A)[-t^2\sigma_0^2/2 + \int(\cos tx - 1)r(dx)]\right\}.$$

Recall that $E|\Lambda(A)|^q < \infty$ if and only if

$$\int_A \int_{|x|>1} |x|^q \rho(s, dx)\lambda(ds) = \lambda(A)2\int_1^\infty x^q r(dx) < \infty.$$

Example 10.7.1 *Moving averages are mixing.*

Assume that λ is the Lebesgue measure on $\mathcal{B}_{\mathbb{R}}$. By the invariance of λ under the action of the shift transformation, if f_0 is Λ-integrable then so is $f_0(\cdot - t)$ for all $t \in \mathbb{R}$.

A symmetric ID process is called a *moving average* if it has the spectral representation

$$\int f_0(t - s)\, d\Lambda(s), \quad t \in \mathbb{R}.$$

In this case we have

$$K(t,s) = \frac{1}{2}\sigma_0^2 t^2 + \int_{-\infty}^{\infty} [1 - \cos(tx)] \, r(dx),$$

$$N_\Psi(f) = \int_{-\infty}^{\infty} K(f(s), s) \, ds$$

$$= \frac{1}{2}\sigma_0^2 \int_{-\infty}^{\infty} f^2(s) \, ds + \int_{-\infty}^{\infty} \int_{-\infty}^{\infty} [1 - \cos(f(s)x)] \, r(dx) \, ds.$$

It is clear that for every $f \in L^\psi(\mathbb{R}, \lambda)$, $N_\Psi(f(\cdot + t)) = N_\Psi(f(\cdot))$ for all $t \in \mathbb{R}$, so that a moving average process is stationary.

Let us show now that every symmetric ID moving average process is mixing or, by Theorem 10.6.1, that

$$N_\Psi(f(\cdot - T) - f(\cdot)) \to 2N_\Psi(f(\cdot)) \quad \text{as } T \to \infty.$$

This, however, follows immediately from the expression defining N_Ψ when f has compact support, and for general g in $L^\psi(\mathbb{R}, \lambda)$ from the fact that

$$g(\cdot)I_{[-c,c]}(\cdot) \to g(\cdot) \quad \text{in} \quad L^\psi(\mathbb{R}, \lambda) \quad \text{as} \quad c \to \infty.$$

An alternative proof establishes likewise condition (iii) of Proposition 9.6.3.

$$\lim_{T \to \infty} E\left(\exp\left\{ i \int_{-\infty}^{\infty} [f(s - T) - f(s)] \, \Lambda(ds) \right\}\right)$$

$$= \left| E \, \exp\left\{ i \int_{-\infty}^{\infty} f(s) \, \Lambda(ds) \right\} \right|^2.$$

Example 10.7.2 *Processes with Poisson, gamma and compound Poisson spectral representations.*

We say that a symmetric ID process **X** has a Poisson (gamma) spectral representation $\int_S f_t \, d\Lambda$ if the random measure Λ has a Poisson (gamma) distribution. In all these cases $\sigma^2(s) = 0$ and $\rho(s, A) = r(A)$ for all $s \in \mathbb{R}$ and $A \in \mathcal{S}$, with $\lambda(A) < \infty$.

The Poisson case corresponds to $r = \delta_{\{-1\}} + \delta_{\{1\}}$. Now, $f \in L^\psi(S, \lambda)$ if and only if $\int_{\{|f|\leq\}} f^2 d\lambda < \infty$ and $\lambda\{|f| > 1\} < \infty$. Moreover, we have

$$K(t, s) = 2(1 - \cos t),$$

$$N_\Psi(f) = 2 \int_S \{1 - \cos[f(s)]\} \, \lambda(ds).$$

Now, let $\{T_t\}_{t \in \mathbb{R}}$ be any group of transformations of $L^\psi(S, \lambda)$ such that for $t, u \in \mathbb{R}$,

$$\int_S \{1 - \cos[tf(s)]\} \, \lambda(ds) = \int_S \{1 - \cos[tT_u f(s)]\} \, \lambda(ds)$$

(one can take, for example, $\{T_t\}_{t \in \mathbb{R}}$ induced by pointwise and measure preserving transformations of $\{S, \mathcal{S}, \lambda\}$). The conditions in Theorems 10.5.1 and 10.6.1 take a more concrete form using the expression of $N_\Psi(f)$. Thus, condition (iv) of Theorem 10.5.1 becomes

$$\lim_{T \to \infty} \frac{1}{T} \int_0^T \left| \int_S \{1 - 2\cos[f(s)] + \cos[(\mathbf{T}_t f - f)(s)]\} \ \lambda(ds) \right| \ dt = 0$$

and similarly condition (ii) of Theorem 10.6.1 becomes

$$\lim_{t \to \infty} \int_S \{1 - \cos[(\mathbf{T}_t f - f)(s)]\} \ \lambda(ds) = 2 \int_S \{1 - \cos[f(s)]\} \ \lambda(ds).$$

The gamma process corresponds to $dr(x) = |x|^{-1} e^{-\theta|x|} dx$ for some $\theta > 0$. In this case, $f \in \mathbf{L}^\psi(S, \lambda)$ if and only if

$$\int_S e^{-\theta/|f|} (e^{\theta/|f|} - 1 - \frac{\theta}{|f|}) f^2 \ d\lambda < \infty$$

and

$$\int \int_{1/|f(s)|}^\infty x^{-1} e^{-\theta x} \ dx \ \lambda(ds) < \infty.$$

We have then

$$K(t, s) = 2 \int_0^\infty [1 - \cos(tx)] x^{-1} e^{-\theta x} \ dx = \ln(1 + \left(\frac{t}{\theta}\right)^2),$$

$$N_\Psi(f) = \int_S \ln[1 + \left(\frac{f}{\theta}\right)^2] \ d\lambda.$$

Thus, the corresponding conditions for ergodicity and mixing can be written in the following way

$$\lim_{T \to \infty} \frac{1}{T} \int_0^T \left| \int_S \left\{ 2\ln\left[1 + \left(\frac{f}{\theta}\right)^2\right] - \ln\left[1 + \left(\frac{\mathbf{T}_t f - f}{\theta}\right)^2\right] \right\} d\lambda \right| \ dt = 0$$

and

$$\lim_{t \to \infty} \int_S \left\{ \ln\left[1 + \left(\frac{\mathbf{T}_t f - f}{\theta}\right)^2\right] \right\} \ d\lambda = 2 \int_S \ln\left[1 + \left(\frac{f}{\theta}\right)^2\right] d\lambda.$$

Both examples considered so far have the finite second moment; $E\Lambda^2(A) < \infty$ for all A with $\lambda(A) < \infty$ (as $\int_1^\infty x^2 r(dx) < \infty$). An example with infinite second moment is the symmetric α–stable case $(0 < \alpha < 2)$ which corresponds to $dr(x) = |x|^{-1-\alpha} dx$. In this case,

$$K(t, s) = |t|^\alpha 2 \int_0^\infty [1 - \cos(y)] y^{-1-\alpha} dy = c_\alpha |t|^\alpha,$$

$$N_\Psi(f) = c_\alpha \int_S |f|^\alpha \, d\lambda,$$

and $L^\psi(S, \lambda) = L^\alpha(S, \lambda)$. For more details in this case see Cambanis, Hardin and Weron (1987), Podgórski (1991) and Podgórski and Weron (1991).

The next process provides a non-stable example with infinite second moment for Λ. Let $dr(x) = p_\alpha(x)dx$, where p_α is the density of a standard symmetric α-stable distribution. We have

$$Ee^{it\Lambda(A)} = \exp\left(-\lambda(A)[\int(1 - e^{itx})p_\alpha(x)dx]\right) = \exp\left(-\lambda(A)(1 - e^{-|t|^\alpha})\right)$$

and $\Lambda(A)$ has the compound Poisson distribution, i.e. the same distribution as $\Lambda(A) = Y_0 + \ldots + Y_N$ where $Y_0 = 0$, $\{Y_i\}_{i=1}^\infty$ are i.i.d. random variables with the standard α-stable distribution and N is a Poisson random variable with mean $\lambda(A)$ and independent of $\{Y_i\}_{i=1}^\infty$. Since for $\alpha < 2$ the second moment of an α-stable random variable does not exist ($\int_1^\infty x^2 p_\alpha(x)dx = \infty$), the second moment of $\Lambda(A)$ does not exist either; in fact, $E|\Lambda(A)|^q < \infty$ if and only if $\lambda(A) < \infty$ and $0 < q < \alpha$. We have

$$K(t, s) = \int_0^\infty [1 - \cos(tx)]p_\alpha(x) \, dx = 1 - e^{-|t|^\alpha},$$

$$N_\Psi(f) = \int_S \left\{1 - e^{-|f|^\alpha}\right\} \, d\lambda.$$

This can be used to simplify the formulas in Theorems 10.5.1 and 10.6.1.

Example 10.7.3 *The α-stable Ornstein–Uhlenbeck process is exact for positive time (or has the K-property for $\mathbb{R}$).*

Let us recall here that the Ornstein–Uhlenbeck process belongs to the class of moving average processes, because

$$X(t) = \int_{\mathbb{R}} e^{-(t-s)} I_{[0,\infty)}(t - s) \, L_\alpha(ds) = \int_{-\infty}^t e^{-t+s} \, L_\alpha(ds), \quad t \geq 0,$$

where $\{L_\alpha(t) : t \in [0, \infty)\}$ is an $S\alpha S$–Lévy motion defined and presented graphically in Section 2.5. The process $\{X(t)\}$ can be also obtained from the Lévy motion by the time change, namely

$$X(t) = e^t \, L_\alpha(e^{-\alpha t}).$$

First, we show that the process $\{X(t) : t \in [0, \infty)\}$ is exact. Since

$$T_t \mathcal{F}_\mathbf{X} = \sigma\{X(s) : \ s \geq t\} = \sigma\{L_\alpha(u) : \ u \leq e^{-\alpha t}\},$$

and the Lévy motion is stochastically continuous with independent increments, it follows from the zero–one law of Blumenthal (1957) for such processes that the σ–field $\cap_{t \geq 0}\sigma\{L_\alpha(u) : u \leq t\}$ is trivial and, consequently, $\cap_{t \geq 0}T_t\mathcal{F}_\mathbf{X}$ is also trivial. According to Definition 9.5.5, the process $\{X(t) : t \in [0, \infty)\}$ is exact. For the proof of this fact in the Gaussian case see Lasota and Mackey (1985). It is clear that if we consider the Ornstein–Uhlenbeck process with real time then it has the K-property.

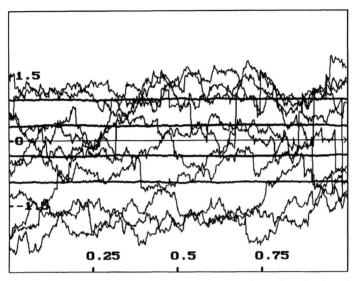

Figure 10.7.1. Computer approximation of the $S_{1.7}(1,0,0)$-valued stationary Ornstein–Uhlenbeck process.

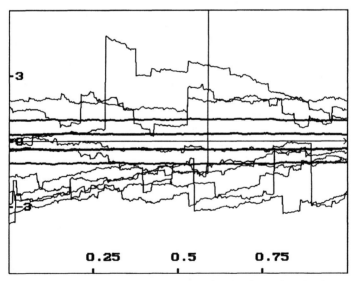

Figure 10.7.2. Computer approximation of the $S_{1.3}(1,0,0)$-valued stationary Ornstein–Uhlenbeck process.

Figures 10.7.1 and 10.7.2 present the visualization of the stationary α–stable Ornstein–Uhlenbeck process $\{X(t)\}$ for $\alpha = 1.7$ and $\alpha = 1.3$, respectively. Both figures show ten typical trajectories of the corresponding stationary Ornstein–Uhlenbeck process $\{X(t)\}$ plotted versus $t \in [0,1]$. The trajectories are represented by thin lines. The two pairs of quantile lines defined by $p_1 = 0.25$ and $p_2 = 0.35$ are approximately parallel indicating the stationarity of the process. In this simulation we have chosen $I = 4000$ and $N = 1000$.

10.8 Random Measures on Sequences of Sets

In this section we investigate the chaotic properties of a simple class of stochastic processes; namely, stationary sequences which can be represented as the *random measures of a stationary sequence of sets*. Although the class of sequences with this representation is not large, it does include examples of sequences which are not symmetric infinitely divisible. Example 10.8.2, taken from Gross and Robertson (1993), is a counter example to the question posed by Cambanis *et al.* (1991) of whether weak mixing and mixing are equivalent for non–Gaussian infinitely divisible processes; in fact, it provides a class of such counter examples which includes symmetric and nonsymmetric α–stable sequences for all $\alpha \in (0,2]$, Poisson sequences, and many others.

The basic set–up is as follows. By $\{\Lambda(\tau^n A)\}_{n \in \mathbf{Z}}$ we mean the spectral representation of a stationary sequence, where A is a set of finite measure in a measure space $(E, \mathcal{E}, \lambda)$, τ is an invertible measure–preserving transformation on $(E, \mathcal{E}, \lambda)$, and Λ is a random measure on $(E, \mathcal{E}, \lambda)$. More specifically, let $(E, \mathcal{E}, \lambda)$ be any σ–finite measure space with nonzero λ; we do not assume any topological structure on E. By $\mathcal{E}_0$ we denote the family of sets in $\mathcal{E}$ with finite measure. Assume that

(i) Λ is an independently scattered random (signed) measure on $(E, \mathcal{E}, \lambda)$;

i.e. Λ is a real–valued stochastic process $\{\Lambda(B)\}_{B \in \mathcal{E}_0}$ on some probability space $(\Omega, \mathcal{F}, P)$ such that whenever $B_1, B_2, \ldots \in \mathcal{E}_0$ are disjoint and $\cup B_i \in \mathcal{E}_0$, the random variables $\Lambda(B_1), \Lambda(B_2), \ldots$ are independent and

$$\Lambda(\cup B_i) = \sum \Lambda(B_i) \qquad a.s.,$$

where convergence of the summation may be conditional. We also assume that

(ii) Λ is *stationary*,

i.e. the distribution of $\Lambda(B)$ depends only on $\lambda(B)$; and

(iii) Λ is *non–degenerate*,

i.e. $\Lambda(B)$ is constant only when $\lambda(B) = 0$. Note that by countable additivity and stationarity, $\Lambda(B) = 0$ whenever $\lambda(B) = 0$; and consequently, we interpret set relations in $(E, \mathcal{E}, \lambda)$ as holding modulo the null sets.

Let $\tau : E \to E$ be a bijection such that $\lambda \circ \tau^{-1} = \lambda \circ \tau = \lambda$; τ is called an *automorphism* (invertible measure–preserving transformation) on $(E, \mathcal{E}, \lambda)$. We assume further that

(iv) there is a set $A \in \mathcal{E}_0$ which *generates* $(E, \mathcal{E})$ under τ,

in the sense that $\cup_{n \in \mathbf{Z}} \tau^n A = E$ and $\sigma\{\tau^n A : n \in \mathbf{Z}\} = \mathcal{E}$. Our goal is to study the sequence $\{\Lambda(\tau^n A)\}_{n \in \mathbf{Z}}$. Thus no generality is lost in assuming that A generates $(E, \mathcal{E})$.

Without loss of generality we also assume that $\{\Lambda(B)\}_{B \in \mathcal{E}_0}$ is defined on $(\Omega, \mathcal{F}, P)$, where $\Omega = \mathbf{R}^{\mathcal{E}_0}$ and $\mathcal{F}$ is the σ–field induced by $\{\Lambda(B)\}_{B \in \mathcal{E}_0}$. Since, the random measure Λ was originally motivated by sequences, we also consider the sub–σ–field

$$\hat{\mathcal{F}} = \sigma\{\Lambda(\tau^n A) : n \in \mathbf{Z}\}.$$

Note that these σ–fields are not generally equal. For instance, $\Lambda(A \cap \tau A)$ is $\mathcal{F}$-measurable but, in general, not $\hat{\mathcal{F}}$-measurable. Abusing notation, we still write "P" for P restricted to $\hat{\mathcal{F}}$.

The transformation τ on $(E, \mathcal{E}, \lambda)$ induces a transformation T on $(\Omega, \mathcal{F}, P)$ by the following relation

$$T\omega(B) = \omega(\tau B) \qquad \forall \omega \in \Omega, \quad B \in \mathcal{E}_0.$$

We write $\hat{T}$ for the restriction of T to $(\Omega, \hat{\mathcal{F}}, P)$. This is the shift transformation on the stationary random sequence $\{\Lambda(\tau^n A)\}$. While our main interest is the transformation $\hat{T}$, it turn out to be convenient to study T in order to study $\hat{T}$.

The transformation T induces the shift operator U_T on $L^2(\Omega, \mathcal{F}, P)$, where

$$U_T f = f \circ T.$$

Similarly, $\hat{T}$ induces $U_{\hat{T}}$ and τ induces U_τ.

If Λ is an infinitely divisible random measure, i.e. if each random variable $\Lambda(B)$ is infinitely divisible, then the sequence $(\Lambda(\tau^n A))$ be infinitely divisible, and similarly if Λ is stable, Gaussian, etc. However, not all such sequences can be represented as above. For instance, observe that if $\Lambda(\cdot)$ is centered Gaussian with variance $\lambda(\cdot)$, then any sequence with the above representation must be nonnegatively correlated.

For convenience, we use the following abbreviations in the next theorem:

$\mathrm{ERG}(\cdot):$	the automorphism $(\cdot)$ is ergodic;
$\mathrm{WMIX}(\cdot):$	the automorphism $(\cdot)$ is weakly mixing;
$\mathrm{WMIX}(\tau, A):$	the sequence $\lambda(A \cap \tau^n A)$ converges to 0 as n approaches infinity outside some set of density 0;
$\mathrm{MIX}(\cdot):$	the automorphism $(\cdot)$ is mixing;
$\mathrm{MIX}(\tau, A):$	the sequence $\lambda(A \cap \tau^n A)$ converges to 0;
$\mathrm{PMIX}(\cdot):$	the automorphism $(\cdot)$ is p–mixing for all $p \geq 1$;
$\overline{\mathrm{LIM}}(\tau, A):$	$\lambda(\limsup(\tau^n A)) = 0$;
$\mathrm{K}(\cdot):$	the automorphism $(\cdot)$ is a K–automorphism.

The following result of Gross and Robertson (1993) is parallel to the results presented in Sections 10.5 and 10.6.

Theorem 10.8.1 *Assume that* $(E, \mathcal{E}, \lambda)$, Λ *A and* τ *satisfy the above assumptions (i)-(iv) and that* T *and* $\hat{T}$ *are induced by* τ.

If $0 < \lambda(E) < \infty$ *then* $\hat{T}$ *is not ergodic (and hence neither is* T *). If* $\lambda(E) = \infty$ *then the following implications hold.*

$$
\begin{array}{ccccc}
& & ERG(\tau) & \Longrightarrow & ERG(T) & \Longleftrightarrow & ERG(\hat{T}) \\
& & \Downarrow & & \Updownarrow & & \Updownarrow \\
WMIX(\tau, A) & \Longleftrightarrow & WMIX(\tau) & \Longleftrightarrow & WMIX(T) & \Longleftrightarrow & WMIX(\hat{T}) \\
\Uparrow & & \Uparrow & & \Uparrow & & \Uparrow \\
MIX(\tau, A) & \Longleftrightarrow & MIX(\tau) & \Longleftrightarrow & MIX(T) & \Longleftrightarrow & MIX(\hat{T}) \\
& & & & \Updownarrow & & \Updownarrow \\
\Uparrow & & & & PMIX(T) & \Longleftrightarrow & PMIX(\hat{T}) \\
& & & & \Uparrow & & \Uparrow \\
\overline{LIM}(\tau, A) & & \Longrightarrow & & K(T) & \Longrightarrow & K(\hat{T})
\end{array}
$$

Remark 10.8.1 *The more significant results shown in the above diagram are as follows: the chaotic properties of a sequence are determined by the behavior of* $\lambda(A \cap \tau^n A)$*; all ergodic sequences with the spectral representation* $(\Lambda(\tau^n A))$ *as described above are weakly mixing; all mixing sequences with this spectral representation are p-mixing for all* $p \geq 1$*. If* $\hat{T}$ *is mixing, weakly mixing, etc., then so is* T*.*

Now, following Gross and Robertson (1993), we construct examples of weakly mixing and mixing automorphisms using the "stacking method". First, however, we give an example related to Theorem 10.8.1.

Example 10.8.1 *There exists a nonergodic automorphism* τ *for which* T *is a K-automorphism.*

Let $E = \mathbb{Z} \times \{1, 2\}$, $\mathcal{E} = 2^E$, let λ be counting measure, and let τ be the shift transformation $\tau\{(x, y)\} = \{(x + 1, y)\}$. Denote

$$A = \{(0, 1), (1, 1), (0, 2)\}.$$

To see that A generates $(E, \mathcal{E})$, observe that $\{(0, 2)\} = A \setminus (\tau A \cup \tau^{-1} A)$ and $\{(0, 1)\} = A \setminus (\{(0, 2)\} \cup \tau A)$. Thus, the hypothesis of Theorem 10.8.1 is satisfied for any random measure which satisfies assumptions (i)-(iii). (Actually, assumption (iii) is not needed to prove that T is a K-automorphism.) Clearly $\limsup(\tau^n A)$ has measure 0. But $\mathbb{Z} \times \{1\}$ is τ-invariant (with infinite measure), so τ is not ergodic.

The next two examples use the "stacking" or "interval-exchange" method of constructing automorphisms. We describe below how a transformation is constructed recursively using "stacks" of subintervals of $[0, \infty)$. We call τ an *infinite rank one* automorphism (by analogy with the finite case) because there is one stack of intervals at each stage. For a more rigorous description of this approach in the finite-measure case see for instance Friedman (1970), Chapter

6. The only difference between our construction of τ described below and the classical cutting and stacking construction is that in our case the measure is infinite.

We take $(E, \mathcal{E}, \lambda)$ to be the half–line $[0, \infty)$ with the Lebesgue measure on the Borel sets. We define stacks of subintervals recursively—at the kth stage we have a stack C_k of height h_k:

$$C_k = (C_k(1), C_k(2), \ldots, C_k(h_k)),$$

where the $C_k(i)$'s are subintervals from $[0, \infty)$ of equal width, which we picture as stacked one above another. We write $\tilde{C}_k = \cup_{i=1}^{h_k} C_k(i)$. In our examples we take

$$C_1 = (A) = ([0, 1)).$$

The stack C_k is constructed from C_{k-1} as follows. Cut C_{k-1} into a given number s_k of subcolumns, each having the same width w_k. On the top of each subcolumn, stack a finite number of disjoint intervals from $[0, \infty) \setminus \tilde{C}_k$ (where the new intervals have the same width as the subcolumns). Let $v(k, l)$ denote the number of intervals stacked on the top of the lth subcolumn in order to construct C_k. These intervals should be chosen consecutively from $[0, \infty) \setminus \tilde{C}_k$, so that no part of $[0, \infty)$ is "skipped". Finally, stack each subcolumn on the top of the one to the left. Thus each stack C_k consists of disjoint intervals of the same width.

The transformation τ_k is defined on $C_k(i), i = 1, \ldots, h_k - 1$, by mapping each interval linearly to the one above it. Clearly, each τ_k is an extension of τ_{k-1}; since intervals are chosen to have equal width, each τ_k is measure–preserving. If $\cup C_k = [0, \infty)$, then $\tau = \lim \tau_k$ is a well–defined automorphism on $(E, \mathcal{E}, \lambda)$.

Let $A = [0, 1), C_1 = (A)$. For $k \geq 1$, define $\eta_k \in \{0, 1\}^{\mathbf{Z}}$ by

$$\eta_k(i) = \begin{cases} 1 & \text{if } 1 \leq i \leq h_k \text{ and } C_k(i) \subset A, \\ 0 & \text{otherwise.} \end{cases}$$

If a sequence has only finitely many 1's, define $a(\cdot)$ to be the position of the right–most 1 in the sequence minus the position of the left–most 1 in the sequence, plus 1. Roughly speaking, $a(\cdot)$ is the "height" of the sequence disregarding leading and trailing 0s. Define

$$\eta_k \cdot \eta_m = \sum_{i \in \mathbf{Z}} \eta_k(i) \eta_m(i).$$

Then $\eta_k \cdot \eta_m$ is the number of positions at which C_k and C_m each have a subinterval of A.

Let S denote the shift operator: $S(\eta(\cdot)) = (\eta(\cdot - 1))$. Define $\eta_k^{(l)}$ to correspond to the part of C_k from the lth subcolumn of C_{k-1}. More precisely, for $l = 1, \cdots, s_k$,

$$\eta_k^{(l)} = S^m \eta_{k-1},$$

where

$$m = (l - 1) h_{k-1} + \sum_{j=1}^{l-1} v(k, j).$$

Remark 10.8.2 *Given the (k-1)-th stack, the parameters s_k and $v(k,l)$, $l = 1, \ldots, s_k$, determine the kth stack. One can see that if*

$$\sum_{k=2}^{\infty} w_k \sum_{l=1}^{s_k} v(k,l) = \infty,$$

then τ is defined on $[0, \infty)$. Also, if $v(k, s_k)$ is greater than $a(\eta_k)$ then $\tau^n A$ "stays in" the stack as long as $0 \leq n < a(\eta_k)$; more precisely,

$$\tau^n A \subset \bigcup_{i=n+1}^{h_k} C_k(i), \qquad 1 \leq n < a(\eta_k).$$

Hence

$$C_k(1) \cap \tau^n A = \emptyset, \qquad 1 \leq n < a(\eta_k).$$

It follows that if $C_1(1) = A$ and $v(k, s_k)$ is greater than $a(\eta_k)$ for each k then

$$C_k(1) = A \setminus \bigcup_{m=1}^{a(\eta_k)-1} \tau^m A, \qquad k \geq 2.$$

This implies that if $C_1(1) = A$ then

$$C_k(1) \in \sigma \left\{ \tau^m A : m \geq 0 \right\} \qquad \forall k \geq 1 \qquad (10.8.1)$$

and $\sigma \left\{ \tau^n A : n \in \mathbb{Z} \right\}$ is the Borel σ-field on $[0, \infty)$.

Example 10.8.2 *There exists an infinite rank one weakly mixing transformation which is not mixing.*

Let $s_k = 2$ for all k; that is, the stack is cut in half at each stage. Take

$$v(k,1) = 0,$$
$$v(k,2) > 2h_{k-1}.$$

By the remarks above, this choice of A and τ satisfies the hypothesis of Theorem 10.8.1 with E the positive half–line and $\mathcal{E}$ the Borel σ–field. We claim that τ is weakly mixing but not mixing.

In fact, τ is ergodic. As we stated at the beginning of this Section, on an infinite measure space ergodicity implies weak mixing. We use the following *characterization of ergodicity*: τ is ergodic if and only if, for any $B_1, B_2 \in \mathcal{E}$ with positive measure, we have $\lambda(B_1 \cap \tau^m B_2) > 0$ for some integer m.

Let x_1 and x_2 be Lebesgue points of B_1 and B_2, respectively. Then there is a $\delta > 0$ such that if J_1 and J_2 are intervals containing x_1 and x_2, respectively, whose lengths are less than δ then

$$\lambda(B_i \cap J_i) > \frac{1}{2}\lambda(J_i), \qquad i = 1, 2.$$

We can choose k large so that there are intervals in the kth stack having this property.

But $\tau^m J_1 = J_2$ for some integer m, by definition of τ. Hence $\tau^m B_1 \cap J_2$ and $B_2 \cap J_2$ both have measure strictly greater than $\lambda(J_2)/2$. Therefore, $\lambda(B_1 \cap \tau^m B_2) > 0$ and τ is ergodic.

To see that τ is not mixing, it is easy to verify by looking at the kth stack that for $n = h_{k-1}$, $k \geq 2$,

$$\lambda(A \cap \tau^n A) = \frac{1}{2}.$$

We claim that τ and A satisfy assumption (iv) of this section. Obviously, $\cup \tau^n A = [0, \infty)$. By the remarks preceding this example we see that $C_k(1)$ is in $\sigma\{\tau^n A : n \geq 0\}$ for all $k \geq 1$, and therefore A generates the Borel σ-field on $[0, \infty)$.

Therefore, by Theorem 10.8.1, if Λ is any random measure satisfying assumptions (i)–(iii) from the beginning of this section then T is weakly mixing but not mixing, and the same is true for $\hat{T}$.

Example 10.8.3 *There exists an infinite rank one mixing transformation.*

Let $A = [0, 1)$, $s_k = k$ (so $w_k = 1/k!$), and let $v(k, l)$'s satisfy

$$v(k, l) > a(\eta_{k-1}) + a(\eta_k^{(1)} + \eta_k^{(2)} + \cdots + \eta_k^{(l)}), \qquad l = 1, 2, \ldots, s_k.$$

By Remark 10.8.2, this choice of A and τ satisfies the hypothesis of Theorem 10.8.1 with E the positive half-line and $\mathcal{E}$ the Borel σ-field. We claim that $\lambda(A \cap \tau^n A) \to 0$.

For any $n > 1$, let $k = k(n)$ be such that

$$a(\eta_{k-1}) < n \leq a(\eta_k).$$

We show that

$$w_k(S^n \eta_k \cdot \eta_k) \to 0 \text{ as } n \to \infty$$

and that this implies $\lambda(A \cap \tau^n A) \to 0$.

Suppose $S^n \eta_k \cdot \eta_k > 0$ for some n, $a(\eta_{k-1}) < n \leq a(\eta_k)$. Then there exist p, q in $\{1, 2, \ldots, k\}$ such that

$$S^n \eta_k^{(p)} \cdot \eta_k^{(q)} > 0.$$

We claim there is only one such pair (p, q).

Observe first that since n is greater than $a(\eta_{k-1})$, p cannot equal q and hence p is strictly less than q.

Consider $S^n \eta_k^{(l)} \cdot \eta_k$, for $l = 1, \ldots, p - 1$. Since $S^n \eta_k^{(p)} \cdot \eta_k^{(q)}$ is nonzero and

$$v(k, q - 1) > a(\eta_k^{(1)} + \eta_k^{(2)} + \cdots + \eta_k^{(p)}),$$

the support of $S^n \eta_k^{(l)}$ lies to the right of the support of $\eta_k^{(q-1)}$ for each such l. But $v(k, p - 1)$ is greater than $a(\eta_{k-1})$, so the support of $S^n \eta_k^{(l)}$ also lies to the left of the support of $\eta_k^{(q)}$. Therefore,

$$S^n \eta_k^{(l)} \cdot \eta_k = 0, \qquad l = 1, \ldots, p - 1.$$

Now consider $S^n \eta_k^{(l)} \cdot \eta_k$ for $l = p+1, \ldots, q$. Since $S^n \eta_k^{(p)} \cdot \eta_k^{(q)}$ is nonzero and $v(k, p)$ is greater than $a(\eta_{k-1})$, the support of $S^n \eta_k^{(l)}$ lies to the right of $\eta_k^{(q)}$ for these values of l. But

$$v(k, q) > a(\eta_k^{(1)} + \eta_k^{(2)} + \cdots + \eta_k^{(q)}),$$

so the support of $S^n \eta_k^{(l)}$ lies to the left of $\eta_k^{(q+1)}$ for these values of l, if $q+1 \leq s_k$. Thus,

$$S^n \eta_k^{(l)} \cdot \eta_k = 0, \qquad l = p+1, \ldots, q.$$

Next, consider $S^n \eta_k^{(l)} \cdot \eta_k$ for $l = q+1, \ldots, s_k$. Since $S^n \eta_k^{(p)} \cdot \eta_k^{(q)}$ is nonzero and for each such l

$$v(k, l) > a(\eta_k^{(1)} + \eta_k^{(2)} + \cdots + \eta_k^{(l)}) + a(\eta_{k-1}),$$

the support of each such $S^n \eta_k^{(l)}$ lies to the left of the support of $\eta_k^{(l+1)}$ (if $l+1 \leq s_k$). Since n is greater than $a(\eta_{k-1})$, the support of $S^n \eta_k^{(l)}$ lies to the left of the support of $\eta_k^{(l)}$. Therefore,

$$S^n \eta_k^{(l)} \cdot \eta_k = 0, \qquad l = q+1, \ldots, s_k.$$

We have shown that

$$S^n \eta_k \cdot \eta_k = S^n \eta_k^{(p)} \cdot \eta_k.$$

But $S^n \eta_k^{(p)} \cdot \eta_k^{(q)} > 0$ and each $v(k, l)$ is greater than $a(\eta_{k-1})$, so $S^n \eta_k^{(p)} \cdot \eta_k^{(l)}$ is 0 except when l equals q. Thus,

$$S^n \eta_k \cdot \eta_k = S^n \eta_k^{(p)} \cdot \eta_k^{(q)}.$$

It is easy to verify that

$$S^n \eta_k^{(p)} \cdot \eta_k^{(q)} \leq \eta_{k-1} \cdot \eta_{k-1} = (k-1)!,$$

so

$$w_k(S^n \eta_k \cdot \eta_k) = \frac{1}{k!}(S^n \eta_k \cdot \eta_k) \leq \frac{1}{k} \to 0 \quad \text{as } n \to \infty$$

(remember that k is a function of n).

Now $v(k, s_k)$ is greater than $a(\eta_k)$, so when the kth stack is shifted by an amount $n \leq a(\eta_k)$, it does not "wrap around" the top, i.e.

$$\lambda(A \cap \tau^n A) = w_k(S^n \eta_k \cdot \eta_k).$$

Therefore, $\lambda(A \cap \tau^n A)$ converges to 0 and, by Theorem 10.8.1, τ is mixing and so are T and $\hat{T}$ for any random measure Λ on $(E, \mathcal{E}, \lambda)$, which satisfies assumptions (i)–(iii). (Actually, assumption (iii) is not needed to prove mixing.)

Remark 10.8.3 *By Theorem 10.8.1, a sufficient condition for T to be a K-automorphism is that* $\limsup \tau^n A$ *have measure 0. This condition is not satisfied by Example 10.8.3. In fact, the tail σ-field $\mathcal{E}^\infty$ is all of $\mathcal{E}$. To see this it is enough to apply τ^n to both sides of equation 10.8.1. We get $C_k(n) \in \mathcal{E}^n$. For $k \geq 2$, let $n = n(k) = h_{k-1} + 1$ and define*

$$B_n = \bigcup_{j=n}^{h_k}(C_k(j) \cap A).$$

Now observe that B_n is in $\mathcal{E}^n$ since for each j, $(C_k(j) \cap A)$ is either $C_k(j)$ or empty. But

$$\lambda(A \cap B_n) = \frac{(k-1)(k-1)!}{k!} \to 1 \quad as \quad k \to \infty.$$

So we have $A \in \mathcal{E}^n$ for every n, and $\mathcal{E}^\infty = \mathcal{E}$. In particular, $\lambda(\limsup \tau^n A) \neq 0$.

Appendix: A Guide to Simulation

Diffusions driven by α–stable Lévy motion. Here we present in detail the computer program STOCH-Lm.c written in the C programming language, solving approximately stochastic differential equations with respect to the α–stable Lévy motion discussed in Section 4.11.

Source code of the program STOCH-Lm.c . Now we give in full extent the source code of this program. Running it (and its several modifications) on the IBM PC we obtained all graphical representations of diffusions reproduced in the book (see e.g. Chapter 7).

```
/* The program STOCH-Lm.c solves a stochastic differential   */
/* equation with the coefficient functions a(t,x), b(t,x),    */
/* visualizing the solution on a given interval [0,T]         */
/* and the histogram and the kernel density estimator for X(T). */

#include  <stdio.h>
#include  <alloc.h>
#include  <math.h>
#include  <stdlib.h>
#include  <conio.h>
#include  <time.h>
#include  <dos.h>
#include  <stdarg.h>
#include  <graphics.h>

#include  "read-f.c"

#define  DIMtr      4003
#define  DIMhi      102
#define  DIMgr      202
#define  DIMqu      12          /* DIMqu < DIMgr !  */
#define  ESC        0x1b
#define  Pi         3.1415926355
#define  sqrt5      2.2360679775
```

315

```
      float   alpha1,sma1,mu1, alpha2,sma2, T;
      int     titer, xstep;
      float   ap,bp,dp, bn;
      int     trmax,himax, Xle[DIMgr], tlgr,trgr;
      float   halpha,tstep,dc,ba, fx,sj,fa,fb;
      float   I_LM[DIMtr], rI[DIMtr], LHI[DIMhi], Xri[DIMgr];
      int     i,j,k,l,jb,jt, xx,yy;
      int     LFR[DIMhi], BBLL[DIMgr], BR, Bdf, BL;
      int     h4,h8,xmin,xmax,ymin,ymax,xmax1;
      float   hh,kk, V,W;

  int     tm,qu, stex,stey,ypi,xpi,xl,xr;
  float   crec,drec,dcrec, pmin[DIMqu],pmax[DIMqu];
  float   trm, eg,fg,hg, gg,ft;

  char    bfr[15];
  int     GraphDriver, GraphMode;
  int     MaxX, MaxY, ErrorCode;
```

/* ================================ */

/* Function Salpha generates the α–stable random variable $S_\alpha(1,0,0)$. */

```
float Salpha(float a)
{ float k,h,s,V,W;
  V=Pi*((0.00001+rand())/32767.9-0.49999);
  k=a*V;
  W=-log((0.00001+rand())/32767.9);
  W/=cos(V-k);
  h=W*cos(V);
  s=fabs(sin(k)*exp(-log(h)/a)*W);
  V=(0.00001+rand())/32767.9;
  if (V<0.5) s=-s;
  return(s);
}
```

/* ================================ */

/* Function kern the computes values of a chosen kernel function. */

```
float kern(float x)
{ float rrr;
  if (fabs(x)>=sqrt5) rrr=0.0;
  else rrr=0.75*(1-x*x/5.0)/sqrt5;
  return(rrr);
}
```

/* ================================ */

/* Function fn computes the values of the constructed kernel estimator. */

```c
float fn(float z)
{  float rr;
   int i;
   rr=0.0;
   for(i=0; i<=trmax; rr+=kern((z-rI[i])/bn), i++);
   rr=rr/bn/(trmax+1.0);
   return(rr);
}
```

```
/* =============================== */
```

/* Function sort_q makes the given sequence nondecreasing. */

```c
void sort_q(float *d,int n)
{  int i,j;
   float V,W;
   if (n<2) return;
   i=0; j=n-1;
   W=d[n/2];
   do
   { for(;(i<j)&&(d[i]<W);i++);
     for(;(j>i)&&(d[j]>W);j--);
     if(i<j)
     { V=d[i]; d[i]=d[j]; d[j]=V;
       i++; j--; }
   } while (i<j);
   if (i==j)
     if (d[i]<W) i++;
   sort_q(d,i);
   sort_q(d+i,n-i);
   return;
} /* end of sort_q */
```

```
/* =============================== */
```

/* A technical procedure. */

```c
void gcvt_outtext(float wx, int wl, int wi, int wj)
{  gcvt(wx,wl,bfr);
   outtextxy(wi,wj,bfr);
}
```

```
/* =============================== */
```

/* A technical procedure for the program read-f.c which edits strings. */

```c
void nazcpy(par)
int par;
{  int i;
```

```
    for(i=0;i<=nzflicz;i++)
      *(*(nazwa+par)+i)=*(nzf+i);
    nzflicz=0;
    free(nzf);
    nzf=calloc(50,sizeof(char));
} /* end of nazcpy */

/* ============================== */

/* Graphics initialization. */

void Initialize(void)
{ GraphDriver = DETECT;
    initgraph( &GraphDriver, &GraphMode, "" );
    ErrorCode = graphresult();
    if (ErrorCode != grOk)
    { printf("Graphics Error:  %s\n", grapherrormsg(ErrorCode));
        exit(1);
    }
    MaxX = getmaxx();
    MaxY = getmaxy();
} /* end of Initialize */

/* ============================== */

/* Input of data defining the problem. */

void data_init_main()
{ char *msgg;
    printf("\n\n DATA FOR A STOCHASTIC DIFFERENTIAL EQUATION\n");
    printf("\n Define the drift function\n");
    printf(" a(t,x) = ");
    do
    { czytf(1); } while(!funk_ok);
    nazcpy(1);

    printf("\n Define the dispersion function\n");
    printf(" b(t,x) = ");
    do
    { czytf(2); } while(!funk_ok);
    nazcpy(2);

    printf("\n Define the index of stability alpha for X(0):\n");
    do
    { printf(" alpha = ");
        scanf("%f",&alpha1);
    } while ((alpha1<=0.0) || (alpha1>2.0) );

    printf("\n Define the parameter sigma for X(0):\n");
```

```
do
{  printf(" sigma = ");
   scanf("%f",&sma1);
}  while ( sma1<=0.0 );

printf("\n Define the real parameter mu for X(0):  \n");
printf(" mu = ");
scanf("%f",&mu1);

printf("\n Define the index of stability alpha for dL(t):\n");
do
{  printf(" alpha = ");
   scanf("%f",&alpha2);
}  while ((alpha2<=0.0) || (alpha2>2.0) );

printf("\n Define the parameter sigma for dL(t):  \n");
do
{  printf(" sigma = ");
   scanf("%f",&sma2);
 }  while ( sma2<=0.0 );

printf("\n The interval of integration is [0,T].  \n");
do
{  printf(" Define T = ");
   scanf("%f",&T);
}  while (T<=0.0);

printf(
"\n The number of subintervals of [0,%5.3f] is titer.\n",T);
do
{  printf(" Define titer = ");
   scanf("%d",&titer);
}  while (titer<1);

printf("\n The size of all statistical samples is trmax.");
printf( "\n Define (not greater then %5d) trmax = ",DIMtr-3);
do
{  scanf("%d",&trmax);
   trmax-=1;
}  while ( (trmax >DIMtr-2) || (trmax<1) );

msgg=
"\n\n If You do not want to see the appropriate "
"\n directions field coming from deterministic part "
"\n of the stochastic equation,"
"\n then press ESC;"
"\n if you do want -- press ANY KEY !\n";
printf(msgg);
```

```
    Bdf=getch();
} /* end of data_init_main() */

/* ================================= */

/* Input of data defining the graphics. */

void data_tra_main()
{ printf("\n\n DATA FOR GRAPHS OF TRAJECTORIES\n");

    do
    { printf(
        "\n Graphs of trajectories should be in [0,T]*[c,d].\n");
        printf(" T = %6.4f.  Define c = ",T);
        scanf("%f",&crec);
        printf(" define d = ");
        scanf("%f",&drec);
    } while (drec<=crec);

    do
    { printf("\n The number of quantils is qu (<=%4d).\n",DIMqu-2);
        printf(" Define qu = ");
        scanf("%d",&qu);
    } while ( (DIMqu-2<qu) || (qu<0) );

    do
    { printf(
        "\n The trajectories will be tm (<=%4d).\n",DIMgr-2*qu-2);
        printf(" Define tm = ");
        scanf("%d",&tm);
    } while ( (DIMgr-2*qu-2<tm) || (tm<1) );

    for (k=1; k<=qu; k++)
    { do
        { printf(
            "\n %3d - pmin from (0.0005,0.5) defines a quantile.",k);
            printf("\n pmin = ");
            scanf("%f",&pmin[k]);
            pmax[k]=1.0-pmin[k];
        } while ( (pmin[k]<0.0005) || (pmin[k]>=0.5) );
    }
} /* end of data_tra_main() */

/* ================================= */

/* Input of data defining density estimators. */

void data_hi_main()
{ printf("\n\n DATA FOR DENSITY ESTIMATORS");
    printf("\n OF CALCULATED X(%5.3f):  \n",T);
```

```
    do
    {  printf("\n Define:  real a = ");
       scanf("%f",&ap);
       printf(" ( b>a !)  real b = ");
       scanf("%f",&bp);
       printf(" positive real d = ");
       scanf("%f",&dp);
    } while ( (ap>=bp) || (dp<=0.0) );

    do
    {  printf(
       "\n Define (<= %4d ) natural himax = ",DIMhi-2);
       scanf("%d",&himax);
    } while ( (himax >DIMhi-2) || (himax<1) );

    printf("\n\n");

    do
    {  printf(
       "\n Define (not greater then 10 ) natural xstep = ");
       scanf("%d",&xstep);
    } while ( (xstep >10) || (xstep<1) );

    do
    {  printf(
       "\n Define positive real parameter bn = ");
       scanf("%f",&bn);
    } while ( bn<=0.0 );
} /* end of data_hi_main */

/* =============================== */

void main()
{
    char *msg;
    msg=
       "\n\n\n The program STOCH-Lm.c solves\n"
       " a general stochastic differential equation\n"
       " with functional coefficients a(t,x), b(t,x),\n"
       " driven by a stable Levy motion.\n"
       " X(0) is a given stable r.v.  \n"
       " The program presents graphically the solution {X(t)},\n"
       " for t from [0,T] and density estimators for X(T). \n\n"
    printf(msg);

/* =============================== */

    data_init_main();
    data_tra_main();
```

```
tstep=T/titer;
halpha=sma2*exp(log(tstep)/alpha2);
trm=0.0;

Initialize();

h8 = textheight( "H" );  h4 = h8/2;
stex=(MaxX+1)/40;  stey=(MaxY+1)/30;
xmax=MaxX-1;  ymax=MaxY-1;
cleardevice();

settextstyle( DEFAULT_FONT, HORIZ_DIR, 2 );
setlinestyle(0,0,1);
setviewport(0,0,MaxX,MaxY,1);
rectangle(0,0,MaxX,MaxY);
xmin+=1;  xmax-=1;
ymin+=1;  ymax-=1;
xpi=xmax-xmin;
ypi=ymax-ymin;
dcrec=drec/(drec-crec);
if ((drec>0) && (crec<0))
{  line(xmin,ymin+ypi*dcrec,xmax,ymin+ypi*dcrec);
   line(xmax-5,ymin+ypi*dcrec+5,xmax,ymin+ypi*dcrec);
   line(xmax-5,ymin+ypi*dcrec-5,xmax,ymin+ypi*dcrec);
}
setlinestyle(0,0,3);
line(xmin,ymin+ypi/4,xmin+6,ymin+ypi/4);
line(xmin,ymin+ypi/2,xmin+6,ymin+ypi/2);
line(xmin,ymin+ypi*3/4,xmin+6,ymin+ypi*3/4);
line(xmin+xpi/2,ymax,xmin+xpi/2,ymax-6);
line(xmin+xpi/4,ymax,xmin+xpi/4,ymax-6);
line(xmin+xpi*3/4,ymax,xmin+xpi*3/4,ymax-6);
gcvt_outtext(T/4.0,4,xmin+xpi/4-stex+h4,ymax-2*stey-2);
gcvt_outtext(T/2.0,4,xmin+xpi/2-stex+h4,ymax-2*stey-2);
gcvt_outtext(T*3/4.0,4,xmin+xpi*3/4-stex+h4,ymax-2*stey-2);
gcvt_outtext(crec+(drec-crec)*3/4.0,4,xmin+3*h4,ymin+ypi/4-7);
gcvt_outtext(crec+(drec-crec)/2.0,4,xmin+3*h4,ymin+ypi/2-7);
gcvt_outtext(crec+(drec-crec)/4.0,4,xmin+3*h4,ymin+ypi*3/4-7);
setlinestyle(0,0,1);

/* ================================ */

if (Bdf!=ESC)
{  gg=(drec-crec)/ypi;
   for ( i=8+(xpi-xpi/16*16)/2; i<xmax; i+=16)
   { eg=T/xpi*i;
       for ( j=7+(ypi-ypi/12*12)/2; j<ymax; j+=12)
       { fg=gg*(ypi-j)+crec;
```

```
            ft=f(1,eg,fg)*T/xpi;
            jb=j-4; jt=j+4;
            BR=0;
            for (l=1; l<12; l++)
            { trgr=ymax-1-(int)((ft*(l-6)+fg-crec)/gg+0.5);
                if ( trgr>jt || trgr<jb ) BR=0;
                else
                { if (BR)
                    { line(i+l-7,tlgr,i+l-6,trgr);
                    tlgr=trgr;
                    }
                    else { BR=1; tlgr=trgr; }
                }
            }
            if (!getpixel(i,j)) line(i,j-3,i,j+3);
        }
    }
}

/* =============================== */

randomize();
tlgr=xmin;

/* =============================== */

/* Simulation of a sample approximating X(0). */

for(i=0; i<=trmax; i++)
{   I_LM[i]=mu1+Salpha(alpha1)*sma1;
    rI[i]=I_LM[i];
}

/* =============================== */

sort_q(rI,trmax+1);
for (j=1; j<=qu; j++)
{   i=(int)(pmin[j]*trmax);
    k=(int)(pmax[j]*trmax);
    eg=rI[i];
fg=rI[k];
    if ( (i>3)&&(i<trmax-3) &&(k>3)&&(k<trmax-3) )
    { eg+=rI[i-1]+rI[i+1]; eg/=3.0;
        fg+=rI[k-1]+rI[k+1]; fg/=3.0;
    }
    xr=ymax-(int)((eg-crec)*ypi/(drec-crec));
    BR=1;
    if (xr<=ymin) { xr=ymin; BR=0; }
```

```
        if (xr>=ymax) { xr=ymax; BR=0; }
        Xle[j+j-2]=xr;
        BBLL[j+j-2]=BR;
        xr=ymax-(int)((fg-crec)*ypi/(drec-crec));
        BR=1;
        if (xr<=ymin) { xr=ymin; BR=0; }
        if (xr>=ymax) { xr=ymax; BR=0; }
        Xle[j+j-1]=xr;
        BBLL[j+j-1]=BR;
   }
   for (i=0; i<=tm-1; i++)
   {  xr=ymax-(int)((I_LM[i]-crec)*ypi/(drec-crec));
      BR=1;
      if (xr<=ymin) { xr=ymin; BR=0; }
      if (xr>=ymax) { xr=ymax; BR=0; }
      Xle[2*qu+i]=xr;
      BBLL[2*qu+i]=BR;
   }

   /* ======== beginning of the main l-loop =========== */
   /* ================================================== */

   for (l=1; l<=titer; l++)
   {  for(i=0; i<=trmax; i++)
      {  fa=f(1,(l-1)*tstep,I_LM[i]);
         fb=f(2,(l-1)*tstep,I_LM[i]);
         I_LM[i]+=fa*tstep+fb*halpha*Salpha(alpha1);
         rI[i]=I_LM[i];
      }
      sort_q(rI,trmax+1);
      for (j=1; j<=qu; j++)
      {  i=(int)(pmin[j]*trmax);
         k=(int)(pmax[j]*trmax);
         eg=rI[i];
fg=rI[k];
         if ( (i>3)&&(i<trmax-3) &&(k>3)&&(k<trmax-3) )
            {  eg+=rI[i-1]+rI[i+1]; eg/=3.0;
               fg+=rI[k-1]+rI[k+1]; fg/=3.0;
            }
         Xri[j+j-2]=eg; Xri[j+j-1]=fg;
      }
      for (i=0; i<=tm-1; Xri[2*qu+i]=I_LM[i], i++);
      trm=l*tstep;
      trgr=xmin+(int)(trm*xpi/T);
      for (i=0; i<=tm+2*qu-1; i++)
      {  if (i<2*qu)
```

```
        setlinestyle(0,0,3);
        else setlinestyle(0,0,1);
        xr=ymax-(int)((Xri[i]-crec)*ypi/(drec-crec));
        BR=1;
        if (xr<=ymin) { xr=ymin; BR=0; }
        if (xr>=ymax) { xr=ymax; BR=0; }
        if ( BBLL[i] || BR ) line(tlgr,Xle[i],trgr,xr);
        BBLL[i]=BR;  Xle[i]=xr;
    }
    tlgr=trgr;
}

/* ========== end of the main l-loop ============== */

/* ================================================== */
getch();
closegraph();

/* ================================================== */

/* Construction of density estimators. */

printf("\n\n Press :");
printf(
    "\n After a wrong guess of parameters for \n")
printf(" density estimators of X(%5.3f)",T);
printf("\n press ESC seeing the graphs - to repeat !\n");

/* ================================= */

do               /* ----- l==ESC ? --------- */
{ data_hi_main();
    V=(bp-ap)/himax;
    rI[trmax+1]=9999999.99;
    k=0;
    for (j=0; j<=himax; j++)
    { W=ap+j*V;
        for (i=0; rI[k]<=W; i++, k++);
        LFR[j]=i; LHI[j]=i/(1.0+trmax);
    }
    for (j=0, k=0; j<=himax; k+=LFR[j], j++);
    LFR[himax+1]=(k<=trmax)?trmax+1-k:0;
    LHI[himax+1]=LFR[himax+1]/(1.0+trmax);

    Initialize();
    xmin=1;
xmax=MaxX-1;
    ymin=1;
ymax=MaxY-1;
```

```
rectangle(xmin,ymin,xmax,ymax);
dc=(ymax-ymin)/dp;
hh=(float)(MaxX-2)/himax;
line(xmax/2+1,ymax,xmax/2+1,ymax-h8);
line(3*xmax/4,ymax,3*xmax/4,ymax-h8);
line(xmax/4+1,ymax,xmax/4+1,ymax-h8);
line(xmin,ymax/2,xmin+h8,ymax/2);
line(xmin,3*ymax/4,xmin+h8,3*ymax/4);
line(xmin,ymax/4,xmin+h8,ymax/4);
gcvt_outtext(ap+(bp-ap)/4.0,4,xmax/4-2*h8, ymax-2*h8);
gcvt_outtext(ap+(bp-ap)/2.0,4,xmax/2-2*h8, ymax-2*h8);
gcvt_outtext(ap+3*(bp-ap)/4.0,4,3*xmax/4-2*h8, ymax-2*h8);
gcvt_outtext(3*dp/4.0,4,2*h8,ymax/4);
gcvt_outtext(dp/2.0,4,2*h8,ymax/2);
gcvt_outtext(dp/4.0,4,2*h8,3*ymax/4);
if (ap<0.0 && bp>0.0)
{ i=(int)(0.5-ap*MaxX/(bp-ap));
   line(i, ymax+3, i, ymin);
   line(i-6, ymin+6, i, ymin);
   line(i+6, ymin+6, i, ymin);
}
xl=1;
for(i=1; i<=himax; i++)
{  xr=1+i*hh;
   k=ymax-(int)(LHI[i]*dc/V+0.5);
   if ( k<=ymax && k>=ymin ) line(xl+1,k,xr-1,k);
   xl=xr;
}
ymin=1; ymax=MaxY-1;
dc=(ymax-ymin)/dp;
xmin=1; xmax=MaxX-1;
ba=(bp-ap)/(xmax-xmin);
setlinestyle(1,0,1);
BL=0;
for (i=4; i<=xmax; i+=xstep)
{  fx=fn(ap+i*ba);
   yy=ymax-(int)(fx*dc+0.5);
   if ( yy>ymax || yy<0 ) BL=0;
   else
   {  if (BL) { line(i-xstep,xx,i,yy); xx=yy; }
      else {BL=1; xx=yy;}
   }
}
l=getch();
if (l==ESC) closegraph();
```

```
   } while (1==ESC);
   closegraph();
/* ================================= */
   exit(1);
} /* end of main */
/* ================================= */

/* Program READ-f.c allows to insert strings defining functions   */
/* of the form f = f(x,t) in the main program STOCH-Lm.c .        */

/* ================================= */

#include<string.h>

#define NUMBR 13
#define cc while(kbhit())getch()

typedef double(*wf)(double);
typedef struct operator
   {  char sym[10];
      int r;
      struct operator *w;
      struct operator *z;
      float vals;
      wf f;
   }
typedef struct symbol
   {  char sb[10];
      wf ff;
   }
typedef char *sgna;
typedef struct operator * (tf0[100]);
sgna *nams;
tf0 *tf;

int coun=0,nzfcoun=0;
int funk_ok;
char a;
char *nzf;
char *nzbuf;

struct symbol lib[] =
   {  {"fabs",fabs},{"sin",sin}, {"asin",asin},
      {"sinh",sinh},{"cos",cos}, {"acos",acos},
      {"cosh",cosh},{"tan",tan}, {"atan",atan},
      {"tanh",tanh},{"exp",exp}, {"ln",log},
      {"log",log10}
```

```
      }
struct operator *expres(struct operator *p);
kbhit(void);
getch(void);
void clrscr(void);
getche(void);
char sgnb(void)
{  char str[2];
   int i=0;
   sscanf(nzbuf,"
   *(nzf+nzfcoun)=*str;
   nzfcoun++;
   while(*(nzbuf+i)!=*str)
   i++;
   *(nzbuf+i)=' ';
   return(a=*str);
}
wf fnd(char buf[10])
{  int i;
   i=0;
   while(i<=NUMBR && strcmp(buf,lib[i++].sb));
   if (i>NUMBR)
   {  gotoxy(17,24);
      printf("Not known function");
   printf("
   while(!kbhit());
   getch();
   funk_ok=0;
   }
   else return(lib[--i].ff);
}
struct operator *factr(struct operator *p)
{  char buf[10];
   int k=0;
   float m;
   memset(buf,0,sizeof(buf));
   if(a!='(' && a!=')' && a!='t' && a!='x' && (a<'0'|| a>'9'))
   {  if (a>='a' && a<='z')
      {  k=0;
         while(a>='a' && a<='z')
         {  buf[k++]=a;
            a=sgnb();
         }
         if (funk_ok)
         {  p->f=fnd(buf);
```

```
                if (funk_ok)
                { strcpy(p->sym,buf);
                  p->r=1;
                  if (a=='(')
                  { a=sgnb();
                     if (a==')')
                     { gotoxy(17,24);
                        printf("argument missing !");
                        while(!kbhit());
                        getch();
                        funk_ok=0;
                     }
                     else
                     { coun++;
                        p->w=calloc(1,sizeof(struct operator));
                        p->w=expres(p->w);
                     }
                  }
                  else
                  { gotoxy(17,24);
                     printf("opening bracket missing !");
                     while(!kbhit());
                     getch();
                     funk_ok=0;
                  }
                }
             }
          }
       }
       else
       { gotoxy(17,24);
          printf("wrong factor !");
          while(!kbhit());
          getch();
          funk_ok=0;
       }
    }
    else
    { switch(a)
       { case '0':
          case '1':
          case '2':
          case '3':
          case '4':
          case '5':
          case '6':
```

```
case '7':
case '8':
case '9':
        p->vals=a-'0';
        a=sgnb();
        while(a>='0' && a<='9')
        {  p->vals=(p->vals)*10+a-'0';
           a=sgnb();
        }
        if (a=='.')
        {  m=0.1;
           a=sgnb();
           while(a>='0' && a<='9')
           {  p->vals+=(a-'0')*m;
              m*=0.1;
              a=sgnb();
           }
        }
        p->r=0;
        break;
case 't':
        p->sym[0]=a;
        a=sgnb();
        p->r=0;
        break;
case 'x':
        p->sym[0]=a;
        a=sgnb();
        p->r=0;
        break;
case '(':
        a=sgnb();
        if(a==')')
        {  gotoxy(17,24);
           printf("wrong expression !");
           while(!kbhit());
           getch();
           funk_ok=0;
        }
        else
        {  coun++;
           p=expres(p);
        }
        break;
deault:
```

```
                          gotoxy(17,24);
                          printf("wrong factor !");
                          while(!kbhit());
                          getch();
                          funk_ok=0;
                  }
          }
      return(p);
}
struct operator *summnd(struct operator *p)
{   struct operator *q;
    p=factr(p);
    if (funk_ok)
    { if(a!=')')
        {   while ((a=='*' || a=='/') && funk_ok)
            {   q=calloc(1,sizeof(struct operator));
                q->w=p;
                q->sym[0]=a;
                q->r=2;
                a=sgnb();
                q->z=calloc(1,sizeof(struct operator));
                q->z=factr(q->z);
                p=q;
            }
        }
    }
    return p;
}
struct operator *wyr(struct operator *p)
{   struct operator *q;
    if (a=='-')
    {   p->r=0;
        p->vals=0;
    }
    else
    p=summnd(p);
    if (funk_ok)
    {   if(a!=')')
        {   while ((a=='+' || a=='-') && funk_ok)
            {   q=calloc(1,sizeof(struct operator));
                q->w=p;
                q->sym[0]=a;
                q->r=2;
                a=sgnb();
                q->z=calloc(1,sizeof(struct operator));
```

```
              q->z=summnd(q->z);
              p=q;
          }
      }
  }
  return p;
}
  struct operator *logi(struct operator *p)
{ struct operator *q;
  p=wyr(p);
  if (funk_ok)
  { if(a!=')')
     { while ((a=='>' || a=='<') && funk_ok)
        { q=calloc(1,sizeof(struct operator));
          q->w=p;
          q->sym[0]=a;
          q->r=2;
          a=sgnb();
          q->z=calloc(1,sizeof(struct operator));
          q->z=wyr(q->z);
          p=q;
        }
     }
  }
  return p;
}
struct operator *and(struct operator *p)
{ struct operator *q;
  p=logi(p);
  if (funk_ok)
  { if(a!=')')
     { while (a=='&' && funk_ok)
        { q=calloc(1,sizeof(struct operator));
          q->w=p;
          q->sym[0]=a;
          q->r=2;
          a=sgnb();
          q->z=calloc(1,sizeof(struct operator));
          q->z=logi(q->z);
          p=q;
        }
     }
  }
  return p;
}
```

```c
struct operator *expres(struct operator *p)
{   struct operator *q;
    p=and(p);
    if (funk_ok)
    {   while(a!=',' && a!=')' && funk_ok)
        {   while (a=='|' && funk_ok)
            {   q=calloc(1,sizeof(struct operator));
                q->w=p;
                q->sym[0]=a;
                q->r=2;
                a=sgnb();
                q->z=calloc(1,sizeof(struct operator));
                q->z=and(q->z);
                p=q;
            }
            if (a!=',' && a!=')')
            {   gotoxy(17,24);
                printf("zly operator");
                while(!kbhit());
                getch();
                funk_ok=0;
            }
        }
    }
    if (funk_ok)
    {   if (a==')')
        {   a=sgnb();
            coun--;
        }
    }
    return p;
}
void funk_del(struct operator *p)
{   if(p!=NULL)
    {   funk_del(p->w);
        funk_del(p->z);
        free(p);
    }
}
void funk_init(void)
{   int i;
    nams=calloc(7,sizeof(sgna));
    for (i=0; i<=6;i++)
    {   *(nams+i)=calloc(50,sizeof(char));
        memset(*(nams+i),0,50*sizeof(char));
```

```
   }
   nzf=calloc(50,sizeof(char));
   tf=calloc(7,sizeof(tf0));
}
void init_del(void)
{  int i;
   for(i=0;i<=6;i++)
   {  funk_del( *(*(tf+i)) );
      free(*(nams+i));
   }
   free(nams);
   free(nzf);
   free(tf);
}
void czytf(par)
int par;
{  int i,j,k,l;
   char zch;
   cc;
   funk_ok=1;
   nzbuf=(char *)malloc(50*sizeof(char));
   for (i=0;i<nzfcoun;i++)
   *(nzf+i)=0;
   nzfcoun=0;
   i=0;
   do
   {  k=wherex();
      l=wherey();
      scanf("
      if (zch!=' ' && zch!='\n')
      {  *(nzbuf+i)=zch;
         i++;
      }
      else gotoxy(k,l);
   }
   while (zch!='\n' || i==0);
   *(nzbuf+i)=',';
   *(nzbuf+i+1)='\0';
   a=sgnb();
   *(*(tf+par))=calloc(1,sizeof(struct operator));
   *(*(tf+par))=expres(*(*(tf+par)));
   if (coun && funk_ok)
   {  gotoxy(17,24);
      printf("error in number of brackets !");
      while(!kbhit());
```

```
        getch();
        coun=0;
        funk_ok=0;
    }
    if (!funk_ok) funk_del( *(*(tf+par)) );
    free(nzbuf);
}
float wart(struct operator *q,float t,float x)
{   float yy;
    switch(q->r)
    {   case 0:
                if (q->sym[0]=='t') q->vals=t;
                else
                    if (q->sym[0]=='x') q->vals=x;
                yy=(q->vals);
                return(yy);
        case 1:
                yy=(q->f)(wart(q->w,t,x));
                    return(yy);
        case 2:
                switch(q->sym[0])
                {   case '+':
                            yy=(wart(q->w,t,x)+wart(q->z,t,x));
                            return(yy);
                    case '-':
                            yy=(wart(q->w,t,x)-wart(q->z,t,x));
                            return(yy);
                    case '*':
                            yy=(wart(q->w,t,x)*wart(q->z,t,x));
                            return(yy);
                    case '/':
                            yy=(wart(q->w,t,x)/wart(q->z,t,x));
                            return(yy);
                    case '>':
                            yy=(wart(q->w,t,x)>wart(q->z,t,x));
                            return(yy);
                    case '<':
                            yy=(wart(q->w,t,x)<wart(q->z,t,x));
                            return(yy);
                    case '&':
                            yy=(wart(q->w,t,x) && wart(q->z,t,x));
                            return(yy);
                    case '|':
                            yy=(wart(q->w,t,x) || wart(q->z,t,x));
                            return(yy);
```

```
                  default:  return(5);
             }
      default:return(5);
   }
}
float f(int par,float t,float x)
{  return(wart(*(*(tf+par)),t,x));
}

/* ================================ */
```

Input of data.   Now, taking as an example the stochastic differential equation (7.3.1) and the graphical visualization of its solution presented in Figures 7.3.2, 7.3.5 in Section 7.3, we explain how to input the initial data and use the program STOCH-Lm.c. We show what kind of information is available on the computer screen and what kind of information the user must provide himself (this is "bold–faced" here).

```
The program STOCH-Lm.c solves
a general stochastic differential equation
with functional coefficients a(t,x), b(t,x),
driven by a stable Levy motion.
X(0) is a given stable r.v.
The program presents graphically the solution {X(t)},
for t from [0,T] and density estimators of X(T).

DATA FOR A STOCHASTIC STABLE DIFFERENTIAL EQUATION

Define the coefficient drift function
    a(t,x) = 4*sin(t)-2*x

Define the coefficient dispersion function
    b(t,x) = 1

Define the index of stability alpha for X(0):
    alpha = 2
Define the parameter sigma for X(0):
    sigma = 1
Define the real parameter mu for X(0):
    mu = 1

Define the index of stability alpha for dL(t):
    alpha = 2
Define the parameter sigma for dL(t):
    sigma = 1

The interval of integration is [0,T].
Define T = 4
```

```
The number of subintervals of [0,4] is titer.
Define titer = 1000

The size of all statistical samples is trmax.
Define (not greater then 4000) trmax = 2000

If You do not want to see the appropriate
directions field coming from deterministic part
of the stochastic equation,
then press ESC,
if you do want -- press ANY KEY !     A

DATA FOR GRAPHS OF TRAJECTORIES
Graphs of trajectories should be contained in [0,T]*[c,d].
T = 4.  Define c = -2         define d = 4

The number of quantile lines is qu (<=10).
Define qu = 3
The number of trajectories is tm (<=194).
Define tm = 5

1 - pmin from (0.0005,0.5) defines a quantile
    pmin = 0.1
2 - pmin from (0.0005,0.5) defines a quantile
    pmin = 0.2
3 - pmin from (0.0005,0.5) defines a quantile
    pmin = 0.3
```

The first part of the program being executed and Fig. 7.3.2 being visible on the computer screen, the user is asked to push any key and start the execution of the second part of the program.

```
DATA FOR DENSITY ESTIMATORS OF CALCULATED X(4):
Define:      real a = -3.0
( b>a !)     real b = 1.0
  positive real d = 1.0
Define (not greater then 100) natural himax = 23
Define (not greater then 10) natural xstep = 4
Define positive real parameter bn = 0.2
```

Running this part of the program the user obtains Fig. 7.3.5 on the computer screen .

A packet of computer programs executable on IBM and compatible PC's is available at the Hugo Steinhaus Center for Stochastic Methods in Science and Technology, Wybrzeże Wyspiańskiego 27, 50–370 Wrocław, Poland.

Bibliography

ADLER, R. J., CAMBANIS, S., and SAMORODNITSKY, G. (1990). On stable Markov processes. *Stoch. Proc. Appl.* **34** 1–17.

ARNOLD, L. (1974). *Stochastic Differential Equations*, Wiley, New York.

ARNOLD, L. and WIHSTUTZ, V. (1982). Stationary solutions of linear systems with additive and multiplicative noise. *Stochastics* **7** 133–155.

ATKINSON, E. N., BARTOSZYŃSKI, R., BROWN, B. W., and THOMPSON, J. R. (1983). Simulation techniques for parameter estimation in tumor related stochastic processes. In *Proceedings of the 1983 Computer Simulation Conference, New York,* 754–757. North Holland, Amsterdam.

BADGAR, W. (1980). In *Mathematical Models as a Tool for the Social Sciences* (B. J. West, ed.) 87–97. Gordon and Breach, New York.

BARNSLEY, M. (1988). *Fractals Everywhere*, Academic Press, Boston.

BARTOSZYŃSKI, R., BROWN, B. W., McBRIDE, C. M., and THOMPSON, J. R. (1981). Some nonparametric techniques for estimating the intensity function of a cancer related nonstationary Poisson process. *Ann. Statist.* **9** 1050–1060.

BEDNAREK, A. R. and ULAM, F., eds. (1990). *Analogies between Analogies, The Mathematical Reports of S. M. Ulam and his Los Alamos Collaborators.* Univ. of California Press, Berkeley.

BENDLER, J. T. (1984). Lévy (stable) probability densities and mechanical relaxation in solids polymers. *J. Stat. Phys.* **36** 625–637.

BERGSTRÖM, H. (1952). On some expansions of stable distributions. *Ark. Mathematicae II* **18** 375–378.

BERLINER, L. M. (1992). Statistics, probability and chaos. *Stat. Science* **7** 69–90.

BICHTELER, K. (1979). Stochastic integrators. *Bull. Am. Math. Soc.* **1** 761–765.

BICHTELER, K. (1981). Stochastic integration and L^p–theory of semimartingales. *Ann. Probab.* **9** 49–89.

BILLINGSLEY, P. (1968). *Convergence of Probability Measures.* Wiley, New York.

BILLINGSLEY, P. (1979). *Probability and Measure.* Wiley, New York.

BLACK, F. and SCHOLES, M. (1973). The pricing of options and corporate liabilities. *J. Political Economy* **81**, 637–659.

BLUMENTHAL, R. M. (1957). An extended Markov property. *Trans. Amer. Math. Soc.* **85** 52–72.

339

BOCHNER, S. (1955). *Harmonic Analysis and the Theory of Probability.* University of California Press, Berkeley.

BRATLEY, P., FOX, B. L., and SCHRAGE, L. E. (1987). *A Guide to Simulation.* Springer, New York.

BREIMAN, L. (1968), (1992). *Probability,* 1st and 2nd editions. Addison-Wesley, Reading.

BRETAGNOLLE, J., DACUNHA–CASTELLE, D., and KRIVINE, J. L. (1966). Lois stable et espaces L^p. *Ann. Inst. H. Poincaré* **B2** 231–259.

BRETON LE, A. and MUSIELA, M. (1988). Filtrage linéaire optimal de processus stables symétriques. *C. R. Acad. Sci. Paris* **307** 47–50.

BROCKWELL, P. J. and BROWN, B. M. (1984). Expansions for the positive stable laws. *Z. Wahrsch. verw. Geb.* **45** 171–194.

BYCZKOWSKI, T., NOLAN, J. P., and RAJPUT, B. (1993). Approximation of multidimensional stable densities. *J. Multivariate Anal.* **46** 13–31.

CAMBANIS, S., HARDIN JR., C. D., and WERON, A. (1987). Ergodic properties of stationary stable processes. *Stochastic Proc. Appl.* **24** 1–18.

CAMBANIS, S., HARDIN JR., C. D., and WERON, A. (1988). Innovations and Wold decomposition of stable sequences. *Probab. Th. Rel. Fields* **79** 1–28.

CAMBANIS, S., LAWNICZAK A., PODGÓRSKI, K., and WERON, A. (1991). Ergodicity and mixing of symmetric infinitely divisible processes. Technical Report No. 346, Center for Stochastic Processes, Department of Statistics, University of North Carolina, Chapel Hill.

CAMBANIS, S. and MAEJIMA, M. (1989). Two classes of self–similar stable processes with stationary increments. *Stoch. Proc. Appl.* **32** 305–329.

CAMBANIS, S., MAEJIMA, M., and SAMORODNITSKY, G. (1992). Characterization of linear and harmonizable fractional stable motions. *Stoch. Proc. Appl.* **42** 91–110.

CAMBANIS, S., NOLAN, J., and ROSINSKI, J. (1990). On the oscillation of infinitely divisible processes. *Stoch. Proc. Appl.* **35** 87–98.

CAMBANIS, S., SAMORODNITSKY, G., and TAQQU, M., eds. (1991). *Stable Processes and Related Topics.* Birkhäuser, Boston.

CAMBANIS, S. and SOLTANI, R. (1984). Prediction of stable processes: spectral and moving average representation. *Z. Wahrsch. verw. Geb.* **66** 593–612.

CASSANDRO, M. and JONA–LASINIO, G. (1978). Critical points behavior and probability theory. *Adv. in Phys.* **27** 919–941.

CHAMBERS, J. M., MALLOWS, C. L., and STUCK, B. W. (1976). A method for simulating stable random variables. *J. Amer. Statist. Assoc.* **71** 340–344.

CHAO, K. and JOANNOPOULOS, J. D. (1992). Ergodicity and dynamical properties of constant temperature molecular dynamics. *Phys. Rev.* **A 45** 7089–7103.

CHATTERJEE, S. and YILMAZ, M. R. (1992). Chaos, fractals and statistics. *Stat. Science* **7** 49–68.

CHIU, S.-T. (1991). Bandwidth selection for kernel density estimation. *Ann. Statist.* **19** 1883–1905.

CIESIELSKI, Z. (1987). Multiple stable stochastic integrals. *Probability Theory and Math. Statist.* vol. I *VNU Sci. Press* Utrecht 363–373.

CIESIELSKI, Z. (1988). Nonparametric polynomial density estimation. *Probab. Math. Stat.* **9** 1–10.

CLINE, D. B. and BROCKWELL, P. J. (1985). Linear prediction of ARMA processes with infinite variance. *Stoch. Proc. Appl.* **19** 281–296.

CORNFELD, I. P., FOMIN, S. V., and SINAI, YA. G. (1982). *Ergodic Theory.* Springer, New York.

CSÖRGŐ, M. and RÉVÉSZ, P. (1981). *Strong Approximations in Probability and Statistics.* Akadémiai Kiadó, Budapest.

DAY, R. (1959). Stable processes with an absorbing barrier. *Trans. Amer. Math. Soc.* **89** 16–24.

DEÁK, I. (1990). *Random Number Generators and Simulation.* Akadémiai Kiadó, Budapest.

DELLACHERIE, C. (1980). Un survol de la théorie de l'intégrale stochastique. *Stoch. Proc. Appl.* **10** 115–144.

DESBOIS, J. (1992). Algebraic areas distributions for two dimensional Lévy flights. *J. Phys. A: Math. Gen.* **25** L755–762.

DEVROYE, L. (1986). *Non–uniform Random Variate Generation.* Springer, New York.

DEVROYE, L. (1987). *A Course in Density Estimation.* Birkhäuser, Boston.

DEVANEY, R. L. (1989). *An Introduction to Chaotic Dynamical Systems.* Addison–Wesley, Reading, Mass.

DEVROYE, L. and GYÖRFI, L. (1985). *Nonparametric Density Estimation: The L_1 View.* Wiley, New York.

DOLÉANS–DADE, C. and MEYER, P. A. (1970). Intégrales stochastic par rapport aux martingales locales. In *Lecture Notes in Math.* **124** 77–107. Springer, New York.

DONEY, R. A. (1987). On the Wiener–Hopf factorization and the distribution of extrema for certain stable processes. *Ann. Probab.* **15** 1352–1362.

DOOB, J. L. (1942). The Brownian movement and stochastic equations. *Ann. of Math.* **43** 351–369.

DOOB, J. L. (1953). *Stochastic Processes*, Wiley, New York.

DUMOUCHEL, W. H. (1973). Stable distributions in statistical inference I. *J. Amer. Statist. Assoc.* **68** 469–482.

DUMOUCHEL, W. H. (1975). Stable distributions in statistical inference II. *J. Amer. Statist. Assoc.* **70** 386–393.

DUMOUCHEL, W. H. (1983). Estimating the stable index α in order to measure tail thickness: A critique. *Ann. Statist.* **11** 1019–1031.

DYM, H. and MCKEAN, H. P. (1976). *Gaussian Processes, Function Theory and the Inverse Spectral Problem.* Academic Press, New York.

ELLIOTT, R. J. (1982). *Stochastic Calculus and Applications.* Springer, New York.

EPLETT, W. J. R. (1986). Approximation theory for the simulation of continuous Gaussian processes. *Probab. Th. Rel. Fields* **73** 159–181.

ERHARD, A. and FERNIQUE, X. (1981). Fonctions aléatoires stables irrégulieres. *C. R. Acad. Sci. Paris* **292** 999–1001.

FAMA, E. (1965). The behavior of stock market prices, *J. Business* **38** 34–105.

FELLER, W. (1966), (1971). *An Introduction to Probability Theory and its Applications*, Vol. 1, 2, 2nd and 3rd editions. Wiley, New York.

FEUERVERGER, A. and MCDUNNOUGH, PH. (1981). On efficient inference in symmetric stable laws and processes. In *Statistics and Related Topics* (M. Csörgő, D. A. Dawson, J. N. K. Rao, and A. K. Md. E. Saleh, eds.) 109–122. North Holand, Amsterdam.

FOMIN, S. V. (1950). Normal dynamical systems. *Ukr. Mat. J.* **2** 25–47.

GARDINER, C. W. (1983). *Handbook of Stochastic Methods for Physics, Chemistry and the Natural Sciences*. Springer, New York.

GARDNER, W. A. (1986). *Introduction to Random Processes with Applications to Signals and Systems*. MacMillan, London.

GARSIA, A.M. (1965). A simple proof of E. Hopf's maximal ergodic theorem. *J. Math. Mech.* **14** 381–382.

GAWRONSKI, W. (1984). On the bell–shape of stable densities. *Ann. Probab.* **129** 230–242.

GINÉ, E. and HAHN, M. G. (1983). On stability of probability laws with univariate stable marginals. *Z. Wahrsch. verw. Geb.* **64** 157–165.

GNEDENKO, B. V. and KOLMOGOROV, A. N. (1954). *Limit Distributions for Sums of Independent Random Variables*. Addison–Wesley, Reading.

GREENWOOD, P. E. (1969). The variation of a stable path is stable. *Z. Wahrsch. verw. Geb.* **14** 140–142.

GRENANDER, U. (1950). Stochastic processes and statistical inference. *Ark. Mat.* **1** 195–277.

GRENANDER, U. (1963). *Probabilities on Algebraic Structures*, Wiley, New York.

GRENANDER, U. (1981). *Abstract Inference*, Wiley, New York.

GROSS, A. (1992a). Ergodic Properties of Some Stationary Infinitely Divisible Stochastic Processes. Ph. D. Thesis, University of California, Santa Barbara.

GROSS, A. (1992b). Some mixing conditions for stationary symmetric stable stochastic processes. Preprint, Institute of Mathematics, The Hebrew University of Jerusalem.

GROSS, A. (1993). Some mixing conditions for stationary symmetric stable stochastic processes. Preprint.

GROSS, A. and ROBERTSON, J. B. (1992). Ergodic properties of stationary Poisson sequences. *J. Comput. Appl. Math.* **40** 163–175.

GROSS, A. and ROBERTSON, J. B. (1993). Ergodic properties of random measures on stationary sequences of sets. *Stoch. Proc. Appl.* **46** 249–265.

GROSS, A. and WERON, A. (1993). On measure preserving transformations and spectral representations of doubly stationary symmetric stable processes. Preprint.

GYÖRFI, L., HÄRDLE, W., SARDA, P., and VIEU, P. (1989). *Nonparametric Curve Estimation from Time Series*. Lecture Notes in Statistics **60**. Springer, New York.

HALL, P. (1980). A comedy of errors: the canonical form for a stable characteristic function. *Bull. London Math. Soc.* **13** 23–27.

HALL, P. and MARRON, J. S. (1989). Lower bounds for bandwidth selection in density estimation. Unpublished manuscript.

HARDIN JR., C. D. (1981). Isometries on subspaces of L^p. *Indiana Univ. Math. Journal* **30** 449–465.

HARDIN JR., C. D. (1982). On the spectral representation of symmetric stable processes. *J. Multiv. Anal.* **12** 385–401.

HARDIN JR., C. D. and PITT, L. (1983). Integral invariants of functions and L^p isometris on groups. *Pacific J. Math.* **106** 293–306.

HARDIN JR., C. D., SAMORODNITSKY, G., and TAQQU, M. S. (1991). Nonlinear regression of stable random variables. *Ann. Appl. Probab.* **1** 583–612.

HERNÁNDEZ, M. and HOUDRÉ, C. (1993). Disjointness results for some classes of stable processes. *Studia Math.* **105** 235–252.

HONEYCUTT, L. (1992). Stochastic Runge–Kutta algorithms. Part I White noise. Part II Coloured noise. *Phys. Rev.* **A 45** 600–610.

HOPF, E. (1937). *Ergodentheorie*. Springer, Berlin.

HUGHES, B. D., SHLESINGER, M. F., and MONTROLL, E. W. (1981). Random walks with self–similar clusters. *Proc. Nat. Acad. Sci. USA* **78** 3287–3291.

IBRAGIMOV, I. A. and CHERNIN, K. E. (1959). On the unimodality of stable laws. *Theor. Probability Appl.* **4** 453–456.

IBRAGIMOV, I. A. and HAS'MINSKII, R. Z. (1981). *Statistical Estimation: Asymptotic Theory*. Springer, Berlin.

IKEDA, N. and WATANABE, S. (1981). *Stochastic Differential Equations and Diffusion Processes*. North Holland, Amsterdam.

IRANPOUR, R. and CHACON, P. (1987). *Basic Stochastic Processes – The Mark Kac Lectures*. MacMillan, London.

ITÔ, K. (1944a). On the ergodicity of a certain stationary processes. *Proc. Imp. Acad. Tokyo* **20** 54–55.

ITÔ, K. (1944b). On the normal stationary process with no histeresis. *Proc. Imp. Acad. Tokyo* **20** 199–202.

ITÔ, K. (1944c). Stochastic integral. *Proc. Imp. Acad. Tokyo* **20** 519–524.

ITÔ, K. (1952). Complex multiple Wiener integral. *Japan J. Math.* **22** 63–86.

ITÔ, K. (1964). The expected number of zeros of continuous stationary Gaussian processes. *J. Math. Kyoto Univ.* **3** 207–216.

JACOD, J. (1979). *Calcul stochastique et problèmes de martingales*. Lecture Notes in Mathematics **794**, Springer, New York.

JACOD, J. (1993). Random sampling in estimation problems for continuous Gaussian processes with independent increments. *Stoch. Proc. Appl.* **44** 181–204.

JACOD, J. and SHIRYAEV, A. (1987). *Limit Theorems for Stochastic Processes.* Springer, Berlin.

JAKUBOWSKI, A. (1988). Tightness criteria for random measures with application to the principle of conditioning in Hilbert spaces. *Probab. Math. Stat.* **9** 95–114.

JAKUBOWSKI, A., MÉMIN, J., and PAGES, G. (1989). Convergence en loi des suites d'intégrales stochastique sur l'espaces $I\!D^1$ de Skorohod. *Probab. Th. Rel. Fields.* **81** 111–137.

JAMES, F. (1990). A Review of pseudorandom number generators. *Comp. Physics Comm.* **60** 329–344.

JANICKI, A. (1990). Computer simulation and visualization of linear stochastic differential equations. In *Stochastic Methods in Experimental Sciences, Proceedings of the 1989 COSMEX Meeting, Szklarska Poreba, Poland 1989* (W. Kasprzak and A. Weron eds.) 210–216. World Scientific, Singapore.

JANICKI, A. and KOKOSZKA, P.S. (1991). The rate of convergence of LePage type series to finite dimensional distributions of Lévy motion. Preprint.

JANICKI, A. and KOKOŚZKA, P.S. (1992). Computer investigation of the rate of convergence of LePage series to α–stable random variables. *Statistics* **23** 365–373.

JANICKI, A., PODGÓRSKI, K., and WERON, A. (1992). Computer simulation of α–stable Ornstein–Uhlenbeck stochastic processes. In *Stochastic Processes, A Festschrift in Honour of Gopinath Kallianpur* (S. Cambanis, J. K. Gosh, R. Karandikar, and P. K. Sen, eds.) 161–170. Springer, New York.

JANICKI, A. and WERON, A. (1994). Can one see α–stable variables and processes ?. *Stat. Sci.* **9** 109–126.

JONA–LASINIO, G. (1975). The renormalization group: a probabilistic view. *Il. Nuovo Cim.* **26** 99–137.

JUREK, Z. J. (1990). On Lévy (spectral) measures of integral form on Banach spaces. *Probab. Math. Stat.* **11** 139–148.

KAC, M. (1959). *Probability and Related Topics in Physical Sciences.* Interscience, New York.

KAKUTANI, S. (1940). Ergodic theorems and a Markov process with a stable distribution. *Proc. Imp. Acad. Tokyo* **16** 49–54.

KALLIANPUR, G. (1980). *Stochastic Filtering Theory.* Springer, New York.

KANTER, M. (1973). The L^p norm of sums of translates of a function. *Trans. Amer. Math. Soc.* **179** 35–47.

KANTER, M. (1975). Stable densities under change of scale and total variation inequalities. *Ann. Probab.* **31** 697–707.

KARATZAS, I. and SHREVE, S. E. (1988). *Brownian Motion and Stochastic Calculus.* Springer, New York.

KARLIN, S. and TAYLOR H. M. (1975). *A First Course in Stochastic Processes,* 2nd edition. Academic Press, New York.

KASAHARA, Y. and MAEJIMA, M. (1988). Weighted sums of i.i.d. random variables attracted to integrals of stable processes. *Probab. Th. Rel. Fields.* **78** 75–96.

KASAHARA, Y., MAEJIMA, M., and VERVAAT, W. (1988). Log–fractional stable processes. *Stoch. Proc. Appl.* **30** 329–339.

KASAHARA, Y. and WATANABE, S. (1986). Limit theorems for point processes and their functionals. *J. of Math. Soc. Japan* **38** 543–574.

KASAHARA, Y. and YAMADA, K. (1991). Stability theorem for stochastic differential equations with jumps. *Stoch. Proc. Appl.* **38** 13–32.

KEANE, M. S. (1991). Ergodic theory and subshifts of finite type. In *Ergodic Theory, Symbolic Dynamics and Hyperbolic Spaces* (T. Bedford, M. Keane and C. Series, eds.) 35–704. Oxford University Press.

KELLA, O. and WHITT, W. (1991). Queues with server vacations and Lévy Processes with secondary jump input. *Ann. Appl. Probab.* **1** 104–117.

KESTEN, H. and SPITZER, F. (1979). A limit theorem related to a new class of self similar processes. *Z. Wahrsch. verw. Geb.* **50** 5–25.

KLAFTER, J., BLUMEN, A., ZUMOFEN, G., and SHLESINGER, M. F. (1990). Lévy walk approach to anomalous diffusion. *Physica* **A 168** 637–645.

KLAFTER, J., SHLESINGER, M. F., ZUMOFEN, G., and BLUMEN, A. (1992). Scale invariance in anomalous diffusion. *Phil.Mag.* **B 65** 755–765.

KLOEDEN, P. E. and PLATEN, E. (1992). *Numerical Solution of Stochastic Differential Equations.* Springer, Berlin.

KLOEDEN, P. E., PLATEN, E., and SCHURZ, H. (1993). *Numerical Solution of Stochastic Differential Equations through Computer Experiments.* Springer, Berlin.

KNUTH, D. E. (1981). *The Art of Computer Programming: Volume 2: Seminumerical Algorithms*, 2nd edition. Addison–Wesley, Reading.

KOKOSZKA, P.S. (1990) Path properties of certain infinitely divisible processes, Ph. D. Thesis. Technical University, Wrocław.

KOKOSZKA, P.S. and PODGÓRSKI, K. (1992). Ergodicity and weak mixing of semistable processes. *Probab. Math. Stat.* **13** 239–244.

KOKOSZKA, P.S. and TAQQU, M.S. (1993). A characterization of mixing processes of type G. Preprint.

KÔNO, N. and MAEJIMA, M. (1991). Self–similar stable processes with stationary increments. In *Stable Processes and Related Topics)* (S. Cambanis, G. Samorodnitsky and M. Taqqu eds.) 275–295. Birkhäuser, Boston.

KRENGEL, U. (1985). *Ergodic Theorems.* de Gruyter, Berlin, New York.

KUELBS, M. (1973). A representation theorem for symmetric stable processes and stable measures. *Z. Wahrsch. Verw. Gebite* **26** 259–271.

KUNITA, H. and WATANABE, S. (1967). On square integrable martingales. *Nagoya Math. J.* **30** 209–245.

KURTZ, T. G. and PROTTER, P. (1991). Weak limit theorems for stochastic integrals and stochastic differential equations. *Ann. Probab.* **19** 1035–1070.

KURTZ, T. G. and PROTTER, P. (1992). Wong–Zakai corrections, random evolutions and simulation schemes for SDE's. Preprint.

KWAPIEŃ, S. and WOYCZYŃSKI, W. A. (1992). *Random Series and Stochastic Integrals – Single and Multiple.* Springer, New York.

LACEY, H.E. (1974). *Isometric Theorem of Classical Banach Spaces.* Springer, Berlin.

LALLEY S. P. (1991). Probabilistic methods in certain counting problems of ergodic theory. In *Ergodic Theory, Symbolic Dynamics and Hyperbolic Spaces* (T. Bedford, M. Keane and C. Series, eds.) 35–704. Oxford University Press.

LAMPERTI, J. (1958). On the isometries of certain function spaces. *Pacific J. Math.* **8** 459–466.

LAMPERTI, J. (1962). Semi–stable stochastic processes. *Trans. Amer. Math. Soc.* **104** 62–78.

LAMPERTI, J. (1972). Semi–stable Markov processes. *Z. Wahrsch. verw. Geb.* **22** 205–225.

LASOTA, A. and MACKEY, M. C. (1985). *Probabilistic Properties of Deterministic Systems.* Cambridge University Press, London.

LEDOUX, M. and TALAGRAND, M. (1991). *Probability Theory in Banach Spaces.* Springer, Berlin.

LEHMANN, E. L. (1990). Model specification: The views of Fisher and Neyman, and later developments. *Statistical Science* **5** 160–168.

LEONOV, V. P. (1960). The use of the characteristic functional and semiinvariants in the theory of stationary processes. *Dokl. Akad. Nauk USSR* **133** 523–526.

LEPAGE, R. (1980), (1989). Multidimensional Infinitely Divisible Variables and Processes. Part I: Stable case. Technical Raport No. 292, Department of Statistics, Stanford University. In *Probability Theory on Vector Spaces IV. (Proceedings, Łańcut 1987)* (S. Cambanis, A. Weron eds.) 153–163. *Lect. Notes Math.* **1391**. Springer, New York.

LEPAGE, R., WOODROOFE, M., and ZINN, J. (1981). Convergence to a stable distribution via order statistics. *Ann. Probab.* **9** 624–632.

LÉVY, P. (1937). *Théorie de l'addition des variables aléatoires,* Gauthier-Villars, Paris.

LÉVY, P. (1948). *Processus stochastiques et mouvement Brownian,* Gauthier-Villars, Paris.

LINDE, W. (1983). *Infinitely divisible and stable measures on Banach spaces,* Teubner Texte, Leipzig.

LINDE, W. (1986). *Probability in Banach spaces – stable and infinitely divisible distributions,* Wiley, New York.

LIPTSER, R. S. and SHIRYAEV, A. N. (1977), (1978). *Statistics of Random Processes I and II.* Springer, New York.

LORZ, U. and HEINRICH, L. (1991). Normal and Poisson approximation of infinitely divisible distribution functions. *Statistics* **22** 627–649.

MAEJIMA, M. (1983). On a class of self–similar processes. *Z. Wahrsch. verw. Geb.* **62** 235–245.

MAEJIMA, M. (1986). A remark on self–similar processes with stationary increments. *Canadian J. Statist.* **14** 81–82.

MAEJIMA, M. (1989). Self–similar processes and limit theorems. *Sugaku Expositions* 2 103-123.

MAKAGON, A. and MANDREKAR, V. (1990). The spectral representation of stable processes: Harmonizability and regularity. *Probab. Th. Rel. Fields* 85 1-11.

MANDELBROT, B. B. (1963), (1972). The variation of certain speculative prices. *J. Business* 36 394-419 and 45 542-543.

MANDELBROT, B. B. (1982). *The Fractal Geometry of Nature.* Freeman, San Francisco.

MANDELBROT, B. B. and VAN NESS, J. V. (1968). Fractional Brownian motions, fractional noises and applications. *SIAM Review* 10 422-437.

MARCUS, D. J. (1983). Non–stable laws with all projection stable.*Z. Wahrsch. verw. Geb.* 64 139-156.

MARCUS, M. B. (1968). Hölder conditions for Gaussian processes with stationary increments. *Trans. Amer. Math. Soc.* 134 29-52.

MARCUS, M. B. (1989). Some bounds for the expected number of level crossings of symmetric harmonizable p–stable processes. *Stoch. Proc. Appl.* 33 217-231.

MARCUS, M. B. and PISIER, G. (1984). Characterization of almost surely continuous p–stable random Fourier series and strongly stationary processes. *Acta Math.* 152 245-301.

MARCUS, M. B. and PISIER, G. (1984). Some results on the continuity of stable processes and the domain of attraction of continuous stable processes. *Ann. Inst. H. Poincaré* 20 171-194.

MARSAGLIA, G. (1977). The squeeze method for generating gamma variates. *Comp. Math. Appl.* 3 321-325.

MARSAGLIA, G. and ZAMAN, A. (1991). A new class of random number generators. *Ann. Appl. Probab.* 1 462-480.

MARUYAMA, G. (1949). The harmonic analysis of stationary stochastic processes. *Mem. Fac. Sci. Kyusyu Ser. Mat.* IV 49-106.

MARUYAMA, G. (1970). Infinitely divisible processes. *Probab. Th. Appl.* 15 3-23.

MASRY, E. and CAMBANIS, S. (1984). Spectral density estimation for stationary stable processes. *Stoch. Proc. Appl.* 18 1-31.

MCKEAN, H. P. (1969). *Stochastic Integrals.* Academic Press, New York.

MCGILL, P. (1989). Computing the overshoot of a Lévy process. In *Stochastic Analysis, Path Integration and Dynamics* (K. D. Elworthy and J–C. Zambrini, eds.) 165-196. North–Holland, Amsterdam.

MÉTIVIER, M. (1982). *Semimartingales: A course on Stochastic Processes*, de Gruyter, Berlin.

MIJNHEER J. L. (1975). Sample paths properties of stable processes. *Math. Centre Tracts* 59, Amsterdam.

MILSTEIN, G. N. (1978). A method of second–order accuracy integration of stochastic differential equations. *Theor. Prob. Appl.* 23 396-401.

MITTNIK, S. and RACHEV, S. T. (1989). Stable distribution for asset returns. *Appl. Math. Lett.* **3** 301–304.

MITTNIK, S. and RACHEV, S. T. (1991). Modeling asset returns with alternative stable distributions. Stony Brook Working Papers, Department of Economics, SUNY, Stony Brook.

MITRA, S. S. (1982). Stable laws with index 2^{-n}, *Ann. Probab.* **10** 857–859.

MODARRES, R. and NOLAN, J. P. (1992). A method for simulating stable random vectors. Preprint.

MONTROLL, E. W. and BENDLER, J. T. (1984). On Lévy (or stable) distributions and the Williams–Watts model of dielectric relaxation. *J. Stat. Phys.* **34** 129–162.

MONTROLL, E. W. and SHLESINGER, M. F. (1983a). Maximum entropy formalism, fractals, scaling phenomena and 1/f noise: A tail of tails. *J. Stat. Phys.* **32** 209–230.

MONTROLL, E. W. and SHLESINGER, M. F. (1983b). On the wedding of certain dynamical processes in disordered complex materials to the theory of stable Lévy distribution functions. *Lecture Notes in Math.* **1035** 109–137. Springer, New York.

MONTROLL, E. W. and SHLESINGER, M. F. (1984). A wonderful world of random walks. In *Studies in Statistical Mechanics*, vol. **11** *Nonequilibrium Phenomena II: From Stochastics to Hydrodynamics* (J. L. Lebowitz and E. W. Montroll, eds.) 1–121. North Holland, Amsterdam.

MUSIELAK, J. (1983). *Orlicz Spaces and Modular Spaces. Lecture Notes in Math.* **1034**. Springer, New York.

VON NEUMANN, J., RICHTMYER, R. D., and ULAM, S. M. (1947). Statistical methods in neutron diffusion. In *Analogies between Analogies, The Mathematical Reports of S. M. Ulam and his Los Alamos Collaborators* (A. R. Bednarek and F. Ulam, eds.) (1990). Univ. of California Press, Berkeley.

NEWTON, D. (1968). On a principal factor system of a normal dynamical system. *J. London Math. Soc.* **43** 275–279.

NEWTON, H. J. (1988). *TIMESLAB: A Time Series Analysis Laboratory.* Wadsworth, Pacific Grove.

NICOLIS, G., PIASECKI, J., and MCKERNAN, D. (1992). Toward a probabilistic description of deterministic chaos. in *From Phase Transitions to Chaos. Topics in Modern Statistical Physics* (G. Györgyi, I. Kondor, L. Sasvári, and T. Tél eds.) 349–362. World Scientific, Singapore.

NOETHER, E. N. (1967). *Elements of Nonparametric Statistics.* Wiley, New York.

NOLAN, J. P. (1989). Continuity of symmetric stable processes. *J. Multivar. Anal.* **29** 84–93.

ORNSTEIN, D.S. (1960). On invariant measures. *Bull. Amer. Math. Soc.* **66** 297–300.

PARDOUX, E. and TALAY, D. (1985). Discretization and simulation of stochastic differential equations. *Acta Applicandae Math.* **3** 23–47.

PARTHASARATHY, K. R. (1967). *Probability Measures on Metric Spaces.* Academic Press, New York and London.

PAULAUSKAS, V. I. (1982). Convergence to stable laws and their simulation. *Litovsk. Mat. Sb.* **22** 146–156.

PITTEL, B., WOYCZYŃSKI, W. A., and MANN, J. A. (1990). Random tree–type partitions as a model for acyclic polymerization: Holmstark (3/2 stable) distribution of the supercritical gel. *Ann. Probab.* **18** 319–341.

PŁONKA, A. (1986). *Time–Dependent Reactivity of Species in Condensed Media. Lecture Notes in Chemistry* **40** 1–151. Springer, New York.

PŁONKA, A. (1991). Developments in dispersive kinetics. *Prog. Reaction Kinetics* **16** 157–333.

PODGÓRSKI, K. (1991). Ergodic properties of stable processes. Ph. D. Thesis. Technical University, Wrocław.

PODGÓRSKI, K. (1992). A note on ergodic symmetric stable processes. *Stoch. Proc. Appl.* **43** 355–362.

PODGÓRSKI, K. and WERON, A. (1991). Characterization of ergodic stable processes via the dynamical functional. In *Stable Processes and Related Topics* (S. Cambanis, G. Samorodnitsky and M. Taqqu eds.) 317–328. Birkhäuser, Boston.

PROTTER, P. (1990). *Stochastic Integration and Differential Equations – A New Approach.* Springer, New York.

RACHEV, S. T. and SAMORODNITSKY, G. (1993). Option pricing formula for speculative prices modelled by subordinated stochastic processes. Preprint.

RAJPUT, B. S. and RAMA–MURTHY, K. (1987). Spectral representations of semistable processes and semistable laws. *J. Multiv. Anal.* **21** 139–157.

RAJPUT, B. S. and ROSINSKI, J. (1989). Spectral representations of infinitely divisible processes. *Probab. Th. Rel. Fields.* **82** 451–488.

REVUZ, D. and YOR, M. (1991). *Continuous Martingales and Brownian Motion.* Springer, New York.

RICE, J. A. (1990). *Mathematical Statistics and Data Analysis.* Wadsworth, Pacific Grove.

ROCHLIN, V. A. (1964). Exact endomorphisms of Lebesgue space. *Amer. Math. Soc. Transl. (2)* **39** 1–36.

ROLSKI, T. and SZEKLI, R. (1991). Stochastic ordering and thinning of point processes. *Stoch. Proc. Appl.* **37** 299–312.

ROOTZÉN, H. (1980). Limit distributions for the error in approximations of stochastic integrals. *Ann. Probab.* **8** 241–251.

ROSINSKI, J. (1989). On path properties of certain infinitely divisible processes. *Stoch. Proc. Appl.* **33** 73–87.

ROSINSKI, J. (1990). On series representations of infinitely divisible random vectors. *Ann. Probab.* **18** 405–430.

ROSINSKI, J. (1992). On uniqueness of the spectral representation of stable processes. Preprint, University of Tennessee, Knoxville.

ROSINSKI, J. and WOYCZYŃSKI, W. A. (1986). On Itô stochastic integration with respect to p–stable motion: Inner clock, integrability of sample paths, double and multiple integrals. *Ann. Probab.* **14** 271–286.

RYBACZUK, M. and WERON, K. (1989). Linearly coupled quantum oscilators with Lévy stable noise. *Physica A* **160** 519–526.

SAMORODNITSKY, G. and TAQQU, M. S. (1990). $1/\alpha$–self–similar α–stable processes with stationary increments. *J. Multiv. Anal.* **35** 308–313.

SAMORODNITSKY, G. and TAQQU, M. S. (1994). *Stable non–Gaussian Random Processes: Stochastic Models with Infinite Variance.* Chapman & Hall, London.

SATO, K. (1991). Self–similar processes with independent increments. *Probab. Th. Rel. Fields* **89** 285–300.

SCHER, H. and MONTROLL, E. W. (1975). Anomalous transit time dispersion in amorphous materials. *Phys. Rev.* **12B** 2455–2477.

SCHER, H., SHLESINGER, M. F., and BENDLER, J. T. (1991). Time–scale invariance in transport and relaxation. *Physics Today* Jan. 26–34.

SCHERTZER, D. and LOVEJOY, S. (1990). Nonlinear variability in geophysics: Multifractal simulation and analysis. In *Fractal's Physical Origin and Properties* (L. Pietronero ed.) 49–79. Plenum Press, New York.

SCHILDER, M. (1970). Some structure theorems for the symmetric stable laws, *Ann. Math. Statist.* **41** 412–421.

SCHMEISER, B. and LAL, R. (1980). Squeeze methods for generating gamma variates. *J. Amer. Statist. Assoc.* **75** 679–682.

SCHNEIDER, W. R. (1986). Stable distributions: Fox function representation and generalization. *Lecture Notes in Phys.* **262** 497–511.

SESHADRI, V. and WEST, B. J. (1982). Fractal dimensionality of Lévy processes. *Proc. Natl. Acad. Sci. USA* **79** 4501–4505.

SHIRYAEV, A. N. (1984). *Probability,* Springer, New York.

SHUMWAY, R. H. (1988). *Applied Statistical Time Series Analysis.* Prentice Hall, Englewood Cliffs.

SŁOMIŃSKI, L. (1989). Stability of strong solutions of stochastic differential equations. *Stoch. Proc. Appl.* **31** 173–202.

SOBCZYK, K. (1991). *Stochastic Differential Equations with Applications to Physics and Engineering.* Kluwer, Dordrecht.

STEIN, P. R. and ULAM, S. M. (1963). Non–linear transformation studies on electronic computers. In *Analogies between Analogies, The Mathematical Reports of S. M. Ulam and his Los Alamos Collaborators* (A. R. Bednarek and F. Ulam, eds.) (1990). Univ. of California Press, Berkeley.

STRASSER, H. (1985). *Mathematical Theory of Statistics.* De Gruyter, Berlin.

SURGAILIS, D., ROSINSKI, J., MANDREKAR, V., and CAMBANIS, S. (1992). Stable generalized moving averages. Center for Stochastic Processes, Tech. Rept. No. 365, UNC, Chapel Hill.

TAKASHIMA, K. (1989). Sample path properties of ergodic self–similar processes. *Osaka J. Math.* 159–189.

TAKAYASU, H. (1984). Stable distribution and Lévy processes in fractal turbulence. *Prog. Theor. Phys.* **72** 471–479.

TAKENAKA, S. (1991). Integral–geometric construction of self–similar stable processes. *Nagoya Math. J.* **123** 1–12.

TALAY, D. (1983). Resolution trajectorielle et analyse numérique des équations différentielles stochastiques. *Stochastics* **9** 275–306.

TALAY, D. (1986). Discrétisation d'une équation différentielle stochastique et calcul approché d'espérance de fonctionnelles de la solution. *Math. Modeling Numer. Anal.* **20** 141–179.

TALAY, D. and TUBARO, L. (1990). Expansion of the global error for numerical schemes solving stochastic differential equations. *Stoch. Meth. Appl.* **8** 483–509.

TALAGRAND, M. (1990). Characterization of almost surely continuous 1–stable random Fourier series and strongly stationary processes. *Ann. Probab.* **18** 85–91.

TAPIA, R. A. and THOMPSON, J. R. (1978). *Nonparametric Probability Density Estimation.* John Hopkins University Press.

TAQQU, M. (1978). A representation for self–similar processes. *Stoch. Proc. Appl.* **7** 55–64.

TAQQU, M. (1986). A bibliographical guide to self–similar processes and long–range dependence. In *Dependence in Probability and Statistics* (E. Eberlein and M. S. Taqqu, eds.) 137–162. Birkhäuser, Boston.

THOMPSON, J. R. and TAPIA, R. A. (1990). *Nonparametric Probability Density Estimation, Modeling, and Simulation.* SIAM, Philadelphia.

TOTOKI, H. (1964). The mixing properties of Gaussian flows. *Mem. Fac. Sci. Kyusyu Ser. Mat.* **18** 136–139.

TUNALEY, J. K. E. (1972). Conduction in a random lattice under a potential gradient. *J. Appl. Phys.* **43** 4783–4786.

ULAM, S. M. (1980). A mathematical physicist looks at computing. In *Science, Computers and People, A collection of Essays from the Tree of Mathematics* (G.-C. Rota and M. Reynolds, eds.) (1986). Birkhäuser, Boston.

ULAM, S. M. and VON NEUMANN, J. (1947). On combination of stochastic and deterministic processes. *Bull. Am. Math. Soc.* **53** 1120–1132.

URBANIK, K. and WOYCZYŃSKI, W. A. (1967). Random integrals and Orlicz spaces. *Bull. Acad. Polon. Sci.* **15** 161–169.

VERVAAT, W. (1985). Sample path properties of self–similar processes with stationary increments. *Ann. Probab.* **13** 1–27.

WALTERS, P. (1982). *An Introduction to Ergodic Theory.* Springer, Berlin.

WANG, K. G. (1992). Long–time correlation effects and based anomalous diffusion. *Phys. Rev.* **A 45** 833–837.

WANG, X. J. (1992). Dynamical sporadicity and anomalous diffusion in the Lévy motion. *Phys. Rev.* **A 45** 8407–8417.

WEISS, G. H. and RUBIN, R. J. (1983). Random walks: Theory and selected applications. *Adv. Chem. Phys.* **52** 363–432.

WERON, A. (1984). Stable processes and measures: A survey. In *Probability Theory on Vector Spaces III* (D. Szynal, A. Weron, eds.) 306–364. *Lecture Notes in Mathematics* **1080**, Springer, New York.

WERON, A. (1985). Harmonizable stable processes on groups: spectral, ergodic and interpolation properties. *Z. Wahr. verw. Geb.* **68** 473–491.

WERON, A. (1993). A remark on disjointness results for stable processes. *Studia Math.* **105** 253–254.

WERON, A. and WERON, K. (1985). Stable measures and processes in statistical physics. In *Lecture Notes in Math.* **1153** 440–452. Springer, New York.

WERON, K. (1986). Relaxation in glassy materials from Lévy stable distributions. *Acta Phys. Polon.* **A 70** 529–539.

WERON, K. (1991). A probabilistic mechanism hidden behind the universal power law for dielectric relaxation: general relaxation equation. *J. Phys. Condens. Matter.* **3** 9151–9162.

WERON, K. and JURLEWICZ, A. (1993). Two forms of self–similarity as a fundamental feature of the power–law dielectric response. *J. Phys. A: Math. Gen.* **26** 395–410.

WEST, B. J. and SESHADRI, V. (1982). Linear systems with Lévy fluctuations. *Physica* **A 113** 203–216.

YAMADA, K. (1989). Limit theorems for jump shock models. *J. Appl. Probab.* **27** 793–806.

YAMADA, T. (1976). Sur l'approximation des équations différentielles stochastiques. *Z. Wahr. verw. Geb.* **36** 133–140.

ZOLOTAREV, V. M. (1966). On representation of stable laws by integrals. *Selected Translations in Mathematical Statistics and Probability* **6** 84–88.

ZOLOTAREV, V. M. (1986). *One-dimensional Stable Distributions.* Translations of Mathematical Monographs of AMS, Vol. 65, Providence.

Index

ISBN 0-8247-8882-6